Troubleshooting and Repairing Diesel Engines

Troubleshooting and Repairing Diesel Engines

Fourth Edition

Paul Dempsey

New York   Chicago   San Francisco   Lisbon   London   Madrid
Mexico City   Milan   New Delhi   San Juan   Seoul
Singapore   Sydney   Toronto

The *McGraw·Hill* Companies

1 2 3 4 5 6 7 8 9 0   DOC/DOC   0 1 3 2 1 0 9 8 7

ISBN 978-0-07-149371-0
MHID 0-07-149371-9

Sponsoring Editor: Larry S. Hager
Production Supervisor: Pamela A. Pelton
Editing Supervisor: Stephen M. Smith
Project Manager: Madhu Bhardwaj
Copy Editor: Ragini Pandey
Proofreader: Anju Pandhari
Indexer: Kevin Broccoli
Art Director, Cover: Jeff Weeks
Composition: International Typesetting and Composition

Printed and bound by RR Donnelley.

McGraw-Hill books are available at special quantity discounts to use as premiums and sales promotions, or for use in corporate training programs. For more information, please write to the Director of Special Sales, McGraw-Hill Professional, Two Penn Plaza, New York, NY 10121-2298. Or contact your local bookstore.

This book is printed on acid-free paper.

About the author

Paul Dempsey is a master mechanic and the author of more than 20 technical books, including *Small Gas Engine Repair* (now in its Second Edition) and *How to Repair Briggs & Stratton Engines* (now in its Fourth Edition), both available from McGraw-Hill. He has also written more than 100 magazine and journal articles on topics ranging from teaching techniques to maintenance management to petroleum-related subjects.

Contents

Foreword

There are several areas that have changed drastically during the last few years with diesel engines and will greatly affect the near future of diesel engine technologics. The highway trucking industry was the first to require these changes to meet federal EPA emissions guidelines for diesel engines back in the late 1980s. In the mid-1990s these same guidelines were required of the off-highway heavy equipment industry. Now even areas not affected in the past such as the marine, petroleum, and agricultural industries have come under these new requirements. They will change these industries in the same way they have previously changed the trucking and heavy equipment industries. During the last 20 years only certain engine horsepower sizes or industries have come under these federal guidelines. However, the 2007, 2010, and 2012 emissions guidelines will cover and affect all horsepower sizes and industries. Additionally, in most areas the current technologies to meet the 2007 guidelines will not completely meet the 2010 and 2012 requirements without additional technological changes or improvements.

These technological changes are inevitable and future technician training needs will be a reality. This is where diesel engine course books like *Troubleshooting and Repairing Diesel Engines* can help the technician stay current with these changing technologies. To show how rapidly these changes have taken place, information of some past and current examples of those areas affected are mentioned.

Since the inception of the EPA guidelines for diesel engines back in the 1980s, most major engine manufacturers have meant the following reductions. Engine particulates have been reduced by 90% and nitrous oxides by nearly 70%. Added to the equation in the 1990s was noise pollution, with reductions required in engine noise levels from 83 to 80 decibels. Although this doesn't seem like much, it is equal to a 50% noise energy reduction. Add to that the effects of the reduction in fuel sulfur in diesel fuels from 5% to 0.5% to 0.05% (in ppm, 5000 to 500 to 50). Sulfur being the lubricating element in diesel fuels has required many changes to fuel system components.

The increased requirements to meet federal EPA emissions have made it extremely important to develop components that can survive these changes. The number of changes that have been made to diesel engines to meet these requirements would be too numerous to mention at this point. However, some of the more interesting areas that have greatly changed due to these requirements include lubrication requirements, fuel system components, and the use of electronic system controls and diagnostics. Those other areas not discussed here will be covered later in this book.

As the demands on the diesel engine have increased so have the requirements on such things as the oils and filters used to maintain these engines. Just 15 years ago we were teaching the needs of using American Petroleum Institute (API) categories CD, CE, and CF type oils. In just the last few years the demands placed on the diesel engine have caused the industry to develop new oils while moving through categories CG-4, CH-4, CI-4, and now CJ-4 oils.

Introduced in 1995, CG-4 was developed for severe duty, high-speed four-stroke engines using fuel with less than 0.5% sulfur. CH-4 was introduced in 1998 and CI-4 was introduced in 2002, also for high-speed four-stroke engines, to meet 1998 and 2002 exhaust emissions, respectively. CI-4 was also formulated for use with exhaust gas recirculation (EGR) systems. Introduced in 2006, CJ-4 is for high-speed four-stroke engines to meet 2007 exhaust emissions. These oils are blended to meet the increased temperatures, speeds, and loads being placed on today's engines. The oils must also meet the needs to cool, cushion, clean, and protect, as well as hold damaging fines and soot in solution until the oil is changed or the filter removes these damaging particles. In conjunction with the changes to the oil, filters are being required to be more efficient than ever before.

Developments in fuel systems design over the past 20 years have involved one of the largest number of changes to any single system on diesel engines. Fuel systems design has seen this area progress from the use of rotary-distribution-type pumps, used by many diesel engine manufacturers mostly on their smaller-size engines. Larger engines used some sort of pump-and-line-type system with fuel nozzles to provide the engine with high-pressure fuel delivery. As technologies improved, the individual fuel pump segment transformed into and was used in making the mechanical unit injector a reality. Improvements continued with the development of electronic unit injectors, and then hydraulically actuated electronic unit injectors. Now it seems that all of these changes and improvements to fuel system development have led us back to the use of an old technology, the very high-pressure common-rail fuel system design. All of these fuel systems are discussed in this book.

As an instructor of electrical/electronic systems for the past 15 years I've witnessed the effect of this change on the diesel industry firsthand. The growth and usages along with changes in the area of electronics have allowed for many of the advancements that have taken place with diesel engines and was the only way to meet the EPA guidelines. The large numbers of advances in electronics have also meant that the amount and levels of training required by today's engine technician have increased dramatically.

What has this meant to those affected by the use of all these electronics? For the engineer, fewer moving parts will provide the ability to design systems to meet most of the requirements or demands placed upon them by the EPA or the individual customer. This flexibility allows the engine manufacturer the ability to make faster changes when design dictates. The customer gains the ability to make updates to programmable parameters to match changes in equipment demands. The technician, with help from ECM/ECUs, sensors, programmed software, and diagnostic tooling, can diagnose problems faster and more accurately. This results in an eventual savings for all, but especially to the customer, because overall reliability today has increased greatly and this affects down time and bottom line. Engines today are better than ever and have come a long way in helping to keep the environment clean.

Where is all of this change to engine technologies eventually going to take us? Even with all the advancements today in technology that will someday diagnose the problem and tell what part to change, we will always need the engine technicians to program the parameters, help diagnose the problem, change parts, and verify that the results fixed the original problem. There is already a big shortage of qualified and trained diesel technicians around the world. Additionally, the need is there to require training of potentially 50,000 to 100,000 diesel technicians by 2010. Being a diesel engine technician is a very demanding career and a career that is going to become more and more important in the future. The future challenges and potential rewards for the trained technician are going to create some very interesting situations within the industry in the very near future.

Bob Hoster
Staff Training Consultant
Caterpillar Global Manpower Development

Troubleshooting and Repairing Diesel Engines

1
CHAPTER

Rudolf Diesel

Rudolf Diesel was born of German parentage in Paris in 1858. His father was a self-employed leather worker who, by all accounts, managed to provide only a meager income for his wife and three children. Their stay in the City of Light was punctuated by frequent moves from one shabby flat to another. Upon the outbreak of the Franco-Prussian War in 1870, the family became political undesirables and was forced to emigrate to England. Work was almost impossible to find, and in desperation, Rudolf's parents sent the boy to Augsburg to live with an uncle. There he was enrolled in school.

Diesel's natural bent was for mathematics and mechanics. He graduated as the head of his class, and on the basis of his teachers' recommendations and a personal interview by the Bavarian director of education, he received a scholarship to the prestigious Polytechnikum in Munich.

His professor of theoretical engineering was the renowned Carl von Linde, who invented the ammonia refrigeration machine and devised the first practical method of liquefying air. Linde was an authority on thermodynamics and high-compression phenomena. During one of his lectures he remarked that the steam engine had a thermal efficiency of 6–10%; that is, one-tenth or less of the heat energy of its fuel was used to turn the crankshaft, and the rest was wasted. Diesel made special note of this fact. In 1879 he asked himself whether heat could not be directly converted into mechanical energy instead of first passing through a working fluid such as steam.

On the final examination at the Polytechnikum, Diesel achieved the highest honors yet attained at the school. Professor Linde arranged a position for the young diploma engineer in Paris, where, in few months, he was promoted to general manager of the city's first ice-making plant. Soon he took charge of distribution of Linde machines over southern Europe.

By the time he was thirty, Diesel had married, fathered three children, and was recognized throughout the European scientific community as one of the most gifted engineers of the period. He presented a paper at the Universal Exposition held in Paris in 1889—the only German so honored. When he received the first of several citations of merit from a German university, he announced wryly in his acceptance speech: "I am an iceman. . ."

The basis of this acclaim was his preeminence in the new technology of refrigeration, his several patents, and a certain indefinable air about the young man that marked him as extraordinary. He had a shy, self-deprecating humor and an absolute passion for factuality. Diesel could be abrupt when faced with incompetence and was described by relatives as "proud." At the same time he was sympathetic to his workers and made friends among them. It was not unusual for Diesel to wear the blue cotton twill that was the symbol of manual labor in the machine trades.

He had been granted several patents for a method of producing clear ice, which, because it looked like natural ice, was much in demand by the upper classes. Professor Linde did not approve of such frivolity, and Diesel turned to more serious concerns. He spent several years in Paris, working on an ammonia engine, but in the end was defeated by the corrosive nature of this gas at pressure and high temperatures.

The theoretical basis of this research was a paper published by N.L.S. Carnot in 1824. Carnot set himself to the problem of determining how much work could be accomplished by a heat engine employing repeatable cycles. He conceived the engine drawn in Fig. 1-1. Body 1 supplies the heat; it can be a boiler or other heat exchanger. The piston is at position C in the drawing. As the air is heated, it expands in correspondence to Boyle's law. If we assume a frictionless engine, its temperature will not rise. Instead, expansion will take place, driving the piston to D. Then A is removed, and the piston continues to lift to E. At this point the temperature of the air falls until it exactly matches cold surface 2 (which can be a radiator or cooling tank). The air column is now placed in contact with 2, and the piston falls because the air is compressed. Note, however, that the temperature of the air does not change. At B cold body 2 is removed, and the piston falls to A. During this phase the air gains temperature, until it is equal to 2. The piston climbs back into the cylinder.

The temperature of the air, and consequently the pressure, is higher during expansion than during compression. Because the pressure is greater during expansion, the

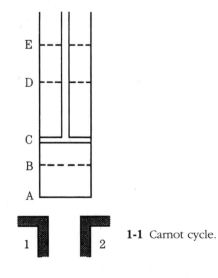

1-1 Carnot cycle.

power produced by the expansion is greater than that consumed by the compression. The net result is a power output that is available for driving other machinery.

Of course this is an "ideal" cycle. It does not take into account mechanical friction nor transfer of heat from the air to the piston and cylinder walls. The infinitesimal difference of heat between 1 and 2 is sufficient to establish a gradient and drive the engine. It would be completely efficient.

In 1892 and 1893 Diesel obtained patent specifications from the German government covering his concept for a new type of *Verbrennungskraftmaschinen*, or heat engine. The next step was to build one. At the insistence of his wife, he published his ideas in a pamphlet and was able to interest the leading Augsburg engine builder in the idea. A few weeks later the giant Krupp concern opened negotiations. With typical internationalism he signed another contract with the Sulzer Brothers of Switzerland.

The engine envisioned in the pamphlet and protected by the patent specifications had these characteristics:

- *Compression of air prior to fuel delivery.* The compression was to be adiabatic; that is, no heat would be lost to the piston crown or cylinder head during this process.
- *Metered delivery of fuel so compression pressures would not be raised by combustion temperatures.* The engine would operate on a *constant-pressure* cycle; expanding gases would keep precisely in step with the falling piston. This is a salient characteristic of Carnot's ideal gas cycle, and stands in contrast to the Otto cycle, in which combustion pressures rise so quickly upon spark ignition that we describe it as a *constant-volume* engine.
- *Adiabatic expansion.*
- *Instantaneous exhaust at constant volume.*

It is obvious that Diesel did not expect a working engine to attain these specifications. Adiabatic compression and exhaust phases are, by definition, impossible unless the engine metal is at combustion temperature. Likewise, fuel metering cannot be so precise as to limit combustion pressures to compression levels. Nor can a cylinder be vented instantaneously. But these specifications are significant in that they demonstrate an approach to invention. The rationale of the diesel engine was to save fuel by as close an approximation to the Carnot cycle as materials would allow. The steam, or *Rankine cycle*, engine was abysmal in this regard; and the *Otto* four-stroke-cycle spark or hot-tub e-ignition engine was only marginally better.

This approach, from the mathematically ideal to materially practical, is exactly the reverse of the one favored by inventors of the Edison, Westinghouse, and Kettering school. When Diesel visited America in 1912, Thomas A. Edison explained to the young inventor that these men worked *inductively*, from the existing technology, and not *deductively*, from some ideal or model. Diesel felt that such procedure was at best haphazard, even though the results of Edison and other inventors of the inductive school were obviously among the most important. Diesel believed that productivity should be measured by some absolute scientific standard.

The first Diesel engine was a single-cylinder four-cycle design, operated by gasoline vapor. The vapor was sprayed into the cylinder near top dead center by means of an air compressor. The engine was in operation in July of 1893. However, it was discovered that a misreading of the blueprints had caused an increase in the size of the chamber. This was corrected with a new piston, and the engine was connected to a pressure gauge. The gauge showed approximately 80 atmospheres before it shattered, spraying the room with brass and glass fragments. The best output of what Diesel called his "black mistress" was slightly more than 2 hp—not enough power to overcome friction and compression losses. Consequently, the engine was redesigned.

The second model was tested at the end of 1894. It featured a variable-displacement fuel pump to match engine speed with load. In February of the next year, the mechanic Linder noted a remarkable development. The engine had been sputtering along, driven by a belt from the shop power plant, but Linder noticed that the driving side of the belt was slack, indicating that the engine was putting power into the system. For the first time the Diesel engine ran on its own.

Careful tests—and Diesel was nothing if not careful and methodical—showed that combustion was irregular. The next few months were devoted to redesigning the nozzle and delivery system. This did not help, and in what might have been a fit of desperation, Diesel called upon Robert Bosch for an ignition magneto. Bosch personally fitted one of his low-tension devices to the engine, but it had little effect on the combustion problem. Progress came about by varying the amount of air injected with the fuel, which, at this time, was limited to kerosene or gasoline.

A third engine was built with a smaller stroke/bore ratio and fitted with two injectors. One delivered liquid fuel, the other a mixture of fuel and air. This was quite successful, producing 25 hp at 200 rpm. It was several times as efficient as the first model. Further modifications of the injector, piston, and lubrication system ensued, and the engine was deemed ready for series production at the end of 1896.

Diesel turned his attention to his family, music, and photography. Money began to pour in from the patent licensees and newly organized consortiums wanting to build engines in France, England, and Russia. The American brewer Adolphus Busch purchased the first commercial engine, similar to the one on display at the Budweiser plant in St. Louis today. He acquired the American patent rights for one million marks, which at the current exchange rate amounted to a quarter of a million dollars—more than Diesel had hoped for.

The next stage of development centered around various fuels. Diesel was already an expert on petroleum, having researched the subject thoroughly in Paris in an attempt to refine it by extreme cold. It soon became apparent that the engine could be adapted to run on almost any hydrocarbon from gasoline to peanut oil. Scottish and French engines routinely ran on shale oil, while those sold to the Nobel combine in Russia operated well on refinery tailings. In a search for the ultimate fuel, Diesel attempted to utilize coal dust. As dangerous as this fuel is in storage, he was able to use it in a test engine.

These experiments were cut short by production problems. Not all the licensees had the same success with the engine. In at least one instance, a whole production run had to be recalled. The difficulty was further complicated by a shortage of trained technicians. A small malfunction could keep the engine idle for weeks, until the customer lost patience and sent it back to the factory. With these embarrassments came the question of whether the engine had been oversold. Some believed that it needed much more development before being put on the market. Diesel was confident that his creation was practical—if built and serviced to specifications. But he encouraged future development by inserting a clause in the contracts that called for pooled research: the licensees were to share the results of their research on Diesel engines.

Diesel's success was marred in two ways. For one, he suffered exhausting patent suits. The Diesel engine was not the first to employ the principle of compression ignition; Akroyd Stuart had patented a superficially similar design in 1890. Also, Diesel had a weakness for speculative investments. This weakness, along with a tendency to maintain a high level of personal consumption, cost Rudolf Diesel millions. His American biographers, W. Robert Nitske and Charles Morrow Wilson, estimate that the mansion in Munich cost a million marks to construct at the turn of the century.

The inventor eventually found himself in the uncomfortable position of living on his capital. His problem was analogous to that of an author who is praised by the critics but who cannot seem to sell his books. Diesel engines were making headway in stationary and marine applications, but they were expensive to build and required special service techniques. True mass production was out of the question. At the same time, the inventor had become an international celebrity, acclaimed on three continents.

Diesel returned to work. After mulling a series of projects, some of them decidedly futuristic, he settled on an automobile engine. Two such engines were built. The smaller, 5-hp model was put into production, but sales were disappointing. The engine is, by nature of its compression ratio, heavy and, in the smaller sizes, difficult to start. (The latter phenomenon is due to the unfavorable surface/volume ratio of the chamber as piston size is reduced. Heat generated by compression tends to bleed off into the surrounding metal.) A further complication was the need for compressed air to deliver the fuel into the chamber. Add to these problems precision machine work, and the diesel auto engine seemed impractical. Mercedes-Benz offered a diesel-powered passenger car in 1936. It was followed by the Austin taxi (remembered with mixed feelings by travelers to postwar London), by the Land Rover, and more recently, by the Peugeot. However desirable diesel cars are from the point of view of fuel economy and longevity, they have just recently become competitive with gasoline-powered cars.

Diesel worked for several months on a locomotive engine built by the Suizer Brothers in Switzerland. First tests were disappointing, but by 1914 the Prussian and Saxon State Railways had a diesel in everyday service. Of course, most of the world's locomotives are diesel-powered today.

Maritime applications came as early as 1902. Nobel converted some of his tanker fleet to diesel power, and by 1905 the French navy was relying on these engines for their submarines. Seven years later, almost 400 boats and ships were propelled solely or in part by compression engines. The chief attraction was the space saved, which increased the cargo capacity or range.

In his frequent lectures Diesel summed up the advantages of his invention. The first was efficiency, which was beneficial to the owner and, by extension, to all of society. In immediate terms, efficiency meant cost savings. In the long run, it meant conserving world resources. Another advantage was that compression engines could be built on any scale from the fractional horsepower to the 2400-hp Italian Tosi of 1912. Compared to steam engines, the diesel was compact and clean. Rudolf Diesel was very much concerned with the question of air pollution, and mentioned it often.

But the quintessential characteristic, and the one that might explain his devotion to his "black mistress," was her quality. Diesel admitted that the engines were expensive, but his goal was to build the best, not the cheapest.

During this period Diesel turned his attention to what his contemporaries called "the social question." He had been poor and had seen the effects of industrialization firsthand in France, England, and Germany. Obviously machines were not freeing men, or at least not the masses of men and women who had to regulate their lives by the factory system. This paradox of greater output of goods and intensified physical and spiritual poverty had been seized on by Karl Marx as the key "contradiction" of the capitalistic system. Diesel instinctively distrusted Marx because he distrusted the violence that was implicit in "scientific socialism." Nor could he take seriously a theory of history whose exponent claimed it was based on absolute principles of mathematical integrity.

He published his thoughts on the matter under the title *Solidarismus* in 1903. The book was not taken seriously by either the public or politicians. The basic concept was that nations were more alike than different. The divisions that characterize modern society are artificial to the extent that they do not have an economic rationale. To find solidarity, the mass of humanity must become part owners in the sources of production. His formula was for every worker to save a penny a day. Eventually these pennies would add up to shares or part shares in business enterprises: Redistributed, wealth and, more important, the sense of controlling one's destiny would be achieved without violence or rancor through the effects of the accumulated capital of the workers.

Diesel wrote another book that was better received. Entitled *Die Enstehung des Dieselmotors*, it recounted the history of his invention and was published in the last year of his life.

For years Diesel had suffered migraine headaches, and in his last decade, he developed gout, which at the end forced him to wear a special oversized slipper. Combined with this was a feeling of fatigue, a sense that his work was both done and undone, and that there was no one to continue. Neither of his two sons showed any interest in the engine, and he himself seemed to have lost the iron concentration of earlier years when he had thought nothing of a 20-hour workday. It is probable that technicians in the various plants knew more about the current state of diesel development than he did.

And the bills mounted. A consultant's position, one that he would have coveted in his youth, could only postpone the inevitable; a certain level of indebtedness makes a salary superfluous. Whether he was serious in his acceptance of the English-offered consultant position is unknown. He left his wife in Frankfort in apparent good spirits and gave her a present. It was an overnight valise, and she was instructed not to open it for a week. When she did, she found it contained 20,000 marks. This was, it is believed, the last of his liquid reserves. At Antwerp he boarded the ferry to Warwick in the company of three friends. They had a convivial supper on board and retired to their staterooms. The next morning Rudolf Diesel could not be found. One of the crew discovered his coat, neatly folded under a deck rail. The captain stopped the ship's progress, but there was no sign of the body. A few days later a pilot boat sighted a body floating in the channel, removed a corn purse and spectacle case from the pockets, and set the-corpse adrift. The action was not unusual or callous; seamen had, and still do have, a horror of retrieving bodies from the sea. These items were considered by the family to be positive identification. They accepted the death as suicide, although the English newspapers suggested foul play at the hands of foreign agents who did not want Diesel's engines in British submarines.

2
CHAPTER

Diesel basics

At first glance, a diesel engine looks like a heavy-duty gasoline engine, minus spark plugs and ignition wiring (Fig. 2-1). Some manufacturers build compression ignition (CI) and spark ignition (SI) versions of the same engine. Caterpillar G3500 and G3600 SI natural-gas fueled engines are built on diesel frames and use the same blocks, crankshafts, heads, liners, and connecting rods.

But there are important differences between CI and SI engines that cut deeper than the mode of igniting the fuel.

Compression ratio

When air is compressed, collisions between molecules produce heat that ignites the diesel fuel. The compression ratio (c/r) is the measure of how much the air is compressed (Fig. 2-2).

Compression ratio = swept volume + clearance volume ÷ swept volume

Swept volume = the volume of the cylinder traversed by the piston in its travel from top dead center (tdc) to bottom dead center (bdc)

Clearance volume = combustion chamber volume

Figure 2-3 graphs the relationship between c/r's and thermal efficiency, which reaffirms what every mechanic knows: high c/r's are a precondition for power and fuel economy.

At the very minimum, a diesel engine needs a c/r of about 16:1 for cold starting. Friction, which increases more rapidly than the power liberated by increases in compression, sets the upper limit at about 24:1. Other inhibiting factors are the energy required for cranking and the stresses produced by high power outputs. Diesels with c/r's of 16 or 17:1 sometimes benefit from a point or two of higher compression. Starting becomes easier and less exhaust smoke is produced. An example is the

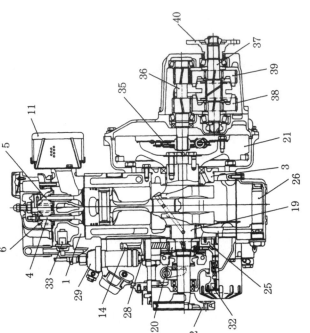

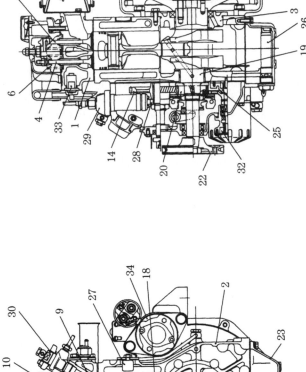

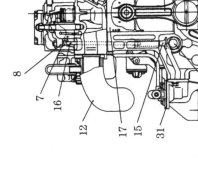

1. Cylinder head
2. Cylinder body
3. Main bearing housing
4. Exhaust valve
5. Intake valve
6. Valve spring
7. Valve rocker arm support
8. Valve rocker arm
9. Precombustion chamber
10. Decompression lever

11. Intake silencer
12. Mixing elbow
13. Camshaft
14. Camshaft gear
15. Tappet
16. Push rod
17. Piston
18. Connecting rod
19. Crankshaft
10. Crankshaft gear

21. Flywheel
22. Crankshaft V-pulley
23. Oil pan
24. Dipstick
25. Lubricating oil pump
26. Lubricating oil inlet pipe
27. Anti corrosion zinc
28. Fuel injection pump cam
29. Fuel injection pump
30. Fuel injection nozzle

31. Fuel feed pump
32. Cooling water pump
33. Thermostat
34. Starter motor
35. Damper disc
36. Input shaft
37. Output shaft
38. Forward large gear
39. Reverse large gear
40. Output shaft coupling

2-1 The Yanmar 1GM10, shown with a marine transmission, provides auxiliary power for small sailboats. The 19.4 CID unit develops 9 hp and forms the basic module for two- and three-cylinder versions.

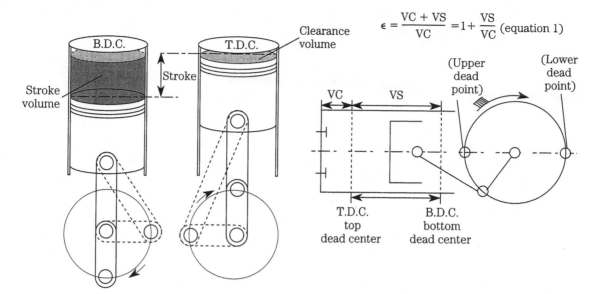

$$\epsilon = \frac{VC + VS}{VC} = 1 + \frac{VS}{VC} \quad (\text{equation 1})$$

2-2 Compression ratio is a simple concept, but one that mathematics and pictures express better than words.

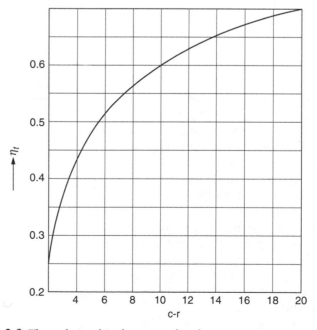

2-3 The relationship between diesel compression ratios and thermal efficiency.

Caterpillar 3208 that has a tendency to smoke and "wet stack," that is, to saturate its exhaust system with unburned fuel. These problems can be alleviated with longer connecting rods that raise the compression ratio from 16.5:1 to 18.2:1.

It should be noted that a compressor, in the form of a turbocharger or supercharger, raises the effective c/r. Consequently, these engines have c/r's of 16 or 17:1, which are just adequate for starting. Once the engine is running, the compressor provides additional compression.

Gasoline engines have lower c/r's—half or less—than CI engines. This is because the fuel detonates when exposed to the heat and pressure associated with higher c/r's. Detonation is a kind of maverick combustion that occurs after normal ignition. The unburned fraction of the charge spontaneously explodes. This sudden rise in pressure can be heard as a rattle or, depending upon the natural frequency of the connecting rods, as a series of distinct pings. Uncontrolled detonation destroys crankshaft bearings and melts piston crowns.

Induction

Modern SI engines mix air and fuel in the intake manifold by way of one or more low-pressure (50-psi or so) injectors. A throttle valve regulates the amount of air admitted, which is only slightly in excess of the air needed for combustion. As the throttle opens, the injectors remain open longer to increase fuel delivery. For a gasoline engine, the optimum mixture is roughly 15 parts air to 1 part fuel. The air-fuel mixture then passes into the cylinder for compression and ignition.

In a CI engine, air undergoes compression before fuel is admitted. Injectors open late during the compression stroke as the piston approaches tdc. Compressing air, rather than a mix of air and fuel, improves the thermal efficiency of diesel engines. To understand why would require a course in thermodynamics; suffice to say that air contains more latent heat than does a mixture of air and vaporized fuel.

Forcing fuel into a column of highly compressed air requires high injection pressures. These pressures range from about 6000 psi for utility engines to as much as 30,000 psi for state-of-the-art examples.

CI engines dispense with the throttle plate—the same amount of air enters the cylinders at all engine speeds. Typically, idle-speed air consumption averages about 100 lb of air per pound of fuel; at high speed or under heavy load, the additional fuel supplied drops the ratio to about 20:1.

Without a throttle plate, diesels breathe easily at low speeds, which explains why truck drivers can idle their rigs for long periods without consuming appreciable fuel. (An SI engine requires a fuel-rich mixture at idle to generate power to overcome the throttle restriction.)

Since diesel air flow remains constant, the power output depends upon the amount of fuel delivered. As power requirements increase, the injectors deliver more fuel than can be burned with available oxygen. The exhaust turns black with partially oxidized fuel. How much smoke can be tolerated depends upon the regulatory climate, but the smoke limit always puts a ceiling on power output.

To get around this restriction, many diesels incorporate an air pump in the form of an exhaust-driven turbocharger or a mechanical supercharger. Forced induction can double power outputs without violating the smoke limit. And, as far as turbochargers are concerned, the supercharge effect is free. That is, the energy that drives the turbo would otherwise be wasted out the exhaust pipe as heat and exhaust-gas velocity.

The absence of an air restriction and an ignition system that operates as a function of engine architecture can wrest control of the engine from the operator. All that's needed is for significant amounts of crankcase oil to find its way into the combustion chambers. Oil might be drawn into the chambers past worn piston rings or from a failed turbocharger seal. Some industrial engines have an air trip on the intake manifold for this contingency, but many do not. A runaway engine generally accelerates itself to perdition because few operators have the presence of mind to engage the air trip or stuff a rag into the intake.

Ignition and combustion

SI engines are fired by an electrical spark timed to occur just before the piston reaches the top of the compression stroke. Because the full charge of fuel and air is present, combustion proceeds rapidly in the form of a controlled explosion. The rise in cylinder pressure occurs during the span of a few crankshaft degrees. Thus, the cylinder volume above the piston undergoes little change between ignition and peak pressure. Engineers, exaggerating a bit, describe SI engines as "constant volume" engines (Fig. 2-4).

Compared to SI, the onset of diesel ignition is a leisurely process (Fig. 2-4). Some time is required for the fuel spray to vaporize and more time is required for the spray to reach ignition temperature. Fuel continues to be injected during the delay period.

Once ignited, the accumulated fuel burns rapidly with correspondingly rapid increases in cylinder temperature and pressure. The injector continues to deliver fuel through the period of rapid combustion and into the period of controlled combustion that follows. When injection ceases, combustion enters what is known as the afterburn period.

The delay between the onset of fuel delivery and ignition (A–B in Fig. 2-5) should be as brief as possible to minimize the amount of unburnt fuel accumulated in the cylinder. The greater the ignition lag, the more violent the combustion and resulting noise, vibration, and harshness (NVH).

Ignition lag is always worst upon starting cold, when engine metal acts as a heat sink. Mechanics sometimes describe the clatter, white exhaust smoke, and rough combustion that accompany cold starts as "diesel detonation," a term that is misleading because diesels do not detonate in the manner of SI engines. Combustion should smooth out after the engine warms and ignition lag diminishes. Heating the incoming air makes cold starts easier and less intrusive.

In normal operation, with ignition delay under control, cylinder pressures and temperatures rise more slowly (but to higher levels) than for SI engines. In his proposal of 1893, Rudolf Diesel went one step further and visualized constant pressure expansion: fuel input and combustion pressure would remain constant during the

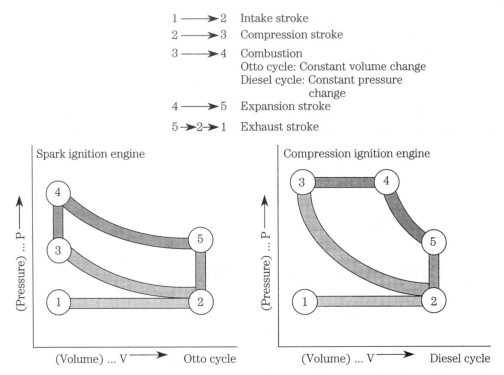

1 ———➤ 2	Intake stroke	
2 ———➤ 3	Compression stroke	
3 ———➤ 4	Combustion	
	Otto cycle: Constant volume change	
	Diesel cycle: Constant pressure	
		change
4 ———➤ 5	Expansion stroke	
5➤2➤1	Exhaust stroke	

2-4 These cylinder pressure/volume diagrams distort reality somewhat, but indicate why SI engines are described as "constant volume" and CI as "constant pressure."

expansion, or power, stroke. He was able to approach that goal in experimental engines, but only if rotational speeds were held low. His colleagues eventually abandoned the idea and controlled fuel input pragmatically, on the basis of power output. Even so, the pressure rise is relatively smooth and diesel engines are sometimes called "constant pressure" devices to distinguish them from "constant volume" SI engines (shown back at Fig. 2-4).

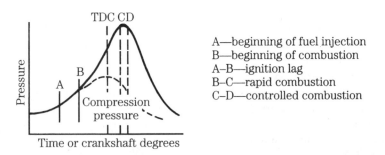

A—beginning of fuel injection
B—beginning of combustion
A–B—ignition lag
B–C—rapid combustion
C–D—controlled combustion

2-5 Diesel combustion and compression pressure rise plotted against crankshaft rotation.

Two- and four-stroke-cycle

CI and SI engines operate on similar cycles, consisting of intake, compression, expansion, and exhaust events. Four-stroke-cycle engines of either type allocate one up or down stroke of the piston for each of the four events. Two-stroke-cycle engines telescope events into two strokes of the piston, or one per crankshaft revolution. In the United States, the term *stroke* is generally dropped and we speak of two- or four-cycle engines; in other parts of the English-speaking world, the preferred nomenclature is two-stroke and four-stroke.

Four-cycle diesel engines operate as shown in Fig. 2-6. Air, entering around the open intake valve, fills the cylinder as the piston falls on the intake stroke. The intake valve closes as the piston rounds bdc on the compression stroke. The piston rises, compressing and heating the air to ignition temperatures.

Injection begins near tdc on the compression stroke and continues for about 40° of crankshaft rotation. The fuel ignites, driving the piston down in the bore on the expansion, or power stroke. The exhaust valve opens and the piston rises on the exhaust stroke, purging the cylinder of spent gases. When the piston again reaches tdc, the four-stroke-cycle is complete, two crankshaft revolutions from its beginning.

Figure 2-7 illustrates the operation of Detroit Diesel two-cycle engines, which employ blower-assisted scavenging. As shown in the upper left drawing, pressurized air enters the bore through radial ports and forces the exhaust gases out through the

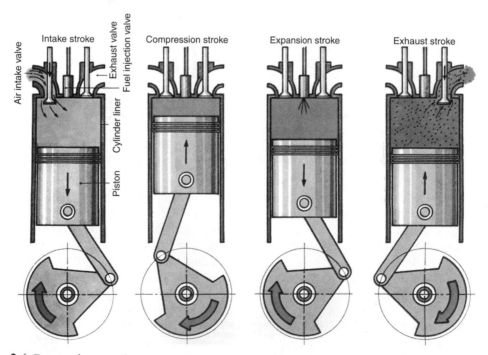

2-6 Four-cycle operation. Yanmar Diesel Engine Co. Ltd.

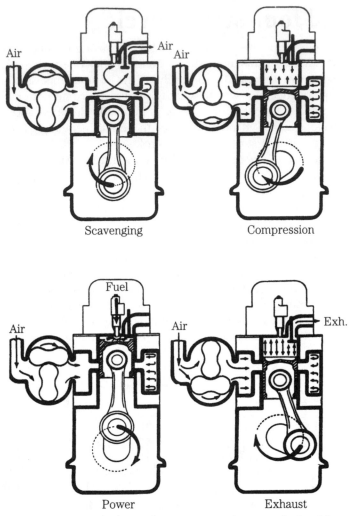

Scavenging

Compression

Power

Exhaust

2-7 Two-cycle operation. The Detroit Diesel engine depicted here employs a Roots-type positive-displacement blower for scavenging.

cylinder without raising its pressure much above atmospheric. The exhaust valve remains open until the ports are closed to eliminate a supercharge effect.

The exhaust valve then closes and the piston continues to rise, compressing the air charge ahead of it. Near tdc, the injector fires, combustion begins, and cylinder pressure peaks as the piston rounds tdc. Expanding gases drive the piston downward. The exhaust valve opens just before the scavenge ports are uncovered to give spent gases opportunity to blow down. These four events—intake, compression, expansion, and exhaust—occur in two piston strokes, or one crankshaft revolution.

Not all two-cycle diesel engines have valves. Combining scavenge air with combustion air eliminates the intake valve, and a port above the air inlet port replaces the exhaust valve. Such engines employ cross-flow or loop scavenging (Fig. 2-8) to

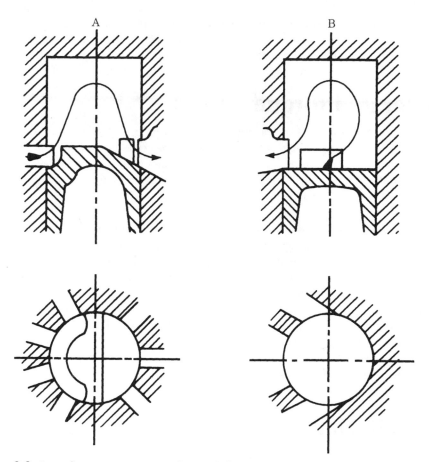

2-8 Cross-flow scavenging employs a deflector on the piston crown to divert the incoming air charge up and away from the exhaust port. Loop scavenging achieves the same effect with angled inlet ports.

purge the upper reaches of the cylinder and to minimize the loss of scavenge air to the exhaust. In the cross-flow scheme, a deflector cast into the piston crown diverts the incoming air stream away from the open exhaust port and into the stagnant region above the piston. The angled inlet ports on loop-scavenged engines produce the same effect.

It is also possible to eliminate the external air pump by using the crankcase as part of the air inlet tract. Piston movement provides the necessary compression to pump the air, via a transfer port, into the cylinder. Not many crankcase-scavenged engines are seen in this country, but the German manufacturer Fichtel & Sachs has built thousands of them.

Because two-cycle engines fire every revolution, the power output should be twice that of an equivalent four-stroke. Such is not the case, principally because of difficulties associated with scavenging. Four-cycle engines mechanically purge exhaust gases, through some 440° of crankshaft revolution. (The exhaust valve opens early during the expansion stroke and closes after the intake valve opens). Two-cycles scavenge in a less positive manner during an abbreviated interval of about 130° Consequently, some exhaust gas remains in the cylinder to dampen combustion.

Power and torque

Horsepower is the ability to perform work over time. In 1782, James Watt, a pioneer developer of steam engines, observed that one mine pony could lift 550 lb of coal one foot in one minute. Torque is the instantaneous twisting force applied to the crankshaft. In the English-speaking world, we usually express torque as pounds of force applied on a lever one foot long.

The two terms are related:

Horsepower = torque × 2pi × rpm. Revolutions per minute is the time component.

Torque = displacement × 4pi × bmep. The latter term, brake mean effective pressure, is the average pressure applied to the piston during the expansion stroke.

High-performance diesels, such as used in European automobiles, develop maximum horsepower at around 5000 rpm. Equivalent SI auto engines can turn almost twice as fast. Since rpm is part of the hp formula, these diesels fall short in the power department. An SI-powered car will have a higher top speed.

But, thanks to high effective brake mean pressures, diesels have the advantage of superior torque. A diesel-powered BMW or Mercedes-Benz easily out-accelerates its gasoline-powered cousins.

Fuel efficiency

High c/r's (or more exactly, large ratios of expansion) give CI engines superior thermal efficiency. Under optimum conditions, a well-designed SI engine utilizes about 30% of the heat liberated from the fuel to turn the crankshaft. The remainder goes out the exhaust and into the cooling system and lubricating oil. CI engines attain thermal efficiencies of 40% and greater. By this measure—which is becoming increasingly critical as fears about global warming are confirmed—diesel engines are the most efficient practical form of internal combustion. (Gas turbines do better, but only at constant speeds.)

Excellent thermal efficiency, plus the volumetric efficiency afforded by an unthrottled intake manifold and the ability to recycle some exhaust heat by turbocharging, translate into fuel economy. It is not unreasonable to expect a specific fuel consumption of 0.35 lb/hp-hr from a CI engine operating near its torque peak. An SI engine can consume 0.50 lb/hp-hr under the same conditions.

The weight differential between diesel fuels (7.6 lb/US gal for No. 2D) and gasoline (about 6.1 lb/US gal) gives the diesel an even greater advantage when consumption is figured in gallons per hour or mile. CI passenger cars and trucks deliver about 30% better mileage than the same vehicles with gasoline engines.

Diesel pickups and SUVs appeal in ways other than fuel economy. Owners of these vehicles tend to become diesel enthusiasts. I'm not sure why, but it probably has something to do with the sheer mechanical presence that industrial products radiate. Earlier generations had the same sort of love affair with steam.

Weight

The Cummins ISB Dodge pickup motor weighs 962 lb and develops 260 hp for a wt/hp ratio of 3.7:1. The 500-hp Caterpillar 3406E, a standard power plant for large (Grade-8) highway trucks comes in at 5.7 lb/hp. The Lugger, a marine engine of legendary durability, weighs 9.6 lb for each of its 120 horses. By comparison, the Chevrolet small block SI V-8 has an all-up weight of about 600 lb and with a bit of tweaking develops 300 hp.

Much of the weight of diesel engines results from the need to contain combustion pressures and heat that, near tdc, peak out at around 1000 psi and 3600°F. And, as mentioned earlier, bmep, or average cylinder pressures, are twice those of SI engines.

There are advantages to being built like a Sumo wrestler. Crankshaft bearings stay in alignment, cylinder bores remain round, and time between overhauls can extend for tens of thousands of hours.

Durability

Industrial diesel engines come out of a conservative design tradition. High initial costs, weight, and moderate levels of performance are acceptable tradeoffs against early failure. The classic diesel is founded on heavy, fine-grained iron castings, liberally reinforced with webbing and aged prior to machining. Buttressed main-bearing caps, pressed into the block and often cross-drilled, support the crankshaft. Pistons run against replaceable liners, whose metallurgy can be precisely controlled. Some of the better engines, such as the Cummins shown in Fig. 2-9, feature straight-cut timing gears, which are virtually indestructible.

Until exhaust gas recirculation became the norm, heavy truck piston rings could, on occasion, go for a million miles between replacements. An early Caterpillar 3176 truck engine was returned to the factory for teardown after logging more than 600,000 miles. Main and connecting-rod bearings had been replaced (at 450,000 and 225,000 miles, respectively) and were not available for examination. The parts were said to be in good condition.

The crankshaft remained within tolerance, as did the rocker arms, camshaft journals, and lower block casting. Valves showed normal wear, but were judged reusable. Connecting rods could have gone another 400,000 miles and pistons for 200,000 miles. The original honing marks were still visible on the cylinder liners.

But Caterpillar was not satisfied, and has since made a series of major revisions to the 3176, including redesigned pistons, rings, connecting rods, head gasket, rocker shafts, injectors, and water pump. Crankshaft rigidity has been improved, and tooling developed to give the cylinder liners an even more durable finish.

2-9 The Cummings ISB employs straight-cut timing gears that, while noisy, are practically indestructible. Gilmer-type toothed timing belts, typical of passenger-car diesels, need replacement at 60,000 miles or less.

Durability is not a Caterpillar exclusive: according to the EPA, heavy-truck engines have an average life cycle of 714,000 miles. Not a few Mercedes passenger cars have passed the three-quarter-million mile mark with only minor repairs.

This is not to say that diesels are zero-defect products. Industrial engines are less than perfect, and when mated with digital technology the problems multiply. Many of the worst offenders are clones, that is, diesels derived from existing SI engines. No one who was around at the time can forget the 1978 Oldsmobile Delta 88 Royale that sheared head bolts, crankshafts, and almost everything in between. Another clone that got off to a bad start was the Volkswagen. Like the Olds, it had problems with fasteners and soft crankshafts. But these difficulties were overcome. Today the VW TDi is the most popular diesel passenger-car engine in Europe, accounting for 40% of Volkswagen's production.

Conventional fuels

Diesel fuel is a middle distillate, slightly heavier than kerosene or jet fuel. Composition varies with the source crude, the refining processes used, the additive mix, and the regulatory climate. ASTM (American Society for Testing Materials) norms for Nos. 1-D and 2-D fuels in the United States are shown in Table 2-1.

Table 2-2 lists characteristics the EPA considers typical for Nos. 1-D and 2-D ULSD sold outside of California, which has its own, more rigorous rules. Note that EPA regulations apply only to sulfur content and to cetane number/aromatic content. Other

Table 2-1. ASTM diesel fuel grades

Grade	Characteristics	Sulfur content
No. 1-D S15 ULSD (ultra-low sulfur diesel)	ULSD is mandatory for use on all 2007-model and later road vehicles. Because of its high volatility No. 1-D ULSD is sometimes substituted for No. 2-D ULSD in cold climates.	15 ppm
No. 1-D S500	Obsolete and in process of phase-out. Damages emission control equipment on 2007 and later model vehicle engines	500 ppm
No. 2-D S15 ULSD	The standard fuel for vehicles and other small, high-speed engines. Produces slightly more power than No. 1-D ULSD	15 ppm
No. 2-D S500	Obsolete and in process phase-out. Damages emission control equipment on 2007 and later model vehicle engines.	500 ppm

fuel qualities, such as lubricity, filterability, and viscosity, are left to the discretion of the refiner. As a general rule, large truck stops provide the best, most consistent fuel.

- Cetane number (CN) and aromatic content refer to the ignition quality of the fuel. U.S. regulations permit 40 CN fuel if the aromatic content does not exceed 35%. In Europe diesel fuel must have a CN of at least 51. Aromatic content expresses the ignition quality of the fuel. High-octane fuels, such as aviation gasoline, have low CNs and barely support diesel combustion. Conversely, ether and amyl nitrate, which detonate violently in SI engines, are widely used as diesel starting fluids.
- API (American Petroleum Institute) gravity is an index of fuel density and, by extension, its caloric value. Heavier fuels produce more energy per injected volume.
- Viscosity also affects performance. Less viscous fuels atomize better and produce less exhaust smoke. But extremely light fuel upsets calibration by leaking past pump plungers. Thick, highly viscous fuels increase delivery pressures and pumping loads.
- Flash point, or the temperature at which the fuel releases ignitable vapors, is a safety consideration.

Table 2-2. ULSD fuel characteristics

	No. 1D	No. 2D
Cetane number	40–54	40–50
Gravity, °API	40–44	32–37
Sulfur, ppm	7–15	7–15
Min. aromatics, %	8	27
Min. flashpoint, °F	120	130
Viscosity, centistokes	1.6–2.0	2.0–3.2

3
CHAPTER

Engine installation

This chapter describes power requirements, mounting provisions, and alignment procedures for installing diesel engines in motor vehicles, stationary applications, and small boats. What I have tried to do here is to provide information that does not have wide currency, but is so critical that it makes or breaks the installation. Vendor catalogs serve for other aspects of the job, such as radiator/keel cooler sizing, selection of anti-vibration mounts, and sound-proofing techniques.

Trucks and other motor vehicles

Normally, installation is a bolt-on proposition, but things become complex when engines or transmissions are not as originally supplied.

Power requirements

Operators often judge a truck's power, or lack of it, by how fast the truck runs. In other words, operators look at maximum rated horsepower available at full governed rpm. But expected road speeds may be unrealistic. For example, numerically low axle ratios can, up to a point, increase top speed, but at the cost of reduced acceleration and less startability, a term that is defined below. Other factors that influence top speed are loaded weight, road conditions, wind resistance (which can double when loads are carried outside of the vehicle bodywork), and altitude. Naturally aspirated engines lose about 3% of their rated power per 1000 ft of altitude above sea level.

The desired cruising speed should be 10% to 20% below rated horsepower rpm, to provide a reserve of power for hill climbing and passing. When fuel economy is a primary consideration, the cruising speed can be set even lower. The power required at cruising speed is the engine's net horsepower.

Other factors to consider are the ability of the vehicle to cope with grades. Startability is expressed as the percentage grade the vehicle can climb from a dead stop. A fully loaded general-purpose truck should be able to get moving up a 10%

grade in low gear. Off-road vehicles should be able to negotiate 20% grades, with little or no clutch slippage. Startability is a function of the lowest gear ratio and the torque available at 800–1000 rpm.

Gradeability is the percentage grade a truck can climb from a running start while holding a steady speed. No vehicle claiming to be self-propelled should have a gradeability of less than 6%. Gradeability depends upon maximum torque the engine is capable of multiplied by intermediate gearing.

Caterpillar and other engine manufacturers can provide assistance for sizing the engine to the particular application. But it's useful to have some understanding of how power requirements are calculated.

The power needed to propel a vehicle is the sum of driveline losses, air resistance, rolling resistance, and grade resistance.

$$\text{driveline losses} = 1 - \text{driveline eff.} \times hp_{air} + hp_{roll} + hp_{grade}$$

driveline eff. = overall efficiency of the driveline, calculated on the assumption that each driven element—main transmission, auxiliary transmission, and rear axle— imposes an efficiency penalty of 4%. Thus, a truck with a single transmission and one driven rear axle would have an overall driveline efficiency of 92% ($0.96 \times 0.96 = 92\%$).

$$hp_{air} = \text{air resistance hp} = (mph^3 \div 375) \times 0.00172 \times \text{modifier} \times \text{frontal area}$$

Without some sort of provision to smooth airflow, the truck has a modifier of 1.0. If an aerodynamic device is fitted, the modifier is 0.60. For purposes of our calculation, frontal area = width in feet × (height in feet − 0.75 ft).

$$hp_{roll} = \text{rolling resistance hp} = GVW \times mph \times Crr$$

where GVW represents the gross vehicle weight in pounds, and Crr represents the rolling resistance. This latter figure depends upon tire type: on smooth concrete, bias-ply tires have a Crr of 17 lb/ton and radial tires 11 lb/ton. Low-profile tires do even better.

$$hp_{grade} = \text{grade hp} = (\text{grade percentage} \times GVW \times mph) \div 37,500$$

Motor mounts

In most instances, the technician merely bolts the engine down to a factory-designed mounting system. But there are times when engine-mounting provisions cannot be taken for granted.

Vehicle engines traditionally use a three-point mounting system, with a single point forward around which the unit can pivot, and with two points at the flywheel housing or transmission. For some engines the forward mount takes the form of an extension, or trunnion, at the crankshaft centerline. A sleeve locates the trunnion laterally, while permitting the engine to rotate. In order to simplify mounting and give more control over resiliency, other engines employ a rigid bracket bolted to the timing cover and extending out either side to rubber mounts on the frame.

Rear mounts normally bolt to the flywheel cover and function to locate the engine fore and aft, while transmitting the torque reaction to the vehicle frame. In order to control vibration, mount stiffness must be on the order of one-tenth of frame stiffness.

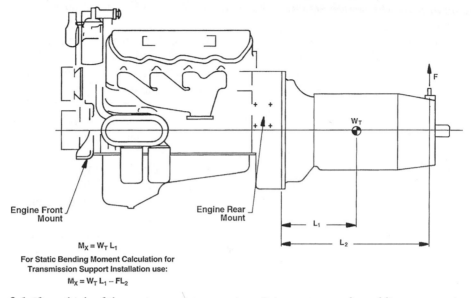

M$_X$ = Static Bending Moment at Flywheel
      Housing/Transmission
L$_1$ = Transmission CG Location
W$_T$ = Weight of Transmission
F = Rear Support Lift
L$_2$ = Rear Support Location

$$M_X = W_T L_1$$

**For Static Bending Moment Calculation for
Transmission Support Installation use:**

$$M_X = W_T L_1 - F L_2$$

3-1 If we think of the motor mounts as springs, it is easy to see that adding a transmission mount reduces the bending forces applied to the bell housing by the weight of the transmission. However for this to happen, the transmission mount must have a lower spring rate than the rear engine mounts. A high spring rate at the transmission neutralizes the rear motor mounts so that the whole weight of the engine and transmission is shared between the front motor mounts and the rear transmission mount. Bending forces increase. And, in practice, frame members adjacent to the transmission mount bend. Courtesy Caterpillar Inc.

On many applications the transmission cantilevers off the engine block without much additional support. The bending moment imposed by the overhung load on the flywheel housing should be calculated and compared against factory specs for the engine. Figure 3-1 illustrates the calculation for a transmission that receives some additional support at the rear with a third mount.

The third mount should have a vertical rate (lb/in. of deflection) considerably lower than the vertical rate of the rear engine mounts. A transmission mount with a higher spring rate than the engine mounts increases the bending moment. In addition, the high spring rate is almost sure to deflect the truck frame and, in the process, generate high forces on the engine/transmission package.

Off-road trucks present special problems, since engines and transmissions are subject to high gravity loads and the potential for frame distortion. Another factor that needs to be taken into consideration is that motor mounts must be able to absorb the torque reactions generated by the ultra-low gear ratios often specified for these vehicles.

Stationary engines

Power requirements

Power requirements for stationary applications can be difficult to calculate. Wherever possible, engine selection should be based upon experience and verified by tests in the field.

Caterpillar rates its industrial engines on a five-tier format based on load- factor duty cycle, annual operating hours, and expected time between overhaul. The load factor is a measure of the actual power output of the engine that, at any particular throttle setting, depends upon load. For example, an engine set to produce 300 hp will produce 50 hp under a 50-hp load, 100 hp under a 100-hp load, and so on. Fuel consumption increases with load demand. The load factor indicates how hard the engine works, and is calculated by comparing actual fuel usage with no-load usage at the throttle setting appropriate for the application.

- Industrial A—100% duty cycle under full load at rated rpm. Applications include pipeline pumping stations and mixing units for oilfield service.
- Industrial B—Maximum 80% duty cycle. Typical applications are oilfield rotary-table drives and drilling-mud, and cement pumps
- Industrial C—Maximum duty cycle 50%, with one hour at full load and speed, followed by one hour at reduced demand. Applications include off-road trucks, oilfield hoisting, and electric power generation for oil rigs.
- Industrial D—Maximum duty cycle not to exceed 10%, with up to 30 minutes of full load and power followed by 1 hour at part throttle. Used for offshore cranes and coiled-tubing drilling units where loads are cyclic.
- Industrial E—Maximum duty cycle not to exceed 5%, with no more than 15 minutes at full power, followed by one hour at reduced load. E-rated engines may need to develop full power at starting or to cope with short-term emergency demands.

All things equal, an oversized engine works at a lower load factor and should run longer between overhauls. Power in reserve also means that overloads can be accommodated without loss of rpm.

Power, the measure of the work the engine performs over time, is only part of the picture. Engines also need to develop torque, an instantaneous twisting force on the flywheel, commensurate with the torque imposed by sudden loads. Load-induced torque slows and, in extreme cases, stalls out the engine. The relationship between engine torque and load torque is known as the torque rise.

$$\text{Torque rise \%} = \frac{(\text{peak torque demand} - \text{rated engine torque} \times 100)}{\text{rated engine torque}}$$

If peak torque demand were twice that of engine torque, the torque-rise percentage would be only 50% and we could expect the engine to stumble and stall under the load. If, on the other hand, engine torque equaled load-induced torque, the engine would absorb the load without protest. However, high-torque-rise engines stress drivelines, mounts, and related hardware. Some compromise must be made.

It should also be noted that not all driven equipment impose sudden torque rises. For example, centrifugal pumps and blowers cannot lug an engine because the efficiency of these devices falls off more quickly with reduced speed than engine torque. Gen-sets run at constant speed and do not require much by way of torque rise. On the other hand, positive-displacement pumps generate high torques when pump output is throttled.

As the engine slows under load, the governor increases fuel delivery. Naturally aspirated engines respond quickly, since the air necessary to burn the additional fuel is almost immediately available. Turbocharged engines exhibit a perceptible lag as the turbo spools up. But naturally aspirated engines have difficulty in meeting emissions limits and, for the same power output, are heavier and more expensive than turbocharged models. Some industrial engines employ a small turbocharger for low-speed responsiveness and a second, larger unit for maximum power. Variable geometry turbocharging (VGT) virtually eliminates lag time.

The black smoke accompanying turbo lag can be reduced with an air/fuel ratio controller. Also known as a smoke limiter, the device limits fuel delivery until sufficient boost is present for complete combustion. Adjustment is a trade-off between transient smoke and engine responsiveness.

The engine manufacturer normally has final say on the mounting configuration that may consist of parallel rails or, in the case of several Caterpillar oilfield engines, a compound base. In the later arrangement the engine bolts on an inner base, which is suspended on springs above the outer base. The technician has the responsibility to see that engine mounts permit thermal growth and that engine and driven element are in dead alignment.

Thermal expansion

Cast iron "grows" less than steel when exposed to heat. The coefficient of expansion is 0.0000055 for cast iron and 0.0000063 for steel. A 94-in.-long iron engine block will elongate 0.083 in. as its temperature increases from 50° to 200° F. Under the same temperature increase, 94-in. steel mounting rails grow 0.089 in. These parts must be free to expand.

Caterpillar 3508, 3512, and 3516 oilfield engines mount on a pair of factory-supplied rails bolted to the oil pan. Standard procedure is to tie the engine down to the rail with a fitted bolt, that is, a bolt inserted with a light push fit into a reamed hole at the right rear corner of the oil pan. This bolt provides a reference point for alignment. Other engine-to-rail mounting bolts fit into oversized holes in the rails to allow for expansion. Clearance-type bolts should be 0.06 in. smaller than the diameter of the holes in the rails (Fig. 3-2). Chocks, used as an installation convenience to position the engine on the rails, must not constrain thermal expansion.

Alignment

Begin by cleaning all mating surfaces to remove rust, oxidation, and paint. If rubber couplings are present, remove them. It may be necessary to fabricate a dial-indicator holder from 1½-in. steel plate that can be bolted down to the machine. While this

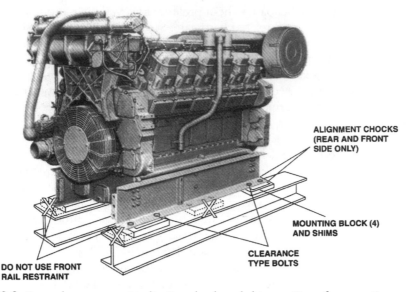

ALIGNMENT CHOCKS
(REAR AND FRONT
SIDE ONLY)

MOUNTING BLOCK (4)
AND SHIMS

CLEARANCE
TYPE BOLTS

DO NOT USE FRONT
RAIL RESTRAINT

3-2 General arrangement indicating chock and shim positions for mounting a stationary engine on rails. Courtesy Caterpillar Inc.

sounds like overkill, many commercially available magnetic indicator holders lack rigidity. Flex in the holder can be detected by the failure of the indicator to return to zero when the shaft is rotated back to the initial measurement position.

Parallel misalignment occurs when the centerlines of the engine crankshaft and driven equipment are parallel, but not in the same plane (Fig. 3-3). Mount a dial indicator on the engine flange with the point against the driven flange. Make several readings while a helper bars over the crankshaft in the normal direction of rotation.

The driven, or load, shaft should, as a general rule, be higher than the engine shaft. Engine main bearings typically have greater clearance than driven-equipment bearings. Until the engine starts, the crankshaft rests on the main-bearing caps, one or two thousandth of an inch below its running height. In addition, some allowance must be made for vertical expansion of the engine at operating temperature, which is nearly always more pronounced than the vertical expansion of the driven machinery.

Figure 3-4 illustrates a method of verifying dial-indicator readings on the outer diameters (ODs) of flanges and other circular objects. Zero the indicator at A in the drawing and, as the flange is rotated, make subsequent readings 90° apart. If the indicating surface is clean and the instrument mount secure, the needle will return to zero in the A position and B + D readings will equal C.

Measurement of angular misalignment can be made with a feeler gauge (Fig. 3-5). Once the problem is corrected by shimming, bolt the flanges together and verify that the crankshaft can move fore and aft a few thousandths of an inch against its thrust bearing.

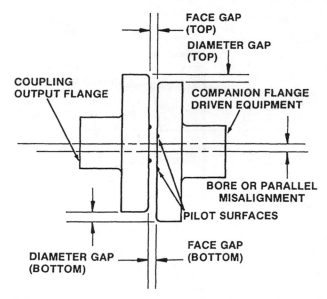

3-3 Parallel misalignment occurs when the centerlines of drive and driven equipment are parallel, but not in the same plane. Courtesy Caterpillar Inc.

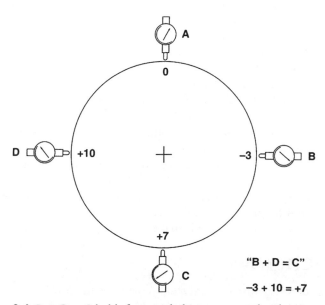

3-4 B + D = C holds for round objects measured with accurate dial indicators. Courtesy Caterpillar Inc.

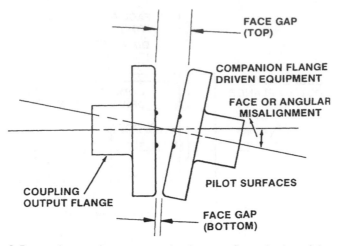

3-5 Angular misalignment nearly always reflects the lay of the shafts. But mis-machined flanges can also contribute to the problem. Courtesy Caterpillar Inc.

Parallel and angular misalignment can originate in driveline hardware, which rarely gets the scrutiny it deserves (Fig. 3-6). Parallel misalignment, called bore runout, refers to the lack of concentricity between the bore of a hub and the shaft centerline. Angular misalignment occurs when the mating face of a flange is not perpendicular with the shaft centerline. This sort of machining error is known as face runout.

Bore runout between the flywheel inner diameter (ID)) and the crankshaft pilot-bearing should be no more than 0.002 in., and no more than 0.005 in. for adaptors that make up to the flywheel. The maximum face runout for driveline components is 0.002 in.

Generators, transmissions, and other close-coupled driven elements that bolt directly to the flywheel housing can also suffer from misalignment. As a check, loosen the bolts securing the unit to the flywheel housing and measure the gap between the parts with a feeler gauge. The interfaces should be within 0.005 in. of parallel.

It is good practice to determine the bending moment of components cantilevered off the flywheel housing, as described previously under truck "Motor mounts."

Use brass shims (steel shims expand as they rust) and tighten the typical four-bolt mounting as shown in Fig. 3-7. Grade-8 hold-down bolts are torqued to specification, an operation that requires a large torque wrench and a torque multiplier in the form of a 3- or 4-ft cheater bar.

Marine engines

Because of the one-off nature of most small-boat construction, the technician is very much on his own when mounting an engine. Figure 3-8 provides an overview of the installation of Yanmar engines on sail and powered boats.

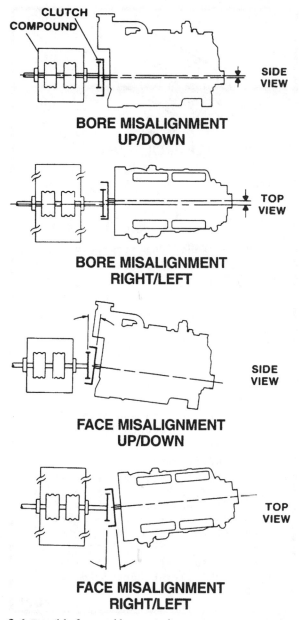

**BORE MISALIGNMENT
UP/DOWN**

**BORE MISALIGNMENT
RIGHT/LEFT**

**FACE MISALIGNMENT
UP/DOWN**

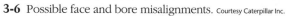

**FACE MISALIGNMENT
RIGHT/LEFT**

3-6 Possible face and bore misalignments. Courtesy Caterpillar Inc.

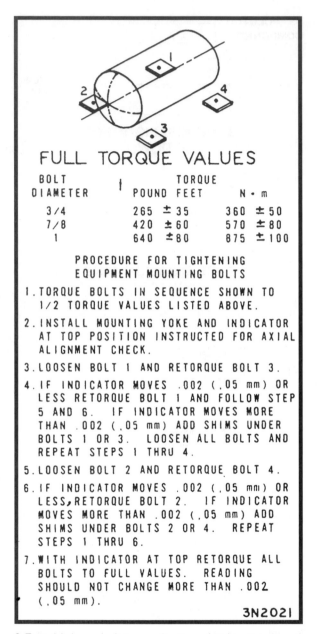

FULL TORQUE VALUES

BOLT DIAMETER	TORQUE POUND FEET	N·m
3/4	265 ± 35	360 ± 50
7/8	420 ± 60	570 ± 80
1	640 ± 80	875 ± 100

PROCEDURE FOR TIGHTENING EQUIPMENT MOUNTING BOLTS

1. TORQUE BOLTS IN SEQUENCE SHOWN TO 1/2 TORQUE VALUES LISTED ABOVE.

2. INSTALL MOUNTING YOKE AND INDICATOR AT TOP POSITION INSTRUCTED FOR AXIAL ALIGNMENT CHECK.

3. LOOSEN BOLT 1 AND RETORQUE BOLT 3.

4. IF INDICATOR MOVES .002 (.05 mm) OR LESS RETORQUE BOLT 1 AND FOLLOW STEP 5 AND 6. IF INDICATOR MOVES MORE THAN .002 (.05 mm) ADD SHIMS UNDER BOLTS 1 OR 3. LOOSEN ALL BOLTS AND REPEAT STEPS 1 THRU 4.

5. LOOSEN BOLT 2 AND RETORQUE BOLT 4.

6. IF INDICATOR MOVES .002 (.05 mm) OR LESS, RETORQUE BOLT 2. IF INDICATOR MOVES MORE THAN .002 (.05 mm) ADD SHIMS UNDER BOLTS 2 OR 4. REPEAT STEPS 1 THRU 6.

7. WITH INDICATOR AT TOP RETORQUE ALL BOLTS TO FULL VALUES. READING SHOULD NOT CHANGE MORE THAN .002 (.05 mm).

3N2021

3-7 Hold-down bolt torque limits and tightening procedure. Courtesy Caterpillar Inc.

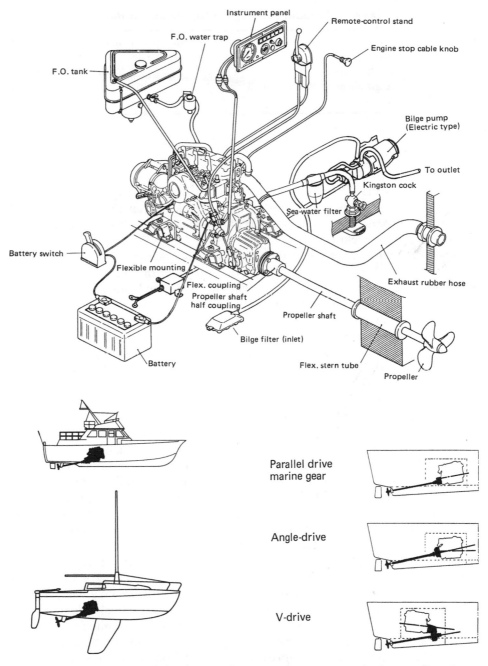

3-8 Engine-room layout and driveline configurations for Yanmar and other small engines.

Table 3-1. Propeller-shaft angles relative to the waterline

Drive type	Recommended angle	Max. permissible angle
Parallel drive	10°	15°
Angle drive	0°–3°	8°
V-drive	5°–8°	15°

Mounts

Inboard engines should be mounted level port-to-starboard and inclined no more than 15° off the horizontal (Table 3-1). Higher crankshaft angles reduce drive efficiency and increase the likelihood of oil starvation for engines with forward mounted oil-pump pickup tubes.

Your supplier will assist in determining propeller diameter, pitch, and style. As installed, the propeller should be at least one propeller diameter below the waterline and permit the engine to achieve maximum rated speed under full load at wide-open throttle. If the engine overspeeds by more than 5%, the propeller is undersized.

Once you have determined the inclination angle at which the engine will be installed, construct the bed, which should have as large a footprint as hull construction permits. Leave room around the engine for servicing and, if possible, arrange matters so that the engine can be removed without structural alterations to the vessel. There must also be sufficient clearance between the flywheel cover/transmission and hull for compression of the flexible motor mounts.

Determine the height of the crankshaft centerline with the engine resting on its mounts. You should be able to get this data from the engine maker. The next and very critical operation is to mark holes for the prop shaft. Some yards use a laser alignment tool, while others rely upon the more traditional approach described here.

Construct a wooden jig, like the one shown in Fig. 3-9, and secure it to the forward engine-room bulkhead. Run a string from the jig to the stern box. Verify that the inclination is correct and that the string is on the vessel centerline. Measure the vertical distance between the string and hull to establish that the transmission, clutch, and other components have sufficient clearance.

Using the string as a guide, mark stern-tube hole locations on the forward and aft sides of stern, or stuffing, box. Drill small holes where indicated on both sides of the box. Pass the string through these holes, which almost certainly will not center in the previously drilled holes. Figure 3-10 illustrates how the final hole dimensions are arrived at. Hole diameter should exactly match stern-tube diameter.

At this point, the crankshaft centerline and stern tube should be in fairly close alignment. Seal the stuffing-box holes with silicone, bolt down the stern tube, and secure the free end of the string to the center of the prop shaft. Using a fabricated marker, align the shaft with the string (Fig. 3-11).

The next step is to fabricate an engine-mounting jig with holes for the string at crankshaft location (Fig. 3-12). Place the jig on the engine bed and adjust the height and position of the mounts as necessary.

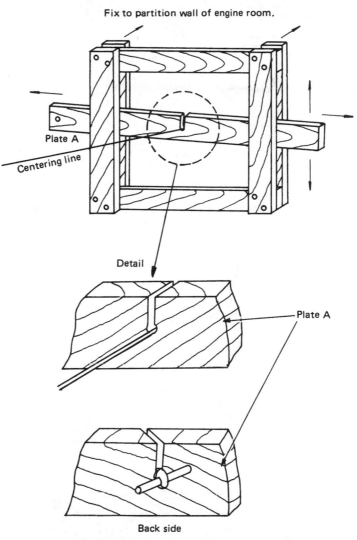

Fix to partition wall of engine room.

Plate A

Centering line

Detail

Plate A

Back side

3-9 An alignment jig mounts on the engine-room forward bulkhead.

Finally, mount the engine and align the prop-shaft flange with the transmission flange as described previously for stationary engines.

Exhaust systems

Wet exhaust systems reduce engine-room heat and simplify installation by allowing oil- and heat-resistant flexible hose to be used aft of the mixing elbow (Fig. 3-13). Basic specifications are:

- Exhaust outlet at transom—at least 6 in. above the loaded draft line.
- Elbow inboard of the transom—at least 14 in. higher than the draft line.

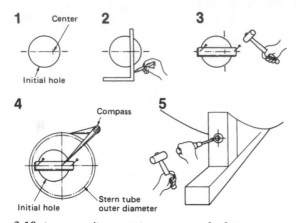

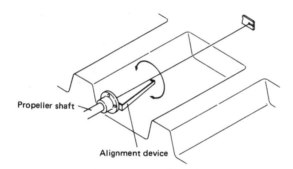

3-10 Accurate alignment is a process of refining approximations: the boat builder drills pilot hole through both faces of the stern box and threads the string through it for a better fix on shaft alignment.

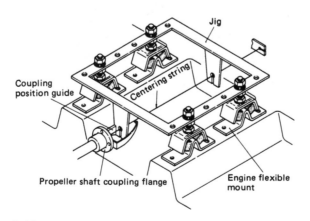

3-11 In preparation for final alignment, a home-made indicator is affixed to one of the flange bolt holes and the shaft centered over the string.

3-12 A jig that replicates the crankshaft-centerline/motor-mount relationship helps to avoid the embarrassment of redrilling engine-mount holes. Once the shaft aligns with the jig, all that remains is to verify flange alignment as described for stationary engines.

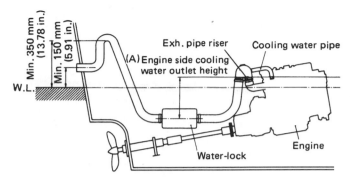

3-13 Exhaust system showing major elements and dimensions above the loaded draft line. The draft line is a convenient marker, but the critical reference is the waterline. Power boats assume a stern-down attitude at speed, which raises the water line. In the application shown, a riser is used to elevate the mixing elbow and the water-injection point 6 in. or so above the waterline.

- Muffler—inboard of the elbow, with the outlet inclined about 8° off the horizontal.
- Water trap—located at the lowest point of the system between the muffler and engine. While a water trap is recommended for all installations, the backpressure developed by the unit can result in turbocharger surge. If this condition occurs, consult the engine manufacturer. It may be necessary to derate the engine by fitting a smaller propeller.
- Cooling-water injection point—6 in. or more above the draft line to prevent water in the hose from flowing back into the engine upon starting and after shutdown. When necessary, a riser can be used to raise the height of the mixing elbow.

4
CHAPTER

Basic troubleshooting

This chapter provides basic troubleshooting information that should help you narrow the source of the problem down to a particular system—fuel, air, starting, or exhaust. Once this is done, turn to the chapter that deals with the system for additional troubleshooting and repair information.

The classic engine was essentially a mechanical device that failed in obvious ways. A few simple tests were enough to establish the source of the difficulty. Newer engines are controlled by a computer. Sensors provide the computer with data on engine and environmental conditions, and actuators carry out the computer's commands, most of which are directed to the fueling system. Failure of any of these components—an erratic sensor, a slow-to-respond actuator, or a loose pin in the wiring harness—affects the way the engine runs. Serious failures, such as loss of the timing signal, shut the engine down.

These electronic engine management systems (EMS) compensate, in part, for their complexity by setting trouble codes when the computer senses something is wrong. The first step in diagnosing computerized engines is to retrieve the trouble codes with the appropriate scan tool. The second step is to gather enough knowledge about the particular system to understand what the trouble codes mean. Chapter 6 describes the functioning of these systems in some detail with emphasis on Caterpillar HEUIs and Ford Power Strokes. In other respects, computer-controlled engines are prone to the same malfunctions as precomputer models. Injector pumps fail, cylinders lose compression, head gaskets leak, and so on.

Again, let me stress that the first step in diagnosing EMS engines is to read the trouble codes. If you merely want to keep an eye on the condition of your engine, a $200 scanner that retrieves the codes and erases them with a push of the button is adequate. If you are serious about repairing EMS engines, you will need to make a serious investment in tooling and documentation as described in Chap. 6. A professional-quality scanner will read active codes—those in effect now—and inactive, or historical, codes that have been set and later erased. The tool will also retrieve events a few seconds before and after each trouble code was set, shut down individual injectors, and report what the real-time data sensors transmit to the computer. Without these capabilities, you will be working blind.

Malfunctions

Common malfunctions are listed below together with probable causes and remedial actions. Fuel-related malfunctions can affect the color of the exhaust smoke, as indicated in Table 4-1. When taken together, engine behavior and smoke color provide a good indication of the source of the problem.

Engine cranks slowly, does not start

Starting system malfunction	Recharge batteries if cranking voltage drops below 9.5V or electrolyte reads less than 1.140 when tested with a hydrometer. Clean battery terminals. Perform starting system check (Chap. 11).
High parasitic loads	Check for binds in driven equipment, overly tight drive belts, shaft misalignments.
Crankshaft viscosity bound	Dark, sticky residue on the dipstick can indicate presence of antifreeze (ethylene glycol) in oil. Have oil analyzed.

Engine cranks normally, does not start

EMS sensor or actuator failure	Retrieve trouble codes. See Chap. 6.
No fuel to injectors	Check for restrictions or air leaks in fuel system, low injector pressure.
Contaminated fuel	Flush fuel system.
Glow-plug failure	Check glow-plug supply circuit, glow-plug control module and individual plugs.
Air inlet restriction	Replace air filter element.
Exhaust restriction	Inspect piping, if necessary perform backpressure test.
Loss of cylinder compression	Test cylinder compression.

Engine starts normally, runs no more than a minute or two, and shuts down

Intermittent EMS-related failure, bad harness connection	Retrieve trouble codes. See Chap. 6.
Air in fuel	Bleed system and check for air leaks.
Fuel return line restricted	Disconnect line to verify flow and remove obstruction.
Air inlet restriction	Replace filter element.
Clogged fuel filter	Replace filter element.

Engine starts normally, misfires

EMS-related failure	Retrieve trouble codes. See Chap. 6.
Air in fuel	Bleed system, check for air leaks.
Air inlet or exhaust system restriction	Replace air filter, check for excessive backpressure.
Clogged fuel filter	Replace filter element.

Malfunctioning injector(s)   Replace injector(s).
Loss of compression in one or   Check cylinder compression.
  more cylinders

Engine fails to develop normal power

EMS-related fault   Retrieve trouble codes. See Chap. 6.
Insufficient fuel supply   Replace filters, check transfer pump output, cap vent, and air in the system.
Contaminated fuel   Verify fuel quality.
Injection timing error   Check injector-pump timing.
High-pressure fuel system malfunction   Inspect high-pressure fuel system for leaks and air entrapment. Verify pump output pressure. If necessary have pump recalibrated.
Air inlet restriction   Change air-filter element.
Exhaust restriction   Inspect exhaust piping, test for excessive back-pressure.
Insufficient turbo boost   Change air-filter element, check boost pressure and for exhaust leaks upstream of the turbo.
Loss of engine compression   Check engine cylinder compression.

Table 4-1. Diagnosis by exhaust smoke color

Black or dark gray smoke Symptom	Probable cause	Remedial action
Smokes under load, especially at high and medium speed. Engine quieter than normal.	Injector pump timing retarded.	Set timing.
Smokes under load, especially at low and medium speed. Engine noisier than normal.	Injector pump timing advanced.	Set timing.
Smokes under load at all speeds, but most apparent at low and medium speeds. Engine may be difficult to start.	Weak cylinder compression.	Repair engine.
Smokes under load, especially at high speed.	Restricted air cleaner.	Clean/replace air filter element.
Smokes under load, noticeable loss of power.	Turbocharger malfunction.	Check boost pressure.
Smokes under load, especially at high and medium speeds. Power may be down.	Dirty injector, nozzle(s).	Clean/replace injectors.
Smokes under load, especially at low and medium speeds. Power may be down.	Clogged/restricted fuel lines.	Clean/replace fuel lines.
Puffs of black smoke, sometimes with blue or white component. Engine may knock.	Sticking injectors.	Repair/replace injectors.

(Continued)

Table 4-1. Diagnosis by exhaust smoke color (*Continued*)

Blue, blue-gray, or gray-white smoke

Symptom	Probable cause	Remedial action
Whitish or blue smoke at high speed and light load, especially when engine is cold. As temperature rises, smoke color changes to black. Power loss across the rpm band, especially at full throttle.	Injector pump timing retarded.	Set timing.
Whitish or blue smoke under light load after engine reaches operating temperature. Knocking may be present.	Leaking injector(s).	Repair/replace injector(s).
Blue smoke under acceleration after prolonged period at idle. Smoke may disappear under steady throttle.	Leaking valve seals.	Replace seals, check valve guides/stems.
Persistent blue smoke at all speeds, loads and operating temperatures.	Worn rings/cylinders.	Overhaul/rebuild engine.
Light blue or whitish smoke at high speed under light load. Pungent odor.	Over-cooling.	Replace thermostat.

Tests

Fuel quality

Take a fuel sample from some convenient point upstream of the filter/water separator. Allow the fuel to settle for a few minutes in a glass container and inspect for cloudiness (an indication of water), algae (jellylike particles floating on the surface), and solids. Placing a few drops of fuel between two pieces of glass will make it easier to see impurities. If fuel appears contaminated, drain and purge the system as described below.

Assuming that the fuel is clean and water-free, the most important quality is its cetane value, which can be determined with fuel hydrometer. A good No. 2 diesel has a specific gravity of 0.840 or higher at 60°F. No. 1 diesel can reduce power outputs by as much as 7% over No. 2 fuel. Blending the two fuels, a common practice in cold climates, results in correspondingly less power.

Fuel system

Most failures involve the fuel system. This system consists of three circuits, each operating at a different pressure.

- Low-pressure circuit. This includes the tank strainer, in-tank pump (on many vehicles), water-fuel separator, filter(s), and lift pump. Pressures vary with the application, but rarely exceed 75 psi.
- Fuel-return circuit. Operating at almost zero pressure, the return line conveys surplus fuel from the injectors back to the tank, filter, or to the inlet side of injector pump. Many of these systems include a restrictor orifice between the fuel-supply line and the return line. The orifice can clog, closing off the fuel return. Air leaks become a matter of concern when fuel is returned to the filter or injector pump, and not to the tank.
- High-pressure circuit. As shown in Fig. 4-1, the high-pressure circuit connects the discharge side of the injector pump with the injectors. For many applications, the connecting plumbing takes the form of dedicated lines from pump to the individual injectors. A more modern approach is to supply the injectors from a common manifold, or rail. The third alternative does away with the external high-pressure circuit by employing unit injectors (UIs). Each UI is a self-contained unit, with its own high-pressure pump operated by an engine-driven camshaft or, in the case of the Hydraulic Electronic Unit Injector (HEUI), by oil pressure. The unit pump system (UPS) is a hybrid, with each injector served by a dedicated pump plunger operating off the engine camshaft. Plunger-to-injector lines are pressurized.

Diesel-Einspritzsysteme von Bosch

VP44
Radialkolben-
Verteiler-
einspritzpumpe

CRS
Common-Rail-
System

UIS
Unit-Injector-
System

UPS
Unit-Pump-
System

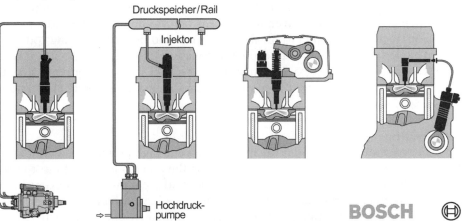

BOSCH

4-1 This drawing illustrates the high-pressure circuits for four modern fuel systems. From a mechanic's point of view, the UI system has much to recommend it. High pressures are confined within the injector bodies and failures tend to be cylinder-specific. (Photo Bosch)

A mechanic has a duty to himself to know the pressures he is dealing with. Most, but not all, older-model engines had fuel systems that operated in the neighborhood of 6000–12,000 psi. But even at 6000 psi, fuel easily penetrates the skin (witness the speckled hands of old diesel mechanics) and often results in blood poisoning. Common-rail and other modern systems generate pressures in excess of 20,000 psi, and the Honda 2.2-L, scheduled for introduction before 2009, develops 30,000 psi. Pressures on this order cut to the bone and, if air bubbles are present, the fuel jets out like water from a hose.

WARNING: Wear eye protection and heavy gloves when working on diesel fuel systems of any vintage. Do not attempt to disable modern ultrahigh-pressure injectors by cracking fuel-line connections.

Hydraulically actuated HEUI injectors shift the high-pressure regime to lube oil supplied at between 800 and 3300 psi, which is still a considerable pressure. These and other electronic injectors (recognized by the presence of wires running to them) pose the risk of electroshock. Injector voltages and amperages can be lethal.

WARNING: Do not disconnect the wiring to electronic injectors while the engine is running.

For fuel to enter the cylinders, the h-p pump must generate enough pressure to unseat the injectors. Pop-off pressure, known more formally as NOP (nozzle opening pressure) or VOP (valve opening pressure), depends upon an unrestricted fuel supply. Air leaks or fuel blockages upstream of the pump, or failure of the pump itself can reduce delivery pressure below NOP. Electronic engine management systems keep close watch on delivery pressure and its effect on other variables, such as the concentration of oxygen in the exhaust. Low delivery pressure triggers one or more trouble codes.

Older, precomputer engines require a more proactive approach to determine if fuel is getting to the cylinders. One technique is to spray starting fluid into the air intake while cranking. If the engine starts, runs for few seconds and stalls, you can be reasonably confident that the problem lies in the fuel system.

CAUTION: Employ starting fluid with discretion. Large amounts of ether in the intake manifold or in the scavenging system on two-cycle engines can result in explosions powerful enough to break piston rings and bend connecting rods.

An alternative approach is to crank the engine over for a 20 or 30 seconds, and crack the fuel-supply line to one or more injectors. Fuel should be present. Note that this technique should be confined to older engines known to generate moderate fuel pressures of 6000 psi or so.

Bleeding

Bleeder screws or Schrader valves are located at high points in the filter assemblies (Fig. 4-2) and in one or more locations on the accessible side of most injector pumps (Fig 4-3). Carefully clean the screws and adjacent areas to prevent foreign matter from entering the fuel circuit. Many engines have a hand-operated bleeder pump as part of the lift pump; others must be primed by cranking. Detroit Diesel supplies a portable hand-pump for engines not equipped with bleeder pumps.

Build a few psi of pressure in the system and, working from the tank forward, loosen the bleeder screws. Tighten the screws as soon as fuel flows in an uninterrupted

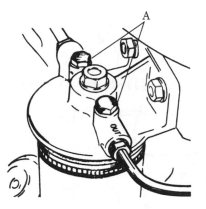

4-2 Fuel filter bleed screws. Lehman Ford Diesel.

stream. Run the engine for at least 20 minutes to ensure that the fuel system is completely purged. If the no-start or hard-start problem recurs after the engine has cooled down, the problem is almost surely an air leak in the low-pressure (injector-pump-supply) circuit.

Low-pressure circuit tests

Failure is most often associated with flow restrictions and air or fuel leaks, although failure of the lift pump is not unknown.

The tests that follow, developed by Ford for light trucks, locate flow restrictions by measuring the pressure drops across individual components while the engine is running. In other words, Ford uses the engine as a test bench. Other manufactures would have the mechanic disassemble the system and test each component separately. If you want to use the Ford approach—which has advantages—establish baseline pressure drops while the system is still healthy and operating normally.

Which gauge will be used depends upon the fuel-filter location relative to the lift pump. If the filter is located on the suction side of the lift pump, a 0–30 in.

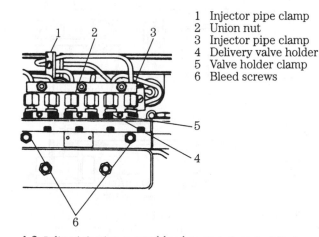

1 Injector pipe clamp
2 Union nut
3 Injector pipe clamp
4 Delivery valve holder
5 Valve holder clamp
6 Bleed screws

4-3 Inline injection pump bleed screws. Ford Industrial Engine and Turbine Operations.

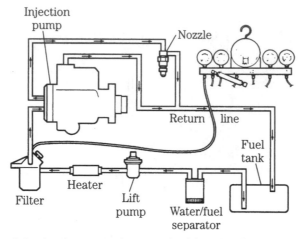

4-4 Filter location relative to the lift pump determines whether a pressure or vacuum gauge is used. In this example, the filter is on the outlet side of the pump and sees positive pressure.

Hg vacuum gauge will be needed. If downstream of the pump, the Ford application calls for 0–15 psi pressure gauge. Other lift pumps develop higher pressures.

1. Connect the appropriate gauge at the output side of the filter (Fig. 4-4). Start the engine and note the gauge reading at 2500 rpm.
2. Connect the gauge to the inlet side of the filter, repeat the test, and compare these readings against the pressure drop for a new filter (Fig 4-5).
3. Use the same procedure for detecting restrictions in the water/fuel separator, heater, and other components.

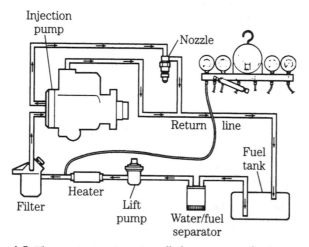

4-5 The next step is to install the gauge at the input side of the filter. The difference between output-side and input-side readings represents the pressure drop across the filter.

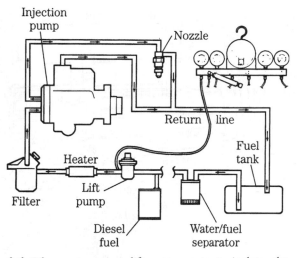

4-6 When measuring lift-pump output, isolate the pump from suction-side pressure drops by providing fuel from a separate source.

The lift pump, also known as the transfer or supply pump, moves fuel from the tank to the injector pump. Figure 4-6 illustrates the Ford approach to measuring lift-pump output pressure. A container of diesel fuel supplies the pump, thus isolating it from possible suction-side restrictions. However, it is usually enough to connect the gauge in parallel with pump output.

Pump pressure is only part of the picture. Manufacturers should, but rarely do, provide volume specifications for this and other pumps. If you're familiar with the engine, some rough idea of output volume can be had by monitoring fuel flow leaving the return line.

Fuel leaks leave a trail, but some mechanics like to check connections with soapy water or the glycerin-based solutions used to detect Freon leaks.

Air can enter on the suction side of the lift pump or when voids develop in the solid column of fuel. The latter condition occurs if the tank runs dry, filter elements are changed or, after shutdown, when fuel cools and contracts in filter canisters.

Entrapped air causes the injector pump to lose prime. Symptoms include reluctance or outright refusal to start, and sudden shutdowns within seconds of starting. Depending upon the amount of air intrusion, the engine may start more or less normally, but refuse to idle.

The time required for air to reach the injector pump gives an indication of the leak source. If the engine dies within seconds of starting, the source of the air must be nearby. Layouts vary, but common practice is to mount the filter and water separator in close proximity to the injector pump. Check these items first. Fuel-return lines that discharge into the filter or suction side of the injector pump present a major leak hazard.

Another technique is to insert a length of clear plastic fuel line of the type used on motorcycles between the lift and injector pumps (Fig. 4-7) or at any location in

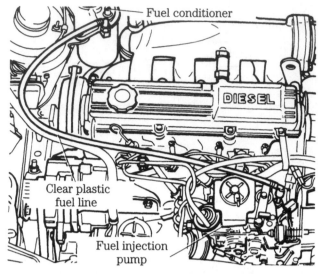

4-7 A length of clear plastic tubing spliced into the fuel line between the lift and injection pumps shows if air is present.

the low-pressure circuit where leaks are suspected. Start the engine and look for bubbles in the fuel stream.

You can also test for air leaks by disconnecting the line at the tank and pump. Plug one end and connect a hand pump or other source of low-pressure air to the free end. Pressurize to no more than 15 psi and apply soapy water or a commercial leak detector solution to the line connections, filter-canister parting surfaces, and other potential leak sources.

CAUTION: Do not pressurize the fuel tank—it might rupture.

Once the leak has been found and repaired, renew o-ring seals on connections, assemble all threaded connections with Loctite 515 Gasket Eliminator, and bleed the system as described above.

High-pressure circuit tests

High-pressure systems fail because of fuel leaks (air leaks are uncommon), malfunctioning injectors, or inadequate pump pressure.

Fuel-pipe unions on engines with pump-fed injectors are a major leak source. Loosening and retightening the union usually corrects the problem. See Chap. 5 for additional information.

Some engines are more prone to leakage than others. For example, hard-used White trucks tend to develop small leaks at the fuel delivery valves, located under the fuel-pipe connections on UTDS, APD, 6BB, T and Q series injector pumps. Replace the o-ring seal and copper ring gasket, and inspect lapped surfaces for channeling. The UTDS pumps in question develop more than 15,000 psi. Crimps or flattened spots in the fuel lines, over-fueling, and clogged injectors will further increase delivery pressure.

WARNING: Wear eye protection and keep hands clear of high-pressure fuel spray. Use a piece of cardboard to detect leaks in hard-to-see places.

Current EM systems focus on overall performance, rather on events occurring within individual cylinders. These systems know when a cylinder goes down or when electronic injectors lose power, but are blind to the roughness and loss of responsiveness associated with skewed spray patterns, fuel dribble, and the like. For that we need more traditional methods.

Symptoms of injector failure include

- Ragged idle or misfiring at certain speeds
- Black, gray, or white smoke
- Vibration and knock

If the injector fails completely, the exhaust runner for that cylinder will be noticeably cooler than the others immediately after startup. Glow-plug resistance, which increases with temperature, will be lower for the affected cylinder. On some engines with remote injector pumps, it is possible to feel the sudden contraction of the fuel pipe as a healthy injector snaps open. The piping to a malfunctioning injector feels inert by comparison. A more direct approach is to disable suspect injectors. This is done by depressing unit-injector cam followers with a pry bar, cracking the fuel-inlet connections on low-pressure pump-fed injectors, or by denying power to solenoid or piezo injectors. A good scan tool will disable electronic injectors safely and on demand. If the injector has failed, taking it out of action has little or no effect upon engine rpm.

The best way of verifying injector performance is to substitute known good injectors and observe the effect on idle, smoothness, and smoke production.

As described in Chap. 5, mechanical injectors can often be repaired in the field. Repairs of electronic injectors are best left to specialists who have the necessary test equipment and parts inventory.

The classic symptom of injector pump failure is low delivery pressure. If the engine is to start, NOP must be attained at cranking speed. If it is to develop maximum torque, the pump must be capable of delivering its full rated pressure and volume. Low or no pump pressure on computer-controlled engines is most often the result of a failed crankshaft position sensor. The throttle position sensor is the next most likely suspect.

Precomputer pumps are, for the most part, rpm-dependent. Test output pressure by connecting a suitable gauge to the test port, usually found at the discharge end of the pump (Fig. 4-8) or at the pump side of the return-line restrictor. Many of these engines also include a key-operated fuel shutoff downstream of the pump. A typical example is shown in Fig. 4-9. And most modern pumps include a bellows or diaphragm-operated altitude-compensator mounted on the pump body (see Chap. 5 for more information). A pin-hole leak in the diaphragm can mimic pump failure.

The governor section of mechanical pumps includes an array of adjustment screws, most of which are sleeping dogs, better left undisturbed. The two that a mechanic needs to know about, set the idle speed and maximum governed speed, usually by restricting the movement of the throttle control lever. The adjustments shown in Fig. 4-10 are fairly standard, although more elaborate designs sometimes split idle speed into two ranges.

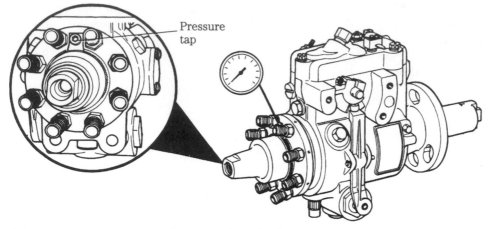

4-8 Internal-pressure test port for a Stanadyne distributor-type pump.

Air inlet system

In its most highly developed form, the air inlet system incorporates a turbocharger and exhaust gas recirculation (EGR).

Air filter

Symptoms of filter restrictions are
- Hard starting
- Black exhaust smoke
- Loss of power
- Failure to reach governed speed

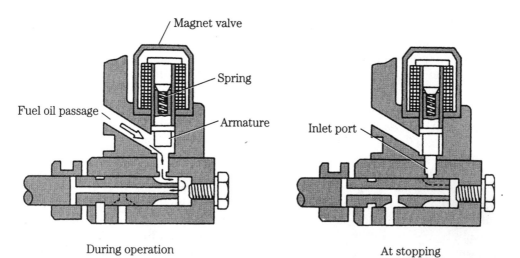

During operation   At stopping

4-9 A solenoid-operated fuel shutoff valve is a standard feature on modern pumps, some of which also incorporate an air shutdown.

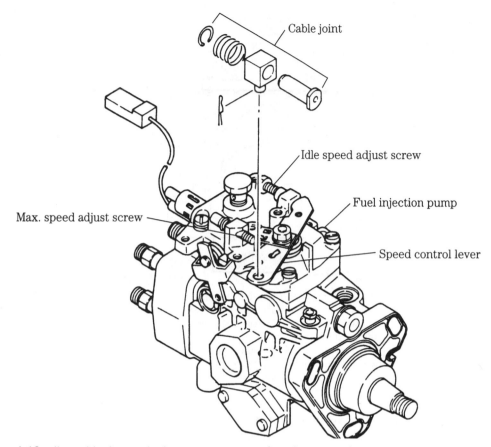

Cable joint

Idle speed adjust screw

Fuel injection pump

Max. speed adjust screw

Speed control lever

4-10 Idle and high-speed adjustment screws as found on several Yanmar pumps.

- Erratic idle (naturally aspirated engines)
- Turbocharger speed surges and possible oil pullover

Service manuals for vehicle and other small engines merely provide a schedule for changing the air filter. If the operating environment varies, as it does on construction sites or during off-road operation, a better approach is to change the element when the pressure drop across it becomes excessive. Some experimentation with new and used filter elements will be necessary to establish a standard. Pressure drops across new filter elements vary over a range of 2–15 in./H_2O. As a general rule, a 50% increase in pressure drop is cause enough to replace the element.

Figure 4-11 illustrates the hookup for measuring pressure drop. The test is made with the engine running at the speed that generates maximum air flow. Air flow peaks at fast idle for naturally aspirated engines and at full power for turbocharged engines.

CAUTION: Do not operate the engine with the air cleaner removed. Unlike spark-ignition (SI) manifolds, which are obstructed by venturis and throttle plates, diesel manifolds open directly (or via expensive compressor wheels) to the combustion chambers. A dropped wing nut or, in the case of turbocharged engines, careless fingers, will have dramatic effects.

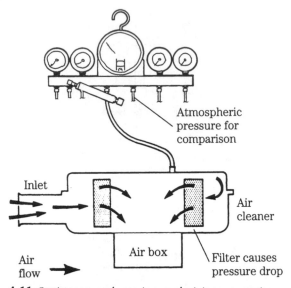

4-11 Stationary and marine technicians sometimes used a manometer—a U-shaped tube, partially filled with water, with one leg open to the atmosphere and calibrated in quarter-inch increments—to measure the pressure drop across the filter. Field mechanics prefer the convenience of a vacuum gauge. The Ford gauge is calibrated in inch/H$_2$O.

Positive crankcase ventilation (PCV)

Positive crankcase ventilation valves require routine replacement, since cleaning is rarely effective. The 2.2-L Isuzu used in several GM vehicles demonstrates how complex these systems can become. The liquid component of blowby gases is filtered out and collected in a holding tank for return to the crankcase. A check valve under the tank isolates the sump from PCV vacuum. If this valve clogs, the oil level rises in the tank and is pulled over into the intake manifold to produce clouds of blue smoke.

Exhaust gas recirculation

Failure of the exhaust gas recirculation (EGR) valve or control circuitry flags trouble codes, increases smoke levels dramatically and costs power. When used with EMS, the EGR valve incorporates a position sensor and an actuator that can be cycled with the appropriate scanner.

Turbocharger

What follows is a quick rundown on some of the more frequently encountered turbo faults. See Chap. 9 for more information about these complex machines.

Turbocharger failure has a number of causes, most of them associated with the demands put on the lubrication system by these devices that, in their most modern form attain as much as 400,000 rpm under peak load (Fig. 4-12). Momentary interruptions in the oil supply, which might occur during initial startup after an oil change

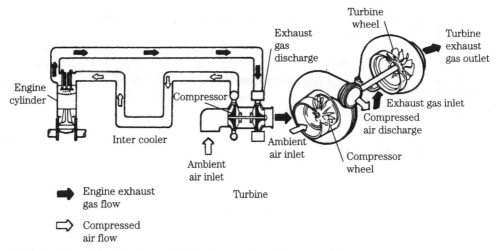

4-12 Gas flow through a Detroit Diesel turbocharger.

or during sudden acceleration immediately after a cold start, destroys the bearings. If the engine is run hard and abruptly shut down, oil trapped in the turbocharger case absorbs heat from the turbine and carburizes into an abrasive. Many EM systems can be programmed to eliminate cold, wide-open-throttle starts and hot shutdowns.

Other problems include carbon and scale accumulations on the wheels, exhaust-side leaks (caused by thermal expansion of the piping), pressure-side leaks (often at the manifold-to-block gasket), and blade-tip erosion. A faulty air filter will "dust" the compressor wheel, giving it a satiny appearance.

Some malfunctions—noise, excessive bearing clearances, cracked exhaust plumbing—obviously involve the turbocharger. A fall off in power, sooty exhaust, low or erratic turbo boost may have causes outside of the unit. Before assuming that the turbocharger is a fault, check the fuel system with particular attention to the injectors, and verify that the pressure drop across the air filter element is within specification. Carefully inspect the turbocharger oil and vacuum lines. And, of course, you need to retrieve the trouble codes. Many EMS blowers incorporate variable geometry, so that boost comes on early during the rpm curve. Malfunctions will flag trouble codes associated with abnormal levels of boost and actuator failure.

Start the engine and listen to the sound the turbocharger makes. With a little practice you will be able to distinguish the shrill sound of air escaping between the compressor and engine and the sound of exhaust leaks, which are pitched at a lower register. If the sound changes in intensity, check for a clogged filer, loose sound-deadening material in the intake duct, and dirt/carbon accumulations on the compressor wheel and housing. A stethoscope will amplify any mechanical noise.

Glow plugs

Slow cold starts and persistent white smoke can be caused by failure of one or more glow plugs on indirect injection (IDI) engines. Test individual plugs with the ignition off and a low-voltage ohmmeter connected between the plug terminal and a good engine ground. Most draw about 2 Ω hot, and nearly 0 Ω cold. If the heating element is open, resistance will be infinite.

An open supply circuit denies power to all glow plugs and makes cold starting virtually impossible. Check for blown fusible links verify that the relay, often a solid-state device with an internal resistance that drops glow-plug voltage to 6V, is functional. EMS glow-plug controllers vary plug on-time with ambient and coolant temperatures. See Chap. 11 for details.

Exhaust backpressure

Excessive backpressure can prevent starting, cost power, increase exhaust temperature, and color the exhaust. Water locks—exhaust pipe risers to block water entry on marine installations—should not be used with turbocharged engines.

Look for damaged mufflers and flattened, crimped, or improperly sized exhaust piping. Trouble codes should be flagged if the EMS senses a clogged particulate trap or catalytic converter.

If excessive backpressure is suspected and the source is not obvious, drill and tap a 11/32 in. hole on a straight section of exhaust piping 6–12 in. downstream of the turbocharger. Tap threads for a 1/8 in. pipe fitting. Connect a manometer and, with coolant temperature normal, apply the maximum possible load to the engine. While backpressure varies with the installation, readings higher than 3 in./Hg are cause for concern.

Engine mechanical

Engine malfunctions fall into four categories: fluid leaks, excessive oil consumption, loss of compression, and bearing wear.

Fuel leaks
External fuel leaks are discussed earlier in this chapter. Fuel contamination of the lube oil (as from failure of Roosa pump seals or HEUI injector o-rings) can be detected by the loss of viscosity and the telltale odor.

Oil leaks
In rough order of frequency, the most common sites of oil leaks are
- Valve cover gasket
- Oil pan gasket
- Timing case gasket
- Front and rear crankshaft seals
- Oil cooler
- Camshaft and oil gallery plugs

Tracing the source of a small leak can be difficult, because oil tends to migrate down and back, toward the rear of the engine. A black light detector or the aerosol powders sold for this purpose help, but the ultimate tactic is to pressurize the crankcase. Connect a low-pressure (5-psi maximum) air line to the dipstick tube. Soap the engine down and look for bubbles. Light foaming at valve cover gaskets is permissible. The rope-type crankshaft seals used in older engines might leak air and still function when the engine runs; lip seals may foam, but should not hemorrhage.

Internal oil leaks past bearings shows up as a loss of pressure, a condition that should be verified with an accurate test gauge made up to the main oil gallery (Fig. 4-13). A pressure drop when loads are applied to a warm engine indicates excessive bearing clearances.

Oil enters the coolant by way of leaks in oil cooler or, more rarely, head gasket. Pressure test the cooler as described in Chap. 8, "Lubrication systems"; the head must be removed to detect leaks across the gasket, which makes the exercise a bit redundant.

The lower, external oil seal on HEUI injectors is all that stands between high-pressure lube oil and the fuel supply. A leak here can empty the sump within a few miles of driving. See Chap. 6 for more information.

Coolant leaks

External leaks are usually visible; identifying internal coolant leak paths requires some detective work. Massive leaks into the combustion chamber produce white smoke at normal engine temperatures. Such leaks can sometimes be seen as coolant flow through the EGR valve.

A burst of steam when the dipstick is touched to a hot exhaust manifold signals the presence of water in the sump. Ethylene glycol coats the stick with brown varnish almost impossible to wipe off. Have the oil tested and, if the test shows the presence of antifreeze, prepare to tear the engine down for cleaning. Large accumulations of coolant in the sump convert the oil to a white, mayonnaise-like emulsion.

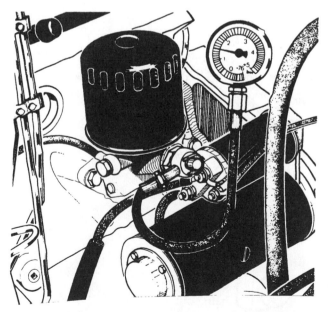

4-13 Verify low oil pressure with an accurate gauge connected to a test port on the main gallery, as shown here on a Peugeot engine.

Excessive oil consumption

Blue smoke is the primary indicator of oil burning, although catalytic converters can disguise the condition. The question facing the mechanic is the source of the oil. Blue smoke at all speeds, hot or cold, points to cylinder/ring wear, which can be verified with tests described below. Smoke during acceleration after periods of idle and immediately upon startup indicates leaking valve seals. Oil in the intake manifold and/or air boxes originates from an upstream source, usually leaking turbo or blower seals. But other possibilities should not be overlooked. Sometimes merely changing the air filter element corrects the problem.

Loss of compression

Disabling one injector at a time will isolate any weak cylinders. If changing out the associated injectors does not restore power, the problem is almost surely compression-related. Loss of compression across all cylinders will be accompanied by high oil consumption and a noticeable reduction in power output.

The cylinder compression gauge (Fig. 4-14) is an essential tool for diesel mechanics. Test procedures vary with make (e.g., some manufacturers specify a cold reading, while others require that the engine be at normal operating temperature). But the general procedure is as follows:

1. Cut off the fuel supply or disconnect the camshaft position sensor on computerized engines. Otherwise, the cylinder under test can fire and destroy the gauge.
2. Disable all cylinders by (depending upon how the gauge is mounted) removing the glow plugs or injectors.

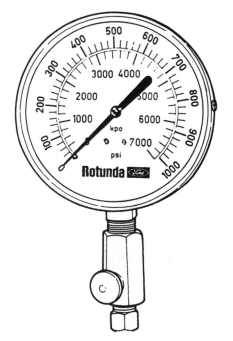

4-14   Ford-supplied cylinder compression gauge.

3. Test each cylinder in turn, allowing about six compression strokes per gauge reading.

4. Check the readings against published specifications and against each other. Variations of 20% or more between cylinders results in noticeable roughness.

WARNING: do not squirt oil into the cylinder in an attempt to determine if the compression loss is due to rings or valves. The cylinder might fire.

Figure 4-15 illustrates a GM leakdown test kit with glow-plug adapter. Unlike competitive testers, the unit does not measure leakdown percentage, but it does indicate the source of air leaks, which is the purpose of the exercise.

1. Remove the radiator cap and warm the engine to operating temperature.

2. Shut off the engine and, using the timing marks, position No. 1 cylinder at top dead center (tdc). Finding tdc for the remaining cylinders is a bit challenging, but can be accomplished with a degree wheel or with the help of the whistle supplied with the kit.

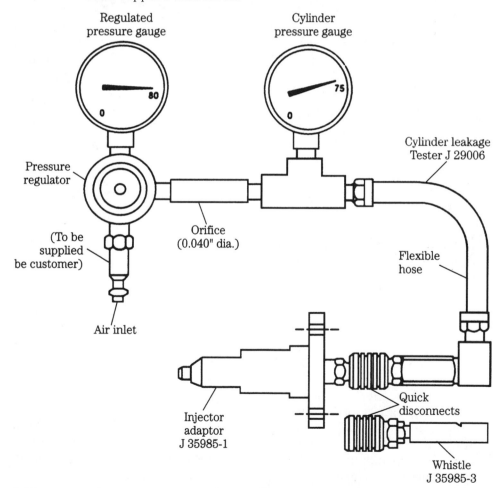

4-15 GM cylinder leakdown tester includes air-line adapters, regulator, gauge, and top dead center (tdc) whistle. The whistle sounds as the piston nears tdc on the compression stroke.

Dipstick or
oil fill:
worn or
sticking
rings

Air inlet:
bad intake
valve

Bubbles in coolant:
warped or cracked
cylinder head or block,
blown head gasket

Leaking from glow
plug hole on
adjacent cylinder:
blown head gasket

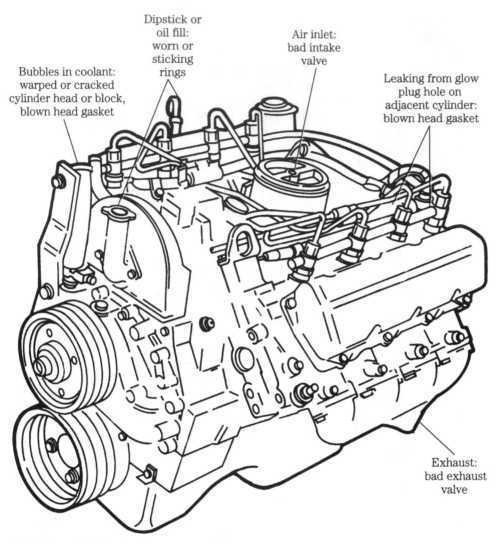

Exhaust:
bad exhaust
valve

4-16   Potential leak sites for compression.

WARNING: if a piston is not at or very close to tdc, the engine will "motor,"
catching a hapless mechanic in the fan or belts.

3. Install the tester and apply about 75 psi of compressed air to it.

4. The significance of the various leak sites is listed in Fig. 4-16.

The amount of blowby past the rings is a rough indicator of cylinder bore/ring
wear. Figure 4-17 illustrates the test apparatus, consisting of a pressure gauge and an
adaptor that replaces the oil filler cap. PCV atmospheric vents must be sealed.
A major limitation of this test is that few manufacturers provide rpm/blowby pressure specifications. The mechanic must establish the norm while the engine is still
healthy.

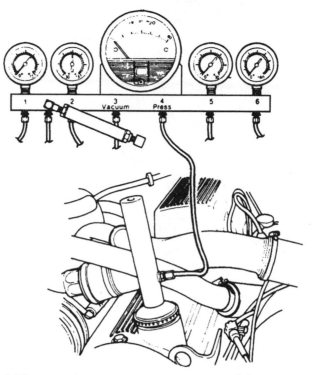

4-17 A crankcase pressure tester is part of the Ford PN 0190002 gauge bar set.

Oil analysis

Short of tearing the engine down for inspection, oil analysis is the best index of wear. The technology, developed initially for diesel locomotives and perfected by the U.S. Navy, can predict engine failure with some exactitude. The oil sample undergoes spectroscopic analysis to identify 16 or more elements with an accuracy of within one part per million. Bearing wear shows up as lead, silver, tin, and aluminum; liner and ring wear as iron and chromium; valve wear as nickel; and bushings as copper. Silicon and aluminum particles suggest air filter failure.

5
CHAPTER

Mechanical fuel systems

Diesel fuel systems were entirely mechanical until the mid-1980s, when manufacturers of vehicle engines turned to computers to give more precise control over fuel delivery and timing. Mechanical systems continue to be specified for many agricultural, construction, and marine applications.

Air blast

Early engines used compressed air to force fuel into the cylinders against compression pressure. Injectors received fuel from a low-pressure pump and air from a common manifold, which was pressurized to between 800 and 1000 psi. The engine camshaft opened the injectors, appropriately enough called "valves," to admit a blast of compressed air and atomized fuel to the cylinders. A throttling valve regulated engine speed by controlling the amount of fuel delivered; air required no metering, since surplus air is always present in diesel cylinders.

Air injection had serious drawbacks. The air/fuel mixture could not penetrate deeply into cylinders that were themselves under 450–500 psi of compression pressure. Attempts to increase engine power by admitting more fuel merely dampened the flame. The compressors were parasitic loads that absorbed 15% and sometimes more of engine output. Nor did air injection make for a compact package: starting and four-stage air-injection compressors accounted for a third of the length 1914-era Krupp submarine engines.[1]

Early common-rail

The need for more compact and silent submarine engines led the British arms maker Vickers to develop airless, or solid, injection. One can make the case that the ultimate consumers of nineteenth- and twentieth-century technology were enemy

---

[1] C.L. Cummins, Jr., "Diesel Engines of 2000 BHP per Cylinder—Pre-1914 Marine Engine Developments," *History of the Internal Combustion Engines,* ed., E.F.C. Somerscales and A. A. Zagotta, p. 19, 1989.

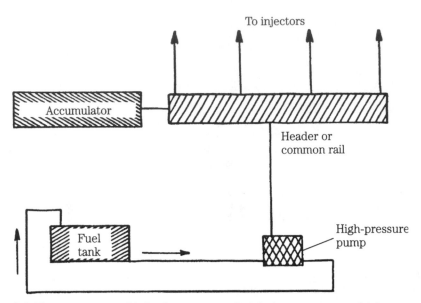

5-1 The common, or third-rail system supplied fuel to cam-operated injectors from a header, pressurized by a remote pump. By the late 1920s, c-r was being phased out by Bosch inline pumps, only to be revived again as the century closed. These new common-rail systems are discussed in the following chapter.

soldiers, sailors, and civilians. At any rate, solid injection was a vast improvement and, after the end of hostilities in 1919, was widely commercialized. Injectors fed from a common rail that was filled with high-pressure fuel and usually fitted with an accumulator to smooth pressure variations generated by injector opening and closing (Fig. 5-1). Adjustable wedges between the camshaft lobes and the injector followers permitted injection duration to vary with speed and load.

Even so, this early form of c-r injection limited engine speed, since increasing the effective width of the wedge by forcing it deeper into contact with the cam also advanced the timing, causing the injector to open earlier. There were also maintenance concerns. According to contemporary accounts, the rails leaked and injectors dribbled fuel throughout the whole operating cycle.

Jerk pump system

The big breakthrough came with the development of the Bosch inline pump, used today in a form almost identical to the first production run of April, 1927. The first pumps were mated with pintle injectors for indirect injection (IDI) applications. Direct injection (DI) multiple-orifice injectors arrived in 1929 and, two years later, Bosch integrated centrifugal governors with the pumps. At this point, the modern diesel engine came into being.

The inline pump consists of a row of individual plungers, one per engine cylinder, operated by the same internal camshaft. High-pressure tubing connects each of the plungers to its injector. The spring-loaded injectors function like pop-off valves to open automatically when a certain pressure threshold is reached. The sudden loss of line pressure gave rise to the term "jerk pump," which while a bit inelegant, is descriptive.

The distributor pump was developed in the early 1960s as a means of reducing the number of extremely precise and expensive plungers. One pumping unit, consisting of one, two, or sometimes three plungers, serves all injectors. After pressurization, the fuel passes through a rotary valve, known as a distributor head, for allocation to the individual injectors. A distributor pump is the hydraulic equivalent of an ignition distributor. Because of their relatively low cost, distributor pumps are standard ware for automobiles and light trucks. The major disadvantage is that the internal cam, which drives the plunger set, depends solely upon fuel for lubrication. Inline pumps lubricate the high-pressure cam faces with motor oil, either from an internal reservoir or from the engine oiling system.

Figure 5-2 illustrates the layout of the jerk-pump system that, in this example, employs an inline pump. A distributor-type pump could be substituted.

A lift pump, shown on the lower left of the drawing, delivers fuel to the filters and from there to the suction side of the inline pump. High-pressure fuel exits the pump through dedicated lines to each injector. A second line made up to each injector recycles surplus fuel back to the tank. The system works at three pressure levels: low pressure, on the order of 30–50 psi, between the lift pump and injector pump, pressure of several thousand psi in the injector piping, and slightly more than zero pressure on the return line.

Inline pumps

The parts breakdown for a Bosch size A Series PE inline pump for six-cylinder engines is shown in Fig. 5-3. Shims (21 and 49) establish camshaft float. Two fuel-delivery adjustments are provided. The basic setting is determined by the height of

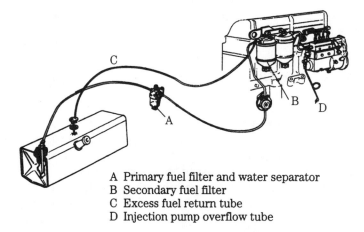

A Primary fuel filter and water separator
B Secondary fuel filter
C Excess fuel return tube
D Injection pump overflow tube

5-2 An inline-pump system as used on Ford-based Lehman marine engines.

5-3 Bosch Series PE pump. This example is found on six-cylinder Chrysler marine engines.

1 Delivery valve holder	21 Adjusting shim	41 Roller bushing
2 Lockplate	22 Tapered roller bearing	42 Screw plug
3 Lockwasher	23 Nut	43 Camshaft
4 Bolt	24 Lockwasher	44 Key
5 Delivery valve spring	25 Stud bolt	45 Lockwasher
6 Delivery valve gasket	26 Screw plug	46 Screw
7 Delivery valve assembly	27 Gasket	47 Center bearing
8 Plunger assembly	28 Oil level gauge	48 Distance ring
9 Air bleeder screw	29 Pinion clamp screw	49 Adjusting shim
10 Gasket	30 Control pinion	50 Tapered roller bearing
11 Air bleeder plug	31 Control sleeve	51 Oil seal
12 Gasket	32 Upper spring seat	52 Bearing cover
13 Pump housing assembly	33 Plunger spring	53 Lockwasher
14 Air bleeder assembly	34 Lower spring seat	54 Screw
15 Gasket	35 Adjusting bolt	55 Control rack
16 Plate cover	36 Locknut	56 Adapter
17 O-ring	37 Tappet	57 Connector ring
18 Gasket	38 Tappet guide	58 Gasket
19 Setscrew	39 Roller pin	59 Connector bolt
20 Distance ring	40 Roller	

the adjustment bolts (35) that thread into the tops of each tappet. In addition, friction clamps (29) permit the control-sleeve pinions (30) to move relative to the rack to compensate for tooth wear and production tolerances. Fuel-delivery adjustments for this and all other injection pumps must be made on a test stand.

The drawing of a single-plunger pump in Fig. 5-4 helps to clarify the relationship between the rack, fuel control gear (plunger pinion), and plunger. Fuel enters through the intake port on the right and exits past the delivery valve at the top of the unit. The obliquely cut groove on the plunger outer diameter (OD) functions in conjunction with the rack to throttle fuel delivery.

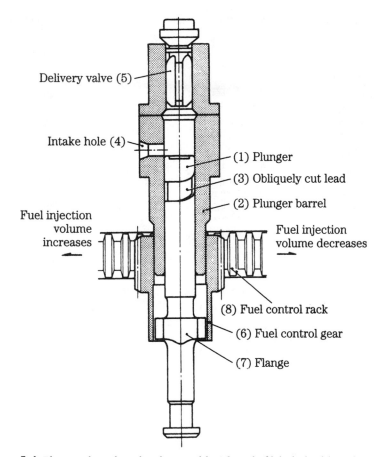

Delivery valve (5)

Intake hole (4)

(1) Plunger

(3) Obliquely cut lead

(2) Plunger barrel

Fuel injection
volume
increases

Fuel injection
volume decreases

(8) Fuel control rack

(6) Fuel control gear

(7) Flange

5-4 Plunger, barrel, and rack assembly. The relief labeled "obliquely-cut lead" is generally called the "helix."

How this is done is shown in Fig. 5-5. At the bottom of the stroke, the plunger uncovers the inlet port. Fuel enters the pressure chamber above the plunger. The plunger rises, initially pushing fuel back out the inlet port. Further movement masks the inlet port. The plunger continues to rise, building pressure on fuel trapped above it. The delivery valve opens, and a few milliseconds later, the injector discharges. Fuel continues to flow until the annular groove milled along the side of the plunger uncovers the inlet port. At this point, pressure bleeds back through the inlet port and injection ceases. Because of the shape of the groove, rotating the plunger opens the inlet port to pressure earlier or later in the plunger stroke.

American Bosch, Robert Bosch, and CAV barrels are drilled with a second port above the inlet port to accept spillage during part-throttle operation. Figure 5-6 illustrates the metering action of a CAV pump. Fuel enters the barrel at A and continues to flow until plunger movement masks the two ports (inlet shown on the left, spill port on the right). At full load, the pressure bleed-down through

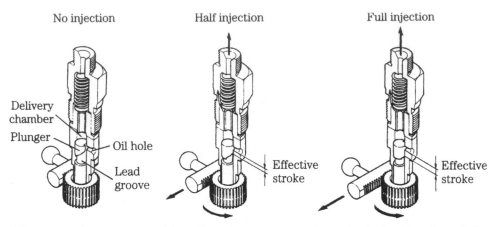

5-5 Constant-beginning, variable-ending plunger action with surplus fuel exiting through the inlet port.

the spill port is delayed until the plunger approaches the end of its stroke, as shown in drawing C. You might wonder why the spill port opens before top dead center (tdc), when a few more degrees of cam movement would raise fuel pressure even more. The reason is that cam-driven plungers behave like pistons, accelerating at mid-stroke and slowing as the dead centers are approached. Opening the spill port early, while plunger velocity and pressure rise are rapid, terminates injection far more abruptly than if the port remained closed until the plunger reached tdc.

In drawing D, the annular groove has rotated to the half-load position. The effects of further rotation are shown at E, which represents idle, and at F, shutdown.

However the porting is arranged, the effective stroke has a constant beginning and a variable ending for pumps this book is concerned with. Some marine and large stationary engines meter fuel delivery at the beginning of the stroke.

Details of the mechanism for transmitting rack movement to the plungers vary with the manufacturer, but always include an adjustment to equalize delivery between plunger assemblies. At a given rack position, each cylinder must receive the same amount of fuel.

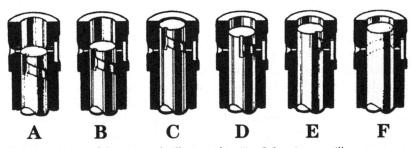

A   **B**   **C**   **D**   **E**   **F**

5-6 A variation of the principle illustrated in Fig. 5-5, using a spill port to vent the unneeded fuel. GM Bedford Diesel

Distributor pumps

The main technical advantage of distributor pumps is that each cylinder receives precisely the same amount of fuel since all cylinders feed from the same plunger or plunger set. And, as pointed out above, reducing the number of plungers, which must be lapped to angstrom tolerances, reduces costs.

Lucas/CAV

Figure 5-7 illustrates a CAV system for medium trucks. The engine-driven feed pump moves fuel from the tank to the filter and regulating valve. Fuel then passes to the transfer pump and, from there, to the inlet side of the distributor pump for pressurization and injection. As explained below, the regulating valve throttles the injection pump during low-speed operation. Surplus fuel returns to the tank through low-pressure lines.

A pair of cam-driven plungers, mounted on the pump rotor and turning with it, generate injector pressure (Fig. 5-8). During the inlet stroke the plungers move outward under pressure from fuel entering through the inlet (or metering) port. As the rotor continues to turn, the inlet port is blanked off by the rotor body and one of the outlet ports is uncovered.

As the outlet port is uncovered, the internal cam, acting through rollers, forces the plungers together. Fuel passes through the port for injection. Fuel that slips by the plungers returns to the filter.

The amount of fuel delivered per plunger stroke is determined by the regulating valve (upper right in Fig. 5-7) that is controlled by the accelerator pedal and the

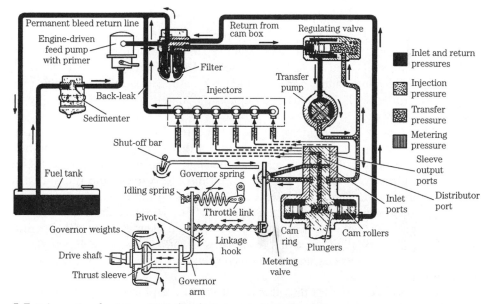

5-7 Schematic of a Lucas-CAV distributor-pump system. GM Bedford Diesel

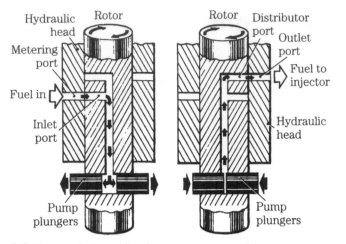

5-8 Opposed-piston distributor pump operation. GM Bedford Diesel

centrifugal governor. At low engine speeds, the regulating valve limits the pressure of fuel going to the transfer pump and into the injection pump. Because the pump plungers are driven apart by incoming fuel, their outward displacement is determined by transfer-pump pressure. As engine speed and/or load increase, the regulating valve increases fuel pressure to force the pump plungers further apart. Unlike Bosch inline pumps, which throttle by varying the effective stroke, the Lucas/CAV unit varies the actual stroke.

Bosch VE

Nearly 50 million VE pumps have been produced since introduction in 1975 with applications ranging from fishing boats to luxury automobiles. Figure 5-9 sketches a basic installation, with filter, gravity-type water separator (sedimentor), fuel lines from the pump and injectors, and a solenoid-operated fuel shutoff (magnet) valve. Normally the fuel shutoff requires battery power to open; for marine applications the valve must be energized to close. This permits the engine to continue to run should the vessel lose electrical power.

The VE turns at half engine speed and is geared to a mechanical (shown in Fig. 5-10) or electronic governor. The rear half of the pump houses the vane-type transfer pump and regulator that supplies the high-pressure section with fuel at pressures ranging from 40 psi to 175 psi at full throttle. Vane-pump pressure also controls the hydraulic timer that advances injection with increased engine speeds.

A tongue-and-groove joint mates the rear half of the two-piece drive shaft with the forward, or plunger, half. This joint, which can be seen clearly in Fig. 5-11, permits the forward half of the shaft to move fore and aft as it rotates under the impetus of the face cam. The cam reacts against the housing through roller bearings, oriented as shown in Fig. 5-12.

A control sleeve regulates the effective stroke of the plunger, which discharges through delivery valves on the distributor head.

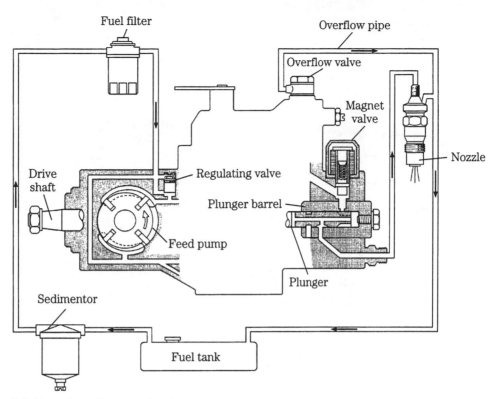

5-9 Typical small-engine distributor-type system. Yanmar Diesel

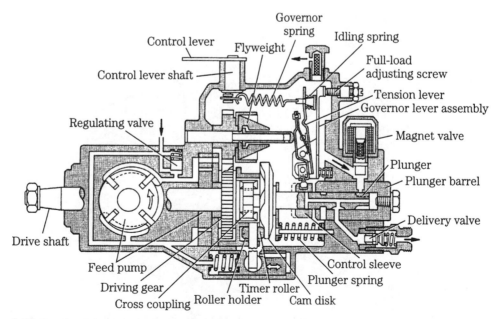

5-10 Bosch VE Series injection pump.

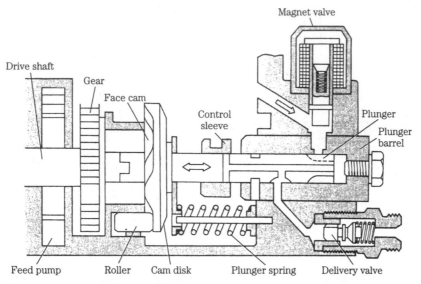

5-11 Cross-sectional view of the VE. Notice the integration and compactness of the design.

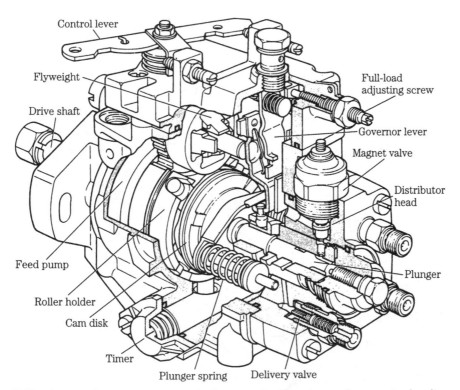

5-12 The VE plunger reciprocates in the pumping function and rotates in the distributor, or fuel-allocation, function. Yanmar Diesel

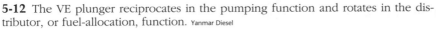

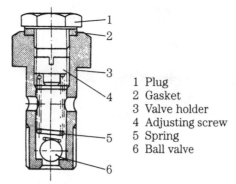

1 Plug
2 Gasket
3 Valve holder
4 Adjusting screw
5 Spring
6 Ball valve

5-13 Fuel overflow valve functions as an automatic bleeder, diverting air and surplus fuel back to tank. Marine Engine Div., Chrysler Corp.

Injector pump service

Bleeding

Because surplus fuel recycles to the tank or filter through fuel overflow valves mounted at high points on pumps and injectors, a half-hour or so of operation is generally sufficient to purge the system of air bubbles (Fig. 5-13). More serious air intrusions require bleeding, using procedures described in Chap. 4 for conventional, pre-computer engines and elaborated upon in Chap. 6 for more modern engines.

Timing

Injection pumps have two provisions for synchronizing fuel delivery with piston movement. Timing marks on the drive gears establish the basic relationship. Elongated mounting-bolt holes, which allow the pump body to be rotated a few degrees, provide the fine adjustment. Reference marks stamped on the pump body and mounting flange enable the adjustment to be replicated (Fig. 5-14).

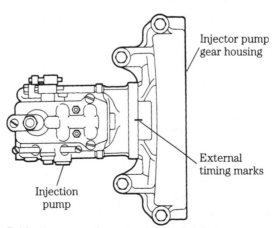

Injector pump gear housing

External timing marks

Injection pump

5-14 Timing marks on a Bosch distributor pump.
Ford Motor Co.

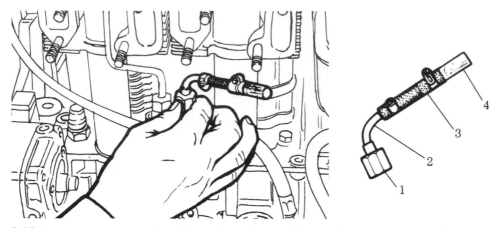

5-15 Transparent section (4) of fuel pipe adapter indicates plunger movement for timing purposes. Lombardini

Before removing a pump, bar the engine over until both valves on No. 1 cylinder close and the timing mark on the harmonic balancer or flywheel aligns with its pointer. This procedure indexes pump-gear timing marks for easy assembly. If the same pump is reinstalled, the reference marks on the pump body and flange should be valid. Substituting another pump puts the marks into question and the engine should be retimed, either statically or dynamically.

Static timing procedures vary enormously, but the purpose of the exercise is to synchronize the onset of fuel delivery with the piston in No. 1 cylinder. Depending upon engine make, model, and application, fuel should begin to flow anywhere from 8–22° btdc as the flywheel is barred over by hand.

Flywheels for small utility engines generally have two marks inscribed on their rims: one representing tdc and the other, always in advance of the first, indicating when fuel should begin to flow from No. 1 delivery valve. A convenient way to monitor fuel flow is to make up an adapter out of a length of clear plastic tubing and a delivery-valve fitting, as shown in Fig. 5-15. The mechanic slowly bars the engine over, while watching for the slightest rise in the fuel level. The onset of fuel movement should occur at the moment the timing mark aligns with its pointer. If tdc is passed and the plunger retreats, the flywheel must be turned back 15° or so to absorb gear lash, and the operation repeated.

Drive gears for Navistar (International) DT358 and its cousins have six timing marks. Which one to use depends upon the engine model and application. In a reversal of traditional practice, certain American Bosch pumps time to the end, rather than the onset, of fuel delivery. The delivery valve, which acts as a check valve, for No. 1 plunger must be disabled before timing.

Timing specifications for distributor pumps are often expressed in thousandths of an inch of plunger movement from bdc. Figure 5-16 illustrates the dial-indicator adapter that replaces the central bolt in the distributor head. Locating bdc—the precise moment when the plunger pauses at the bottom of its stroke—requires patience.

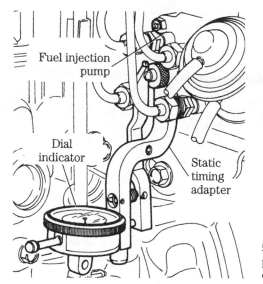

Fuel injection
pump

Dial
indicator

Static
timing
adapter

5-16 For critical applications, plunger travel is measured with a dial indicator. Ford Motor Co.

Once bdc is found, the mechanic zeros the gauge and bars the engine over in the normal direction of rotation to the appropriate crankshaft or harmonic-balancer mark. He then rotates the pump body as necessary to match lift with the published specification.

Dynamic timing, made with a strobe light while the engine ticks over at slow idle, compensates for pump-gear wear and other variables. It is the only way that van and other inaccessible engines can be timed.

The Sun timing light draws power from a wall outlet or, if equipped with an inverter, from the engine's 12-V or 24-V batteries (Fig. 5-17). A transducer clamps over No. 1 fuel line to trigger the strobe when the injector opens and the sudden drop in fuel pressure contracts the line. The instrument also tracks how many crankshaft degrees the timing advances as engine speed increases.

Unit injectors (UIs), which integrate the pump function with injection, are timed as described below.

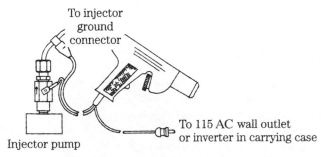

To injector
ground
connector

To 115 AC wall outlet
or inverter in carrying case

Injector pump

5-17 Sun-timing light triggers from pulses in the fuel pipe.

Repair

Other than periodically checking the oil level for in-line pump reservoirs and adding rust inhibitor to the fuel before long-term storage, injector pumps require no special attention. When loss of pressure indicates that the pump has failed, the mechanic farms the unit out to a specialist for cleaning or repair.

While most mechanics would tackle Bosch VE, inline pumps are simple devices that can be opened for cleaning and elementary repairs. But the work requires patience and the highest possible levels of cleanliness. Plungers are lapped to their barrels and must be assembled as found. Nor should these sensitive parts be touched with bare fingers. Wear surgical gloves or use forceps. Tappets and other adjustable parts must be kept with their plungers. When disassembly entails loss of adjustment, the existing adjustments must be scribe marked. If at all possible, obtain a drawing of the pump before you begin.

Inline pumps come apart as follows:

- Unscrew the delivery valves mounted above each barrel and lay them out in sequence on the bench, which should be covered with newspaper to reduce the possibility of contamination.
- Most inline pumps incorporate a side cover for tappet access; others mount the individual plunger assemblies from above with studs. On those with side covers, rotate the camshaft to bring each tappet to the top of its stroke and shim the tappet. Withdraw the plungers from the top with the aid of a hooked wire or expansion forceps. Mark or otherwise identify the plungers for correct assembly.
- Inspect the lapped surfaces with a magnifying glass. Deep scores or pronounced wear marks mean that the pump has come to the end of its useful life. Over-tightening the delivery valves warps the barrels and produces uneven wear patterns. Water or algae in the fuel leave a dull, satiny finish on the rubbing surfaces. Loss of the sharp edges on the helix profile upsets calibration.
- A plunger should fall of its own weight when the barrel is held 45° from vertical. Gummed or varnished plungers can make governor action erratic and accelerate wear on the helixes and rack.
- If you have the specifications, check the cam profile with an accurate (i.e., recently recalibrated) micrometer.
- Examine the rack teeth for wear. Figure 5-18 illustrates, in exaggerated fashion, the wear pattern. The rack should move on its bushings with almost no perceptible side or vertical play. Installation of new bushings requires a factory reamer.

Delivery valves

Injectors feed through delivery valves mounted on the distributor head or above each barrel on inline pumps. These check valves open under pressure to supply fuel to the injectors and close automatically during the suction stroke. Most delivery valves consist of a conical sealing element shaped like an inverted top hat (Fig. 5-19). The extension below the element, known as the piston, has two functions. It centers the valve over its seat and acts as a ram to force fuel back into the pump as the valve

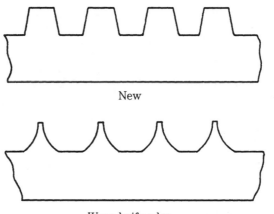

New

Worn-knife edge

5-18 Control-rack teeth bell-mouth in service. GM
Bedford Diesel

closes. This action assures a rapid drop in pressure so that the injector snaps shut without after-dribble.

Delivery valves come in various shapes and sizes. Some replace the conical element with a disc, others hold a certain percentage of pressure in the downstream plumbing after closing. But all work on the same general principle.

Service

Delivery-valve problems are cylinder specific. If the valve sticks open, no fuel passes to the associated injector. Leakage is harder to diagnose. White smoke that persists after all bases have been touched—pump pressure and timing, new injectors, and engine compression—suggests that one or more delivery valves may be at fault. Disassemble one valve at a time, clean the parts thoroughly, and test for leaks by blowing through the outlet port. Older, simpler valves can sometimes be resurfaced.

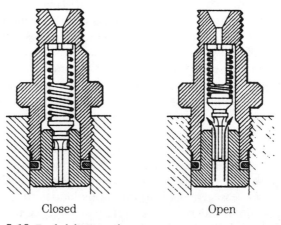

Closed   Open

5-19 Fuel delivery valve operation. GM Bedford Diesel

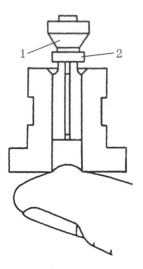

5-20 Delivery valve test. Check valve shown at 1, piston at 2.

Marine Engine Div, Chrysler Corp

Check piston fit by depressing the valve and placing your finger over the inlet port (Fig. 5-20). You should feel the vacuum as the piston falls.

Injectors

Mechanical injectors consist of two main parts—an injector body, or nozzle holder, and a needle (Fig. 5-21). The injector body secures the assembly to the cylinder head, and includes connections for fuel inlet and return lines.

A spring-loaded needle ("nozzle valve") in the drawing, controls fuel flow through the injector. At NOP fuel pressure exerts sufficient force on the raised shoulder near the tip of the needle to overcome spring tension. The needle retracts clear of the orifice and injection commences. Once the needle lifts, its full cross-sectional area—the shoulder and tip—comes under fuel pressure. Consequently, less pressure is required to hold the needle open than to unseat it. This feature holds the nozzle open as fuel pressure drops in response to injection. When the pump plunger descends and the delivery valve closes, fuel pressure drops abruptly. The return spring forces the needle down to seal the orifice and injection ceases.

The smaller the nozzle orifice, the finer the fuel spray and, all things equal, the less ignition lag. But if too small, the orifice limits engine output by restricting fuel flow. Multi-hole nozzles, some with a dozen or more, electrically discharge, machined orifices that satisfy both flow rate and atomization requirements, have become standard on DI engines.

IDI engines employ pintle nozzles, two varieties of which are illustrated in Fig. 5-22. The pintle is an extension of the needle that, when retracted, produces a hollow, cone-shaped spray. Because of the large diameter of the orifice and the shuttle action of the pintle, these nozzles are less susceptible to carbon buildup than multi-hole types.

The throttling pintle introduces fuel in a staged sequence to reduce the sudden rise in pressure and heat during initial combustion. The pintle has a reduced diameter

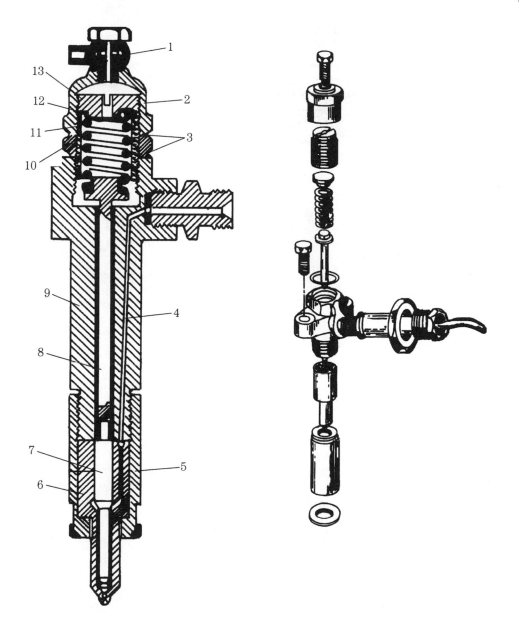

1. Leak off union
2. Body cap nut
3. Seal washers
4. Fuel inlet passage
5. Nozzle cap nut
6. Nozzle
7. Nozzle valve
8. Nozzle valve spindle
9. Injector body
10. Spring cap locknut
11. Nozzle valve spring
12. Spring seat washer
13. Spring cap nut

5-21 Cross-section and exploded view of two CAV injectors. GM Bedford Diesel and Lehman Ford Diesel.

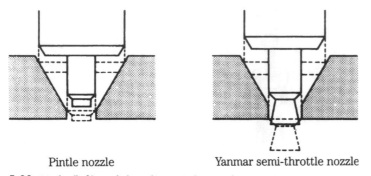

Pintle nozzle   Yanmar semi-throttle nozzle

5-22 Pintle (left) and throttling pintle nozzles. Yanmar Diesel

on the tip that, when seated, extends into the combustion chamber. Yanmar and most other manufacturers use a throttling pintle with the tip in the form of an inverted cone, shown on the right in Fig. 5-22. During the initial phase of injection, the cone lifts so that its base almost fills the orifice. At full lift the cone retracts clear of the orifice for maximum fuel delivery.

Injector service

Regardless of the nozzle type, injectors should
- Discharge cleanly without "after-dribble," which tends to collect in the cylinder between cycles.
- Atomize fuel for rapid combustion.
- Direct the spray into the far reaches of the cylinder, but without impinging against the piston or chamber walls.

Most malfunctions are caused by carbon buildup on the tip, which distorts the spray pattern and can sometimes stick the needle or pintle in the open position. Automatic injectors can be tested with the apparatus shown in Fig. 5-23. Similar tests

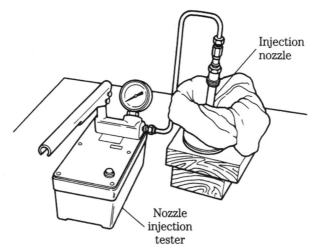

Injection nozzle

Nozzle injection tester

5-23 Ford-supplied injector tester. These tools have multiple uses, including testing common-rail fuel manifolds and HEUI oil galleries for leaks.

can be performed on unit injectors by loading the pump cavity with fuel and applying force to the pump plunger.

WARNING: Wear eye protection and heavy gloves when testing injectors. Fuel spray can cause blindness and blood poisoning.

There are four standard tests—spray pattern, NOP, sealing effectiveness, and chatter.

Spray pattern

Connect the tester as shown and, operating the pump lever in short, rapid strokes, observe the spray pattern. Pintle nozzles should produce a finely divided spray of uniform consistency and penetration (Fig. 5-24). Multi-hole nozzles generate a wide, fan-like pattern (Fig. 5-25). It can be helpful to direct the discharge against a piece of paper and compare the pattern to one produced by a new injector.

Nozzle opening pressure

Slowly depress the pump lever and note the peak gauge reading. Compare this pressure with the manufacturer's NOP specification. Some specifications are fairly broad, and engines may not idle properly unless NOPs cluster at one or the other end of the specification range. Most needle-return springs can be stiffened with shims.

Sealing

Dribble standards have become tighter in recent years. The pintle injector illustrated in Fig. 5-26 should remain fuel tight under 80% of NOP for at least 10 seconds. Note and correct any other source of leakage.

Chatter

Injector opening is normally accompanied by a sharp "pop," which suggests that the needle retracts cleanly. Verify this action as the pump lever is cycled rapidly. However, leaking injectors can be made to chatter and some perfectly good throttling pintle injectors pop open, but refuse to chatter, no matter how adroitly the pump handle is worked.

Timers

Moving fuel through an open injector requires time and, once injected, additional time is needed for the fuel to ignite. As engine speeds increase, less time is available to accommodate these delays, and the start of injection must occur earlier.

The drawing back at Fig. 5-10 shows the VE hydraulic timer in cross-section. Pressure developed by the feed pump rises in an almost linear fashion with engine speed. As pressure increases, the piston moves to the right, compressing its return spring. A roller transfers this motion to the drive cam, turning it few degrees against direction of shaft rotation.

Mechanical timers are similar to SI advance mechanisms, in that they employ centrifugal weights to sense engine speed and advance the cam (Fig. 5-27). Spring tension on the weights determines the cut-in speed and shims fix the travel limit.

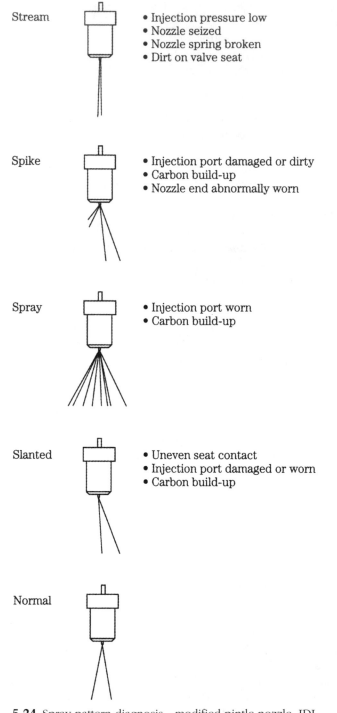

Stream
- Injection pressure low
- Nozzle seized
- Nozzle spring broken
- Dirt on valve seat

Spike
- Injection port damaged or dirty
- Carbon build-up
- Nozzle end abnormally worn

Spray
- Injection port worn
- Carbon build-up

Slanted
- Uneven seat contact
- Injection port damaged or worn
- Carbon build-up

Normal

5-24 Spray pattern diagnosis—modified pintle nozzle, IDI engines.

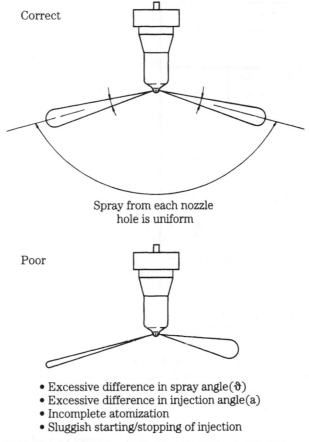

Correct

Spray from each nozzle
hole is uniform

Poor

- Excessive difference in spray angle(ϑ)
- Excessive difference in injection angle(a)
- Incomplete atomization
- Sluggish starting/stopping of injection

5-25 Spray pattern diagnosis—multi-hole nozzle, DI engines.

To check timer operation, connect a timing light to an injector pipe and run the engine up to speed. Injection should advance smoothly without hesitation as the rack is extended. Specifications vary, but most applications require 5–7.5° of total advance.

Diaphragm controls

Injector pumps often include a diaphragm-operated control that reduces fuel delivery at high altitudes and/or increases fuel at wide throttle angles for turbocharged applications. Diaphragms are tested with a vacuum gauge.

Centrifugal governors

Spark-ignition engines have a built-in governor in the form of a throttle valve that, even wide open, limits the amount of air passing into the engine. Diesels have no such limitation, since surplus air is always available for combustion. Engine speed

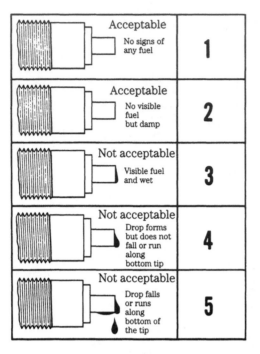

Acceptable — No signs of any fuel	1
Acceptable — No visible fuel but damp	2
Not acceptable — Visible fuel and wet	3
Not acceptable — Drop forms but does not fall or run along bottom tip	4
Not acceptable — Drop falls or runs along bottom of the tip	5

5-26 Dribble test—pintle nozzle, IDI engines.

depends solely upon the amount of fuel delivered. If the same volume of fuel necessary to cope with severe loads were delivered under no-load conditions, the engine would rev itself to destruction. Consequently, all diesel engines need some sort of speed-limiting governor.

Most governors also control idle speed, a task that verges on the impossible for human operators. This is because the miniscule amount of fuel injected during idle exaggerates the effects of rack movement. Automotive applications are particularly critical in this regard. Idle speed must be adjusted for sudden loads, as when the air-conditioner compressor cycles on and off, and when the driver turns the wheel and engages the power-steering pump.

In addition to limiting no-load speed and regulating idle speed, many governors function over the whole rpm band. The operator sets the throttle to the desired speed and the governor adjusts fuel delivery to maintain that speed under load. The degree of speed stability varies with the application. No governor acts instantaneously—the engine slows under load before the governor can react. Course regulation holds speed changes to about 5% over and under the desired rpm, and is adequate for most applications. Fine regulation, of the kind demanded by AC generators, cuts the speed variation by half or less.

Centrifugal governors sense engine speed with flyweights and throttle position with spring tension. Figure 5-28 illustrates the principle: As engine speed increases, the spinning flyweights open to reduce fuel delivery. As the throttle is opened, the spring applies a restraining force on the flyweight mechanism to raise engine speed.

So much for theory. In practice, centrifugal governors demonstrate a level of mechanical complexity that seems almost bizarre, in this era of digital electronics.

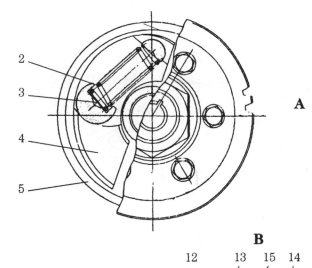

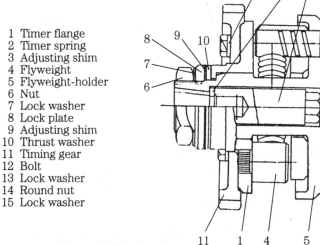

1 Timer flange
2 Timer spring
3 Adjusting shim
4 Flyweight
5 Flyweight-holder
6 Nut
7 Lock washer
8 Lock plate
9 Adjusting shim
10 Thrust washer
11 Timing gear
12 Bolt
13 Lock washer
14 Round nut
15 Lock washer

5-27 Centrifugal fuel timer varies injection timing in response to changes in engine speed.

But millions of these governors are in use and some discussion of repair seems appropriate.

Service

Governor malfunctions—hunting, sticking, refusal to hold adjustments—can usually be traced to binding pivots. In some cases, removing the cover and giving the internal parts a thorough cleaning is all that's necessary. If more work is needed, the governor and pump should be put into the hands of a specialist.

To appreciate why this is so, consider the matter of bushing replacement, which seems like a simple enough job. Using the CAV unit as a model, the critical bushings

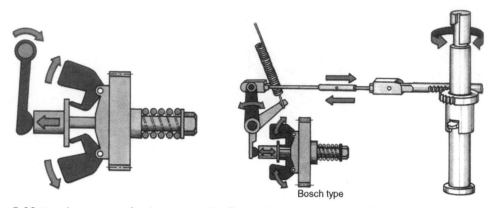

Bosch type

5-28 Bosch-type centrifugal governor. The flyweights react against throttle-spring tension to set the speed of the engine. As loads are encountered, the engine slows, the flyweights exert less force, and the spring extends the rack to restore engine rpm. Yanmar Diesel Engine Co., Ltd.

are those at the flywheel pivots (225 in Fig. 5-29) and the sliding bushing (231) on the governor sleeve. To remove and install the flyweights you will need CAV special tool 7044-8. Bushing press in, but factory reamers are needed to size them properly. And a governor with enough hours on it to wear out bushings will almost surely need recalibration, which can only be accomplished with a factory tooling that correlates piston stroke with rack position.

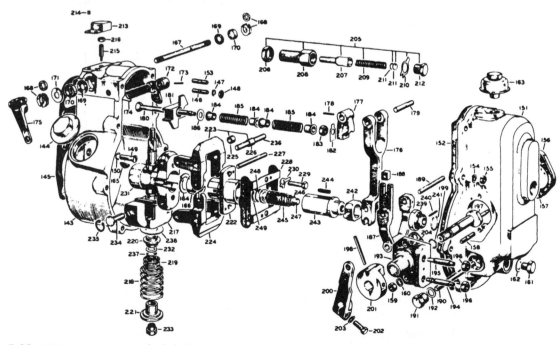

5-29 CAV governor in exploded view.

Pneumatic governors

Pneumatic, or flap valve, governors are less expensive than centrifugal governors and more amenable to repair. The velocity of air moving through the intake manifold is a function of piston speed. The faster the engine runs, the greater the velocity. Pneumatic governors sense this velocity as vacuum developed by a venturi mounted on the air intake.

The venturi restricts the flow of air and, so doing, reduces its pressure. This relationship has a constant of 20, as shown in Fig. 5-30. If we use consistent terms and multiply pressure times velocity at any point in the venturi, the answer is always 20, or something approximating that number. Nor need the venturi be the sort of streamlined restriction shown in the drawing; any impediment accelerates air flow and reduces its pressure.

Pneumatic governors, while by no means the state of the art, are still encountered on marine and industrial engines. The example discussed below is one of the most complex.

MZ

This unit, shown in cross-section in Fig. 5-31, uses a flap, or butterfly, valve (4) as a variable venturi. A tube bleeds vacuum from the edge of the valve to the left or low-pressure side of the diaphragm housing. A second tube brings filtered air at atmospheric pressure to the right side of the housing. The spring-loaded diaphragm (8) separates these two halves of the housing and connects to the fuel rack.

The flap valve is free to pivot in response to incoming air velocity. At low engine speeds the valve positions itself as shown in the drawing. All air entering the engine must accelerate and squeeze past the obstruction created by the edge of the valve and the vacuum-line connection. The diaphragm responds to the resulting depression by moving toward the left, or low-pressure, side of the housing. Fuel delivery is reduced. If the engine decelerates as when encountering a load, vacuum drops and the diaphragm shifts toward the right to increase fuel flow.

At wide-open throttle, the velocity of incoming air pivots the flap valve full open. Sensing the low level of vacuum, the diaphragm shifts to the right, extending the rack

The governor also includes the mechanism shown in Fig. 5-32 that reduces fuel delivery at idle. As the diaphragm moves to its extreme leftward position, it brings

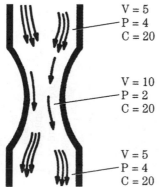

V = 5
P = 4
C = 20

V = 10
P = 2
C = 20

V = 5
P = 4
C = 20

5-30 The venturi constant.

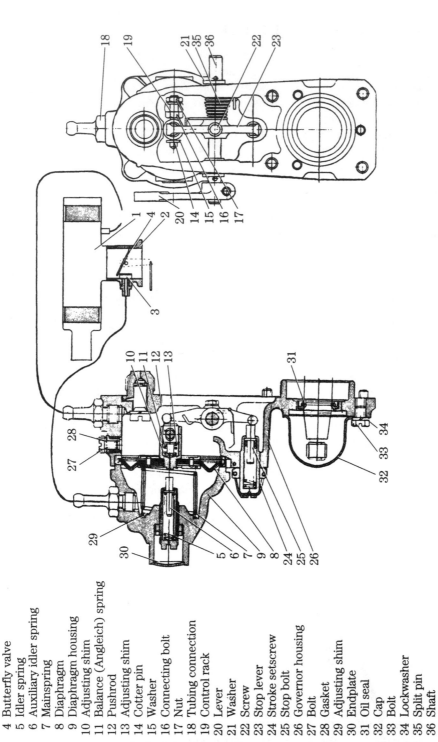

1 Air cleaner
2 Auxiliary venturi
4 Butterfly valve
5 Idler spring
6 Auxiliary idler spring
7 Mainspring
8 Diaphragm
9 Diaphragm housing
10 Adjusting shim
11 Balance (Angleich) spring
12 Pushrod
13 Adjusting shim
14 Cotter pin
15 Washer
16 Connecting bolt
17 Nut
18 Tubing connection
19 Control rack
20 Lever
21 Washer
22 Screw
23 Stop lever
24 Stroke setscrew
25 Stop bolt
26 Governor housing
27 Bolt
28 Gasket
29 Adjusting shim
30 Endplate
31 Oil seal
32 Cap
33 Bolt
34 Lockwasher
35 Split pin
36 Shaft

5-31 MZ centrifugal governor.

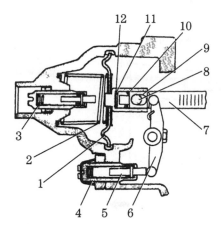

1 Diaphragm
2 Governor spring
3 Auxiliary idler spring
4 Stroke setscrew
5 Stop bolt
6 Stop lever
7 Control rack
8 Connecting bolt
9 Pushrod
10 Adjusting shim
11 Balance (Angleich) spring
12 Adjusting shim

5-32 MZ balance mechanism.

the stop lever (6) to bear against the stop bolt (5). This action compresses the balance spring (11) and displaces the rack (7) toward the no-fuel position. As engine speed increases, the diaphragm moves to the right. The balance spring mechanism moves with it, away from the stop lever. Once clear of the lever, the spring has nothing to react against and is out of the circuit.

Service

The leather diaphragm should hold 500 mm/H$_2$O of vacuum with a leak-down rate of no more than 2 mm/sec. It should be flexible enough to collapse of its own weight when held by the rim. Nissan dealers can supply the special diaphragm oil required to soften the leather.

A screw under the circular end cover reacts with the auxiliary idle spring (3) to set the idle speed. Tightening the screw moves the rack to the right to increase fuel delivery and idle rpm. But the effect extends to all engine speeds (Fig. 5-33). Fuel delivery for best idle over-fuels the engine at high speed and, conversely, a setting that gives best fuel economy at speed, costs low-end power. The adjustment is a compromise.

Other adjustments require a test fixture available to dealer mechanics. For example, the stroke adjustment, shown at 4 in Fig. 5-32, is made to factory specifications with the diaphragm chamber evacuated and the pump turning 500 rpm. To further complicate matters, balance-spring adjustment must be done by comparing its action with a known good unit.

Unit injection

Thus far, we have discussed three types of mechanical fuel systems: air blast, early common-rail, and jerk pump. The last of these systems, and in some ways the most attractive, is unit injection.

Unit injection combines the pumping function with injection. That is, each UI incorporates its own high-pressure pump driven from the engine camshaft. Injection

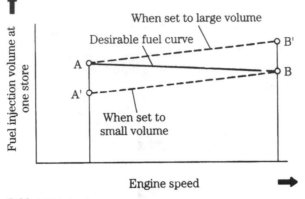

5-33 MZ injection curves.

pressures, often in excess of 20,000 psi, are confined entirely within the injector bodies. The lines that connect the lift pump with the injectors see only moderate pressures, on the order of 50 psi. Consequently, UI systems are virtually leak-proof. Mechanics, at least the ones old enough to remember first-generation Volkswagens and Detroit Diesel two-strokes, like the way UIs localize problems. When a UI fails, only the associated cylinder goes out. And because an engine will run with one or two bad injectors, it becomes its own test stand.

Figure 5-34 illustrates the standard Detroit Diesel fueling circuit. The restrictor draws off a portion of the incoming fuel to cool the injectors before returning to the tank. Two stages of filtering are provided, with the primary filter on the suction side of the fuel pump.

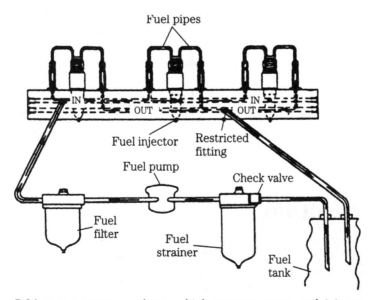

5-34 Unit injectors combine a high-pressure pump and injector nozzle in the same assembly. Detroit Diesel

In addition to the parts found on more conventional injectors, a UI has a mushroom-shaped cam follower, return spring, pump plunger, fuel-supply chamber, and one or more check valves (Fig. 5-35). The rack controls fuel delivery by rotating the plunger as described under "Inline pumps."

Fuel enters the injector and passes to the supply chamber. As the engine cam forces the plunger down, fuel trapped in the supply chamber comes under increasing

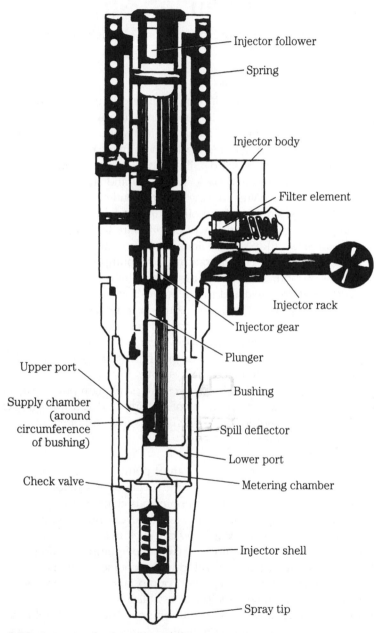

5-35 Cutaway of a Detroit unit injector.

pressure. At NOP, the needle valve lifts and injection begins. The disc-shaped check valve is fail-safe. Should needle fail to seat, the check valve closes to prevent air from entering the fuel supply.

Other than changing out plunger-related parts, unit injectors are not normally considered field-repairable.

Low-pressure system

Regardless of the type of high-pressure system, all engines employ similar low-pressure systems.

Lift pumps

A fuel supply, feed, transfer, or lift (the terms are interchangeable) pump supplies low-pressure fuel to the suction side of the injector pump or unit injectors. Most lift pumps are driven mechanically, although in recent years, electric drive has become popular. Pressure varies with 50 psi as a ballpark figure.

Stanadyne, Lucas/CAV and Bosch distributor-type injector pumps employ an integral vane-type lift pump as pictured back in Fig. 5-9 and subsequent drawings. The sliding vanes rotate in an eccentric housing with the outlet port on the periphery.

Gear-type lift pumps, similar to lube-oil pumps, develop pressure from the tooth mesh and usually incorporate a spring-loaded pressure-limiting valve.

Chrysler-Nissan SD22 and SD33 engines feed through a piston pump driven off the injector pump (Fig. 5-36). The operation of the pump is slightly unorthodox, in that fuel is present on both sides of the piston to eliminate air locks. As the piston retracts, fuel under it is expelled and the inlet check valve opens to admit fuel into the chamber above the piston (Fig. 5-37). As the piston rises, the check valve closes to permit the upper chamber to be pressurized. Fuel under the piston then reverses course to fill the void left by the rising piston. Near tdc the spring-loaded discharge-side valve opens. Spring tension determines pump pressure. The upper chamber can also be pressurized manually to bleed the system.

Small engines are often fitted with low-pressure diaphragm pumps such as the AC unit pictured in Fig. 5-38. The pump consists of a spring-loaded diaphragm activated by the engine camshaft. Inlet- and discharge-side check valves are often interchangeable.

Lift-pump service

Loss of pressure is the definitive symptom of pump failure, although it would be helpful if manufacturers provided output volume specifications. Before condemning the pump, check for air leaks as described in the previous chapter. A length of transparent plastic tubing spiced into the pump output line will reveal the presence of air bubbles.

Pump vanes wear in normal service and can stick in their grooves when the fuel turns resinous after an extended shutdown. Gear-type pumps should require no routine maintenance other than inspection during the course of engine overhaul. Wear tends to localize on the pump cover. When pump failure, as from a sheared drive gear, is suspected, it is sometimes possible to establish that the gears turn by inserting a fine wire into the outlet and cranking the engine. The wire should vibrate from contact with the spinning gears.

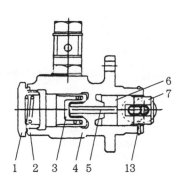

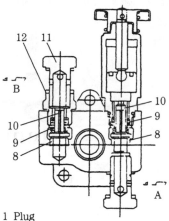

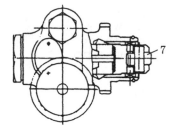

1 Plug
2 Piston spring
3 Piston
4 Feed pump housing
5 Push rod
6 Tappet
7 Roller
8 Check valve seat
9 Check valve spring
10 Check valve
11 Connector bolt
12 Nipple
13 Snap ring
A From the fuel tank
B To the fuel filter

5-36 Piston-type Chrysler-Nissan lift pump.

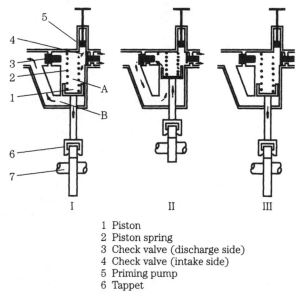

1 Piston
2 Piston spring
3 Check valve (discharge side)
4 Check valve (intake side)
5 Priming pump
6 Tappet
7 Camshaft

5-37 Operation of the unit shown in the previous illustration. Marine Engine Div., Chrysler Corp.

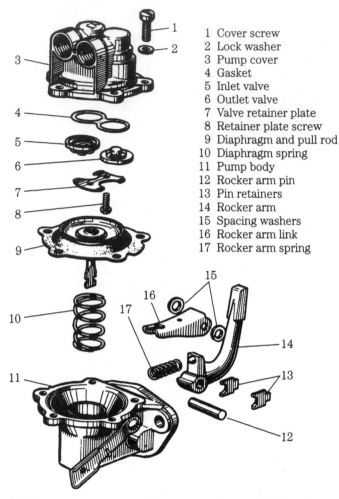

1 Cover screw
2 Lock washer
3 Pump cover
4 Gasket
5 Inlet valve
6 Outlet valve
7 Valve retainer plate
8 Retainer plate screw
9 Diaphragm and pull rod
10 Diaphragm spring
11 Pump body
12 Rocker arm pin
13 Pin retainers
14 Rocker arm
15 Spacing washers
16 Rocker arm link
17 Rocker arm spring

5-38 AC diaphragm-type lift pump used on Bedford trucks.

Check valves are the weak point of reciprocating (piston or diaphragm) pumps. Most check valves can be removed for cleaning or replacement. When this is not possible, a piece of wire inserted into the fuel entry port should unstick the ball, at least temporarily. Diaphragms should be changed periodically and the housing cleaned every 200 hours or so of operation (Fig. 5-39). When installing these pumps, it is good practice to bar the engine over so the pump lever rides on the lower part of the cam.

Fuel filters and water separators

Most engines incorporate three stages of filtration—a screen at the tank and paper-element primary and secondary filters. Filters are rated by the size of particle trapped: a typical installation would employ a 30-micron primary filter and a secondary

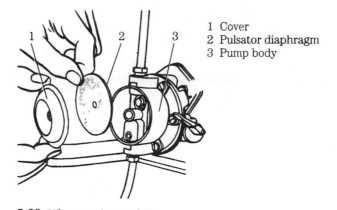

1 Cover
2 Pulsator diaphragm
3 Pump body

5-39 Lift pump inspection. Ford Industrial Engine and Turbine Operations

of between 10 and 2 microns. A water separator is often built into the primary filter (Fig. 5-40), which also may include a connection for fuel return and an electric heater to prevent fuel gelling and waxing in cold climates.

When a filter element is changed, entrapped air must be bled as described in Chapter 4.

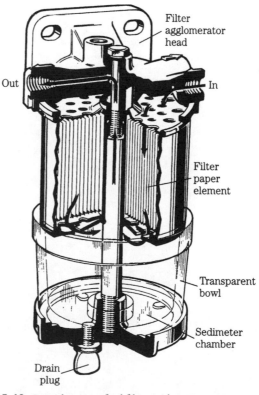

Filter
agglomerator
head

Out

In

Filter
paper
element

Transparent
bowl

Sedimeter
chamber

Drain
plug

5-40 Cartridge-type fuel filter with water separator.

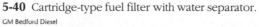

GM Bedford Diesel

6
CHAPTER

Electronic management systems

This chapter describes the operation of four popular engine management systems (EMS)—Bosch, Caterpillar, Ford Power Stroke, and Detroit Diesel. The workings of the onboard computer, electronic injectors, and other key components are discussed, together with basic troubleshooting and some repair procedures. But readers should understand that this material is merely an introduction to these complex systems and that it does not substitute for factory documentation and training.

Background

Before we get down to the nitty-gritty of repair and troubleshooting of specific systems, it is helpful to take a brief look at how electronics have transformed the diesel engine from an industrial workhorse to power for luxury automobiles. Most of this work was done in Europe and primarily by Robert Bosch GmbH.

Modern common-rail systems

Since the 1960s, engineers realized that an updated common-rail system, using computer-controlled injectors and ultra-high fuel pressures, offered revolutionary possibilities for the diesel engine. As shown in Fig. 6-1, the modern c-r system employs a remote pump to pressurize a fuel rail, which functions both as a reservoir and as an accumulator. The rail expands to dampen pump pressure peaks and contracts to stabilize pressure when the injectors open. Since rail pressure is almost constant, fuel can be injected at will, independent of pump plunger movement. When coupled with split-shot electronic injectors, fuel delivery begins early during compression stroke and can be initiated after combustion to light-off carburized particulate traps. In other words, common-rail was an enabling technology for electronic fuel injection.

Two Nippon Denso engineers—Shokei Itoh and Mashiko Miyaki—were responsible for the first commercial version of the system, which appeared on 1995-model

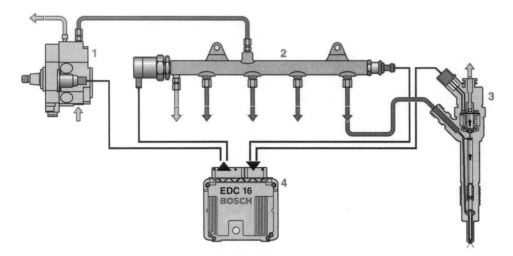

6-1 Basic components of the modern common-rail system are a high-pressure pump (1), common rail (2), electronic injectors (3), and onboard computer (4). Courtesy Bosch

Hino Rising Ranger trucks. Meanwhile Fiat and its subsidiary Magneti Marelli Power Trains were working on c-r for passenger cars and light trucks. By the late 1990s, the prototype was turned over to Robert Bosch GmbH for production. The first-generation 1350-bar (1 bar = 14.51 psi) Bosch c-r system made its debut in 1997 on Alfa Romeo and Mercedes-Benz high-speed touring cars.

Subsequent Bosch developments were rapid:

- 1999—First generation, truck 1480-bar system (Renault).
- 2001—Second generation, passenger-car 1600-bar system (Volvo and BMW).
- 2002—Second generation, truck 1600-bar system (MAN).
- 2003—Third generation, passenger-car 1600-bar system (Audi V-6). Piezo injectors reduced emissions by 20%, boosted power 5%, reduced fuel consumption 3%, and engine noise by 3 dB(A).
- 2006—Fourth generation under development with higher pressures and revised injector geometry.

Since the common rail is not protected by patent (low-pressure versions have been around since the 1920s), the technology has become nearly universal for automotive and light truck engines, manufactured by companies as diverse as Hyundai, Cummins, and Mercedes-Benz. Nor is c-r limited to automotive applications: L'Orange GmbH has prototyped a c-r upgrade for marine engines with cylinder bores as large as 500 mm, and Cummins will soon release a common-rail kit for retrofit to locomotives.

Solenoid-actuated injectors

Figure 6-2 is a schematic of a multi-hole, solenoid-operated injector of the type used on common-rail systems. The upper end of the spring-loaded needle is flat and the large-diameter lower end is shouldered. Hydraulic pressure applied to the flat

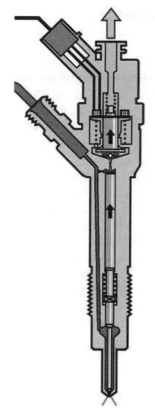

6-2 Bosch electronic injector initiates injection with ball-type spill valve, just visible below the solenoid in the drawing.
Courtesy Bosch

tends to push the needle down, while pressure bearing against the shoulder would raise it. Movement depends upon the balance of pressures acting on these surfaces and upon spring tension.

A ball-type spill valve stands between high-pressure fuel from the rail and the low-pressure circuit that returns fuel to the tank. When at rest, the solenoid is not energized and the armature spring holds the ball valve closed. No fuel flows to the tank, but both needle working surfaces see the same high pressure. The forces generated—lift from the shoulder and down-force from the flattened top—neutralize each other and spring tension keeps the needle seated.

When the solenoid is energized, the armature retracts and the ball valve opens to the low-pressure return line. Because the orifice is quite small, most of the pressure drop is confined to the area above the valve. Pressure in the nozzle cavity remains high. This pressure lifts the needle to commence injection. When de-energized, the solenoid closes the ball valve, pressures equalize and the spring forces the needle down to end injection.

Even though hydraulics do the heavy work, solenoids require large amounts of electric power. To overcome armature inertia, the drag of heavy fuel, and other operating variables, Bosch solenoids receive a 50V, 20A jolt during the opening phase. Once the armature is retracted, battery voltage holds it in place.

Suspect injectors must be farmed out to specialists for cleaning, repair, and re-calibration.

Piezo-actuated injectors

Piezo injectors evolved from the flow-control technology developed by Siemens VDO for ink jet printers. Piezo crystals expand when excited by voltage and, conversely, generate voltage when compressed. The latter feature is used in cigarette and barbecue-pit lighters.

The growth of piezo crystals under voltage is miniscule: a single wafer 80 μm thick elongates about 0.1 μm at 160V. Siemens stacks the wafers to produce a stroke of about 40 μm that is hydraulically amplified to unseat the spill valve. The wafer pack responds within 80 milliseconds of excitation, or three times faster than the best solenoids, and with an actuation delay an order of magnitude faster. As a result, injection can be divided into as many as seven precisely metered shots, each accurate to within less than a milligram of fuel. This translates into 3% more power and 15% better fuel economy than solenoids offer, and the ability to meet Euro 4 standards without particulate traps.

These injectors run on the same high-pressure regime as solenoid types: piezo-actuated common-rail injectors, like the one shown in Fig. 6-3, see peak pressures of around 1600 bar, a figure that will be increased to 2000 bar in the next generation. Piezo unit injectors develop pressures in excess of 2000 bar. Depending upon the application, injector tips have as many as 10 orifices with diameters of less than 0.1 mm.

To date Siemens has built more than 15 million c-r injectors, used on upscale vehicles such as the Audi, Mercedes, Land Rover, Jaguar, and the 2007 Mercedes E320, which will introduce the technology to the North American market. Siemens also manufactures piezo unit injectors for VW and International. The latter operation is based in Blythewood, South Carolina, with all production going to the American diesel maker.

Solenoid injectors from Bosch, Denso, and Delphi continue to find buyers since they cost half as much as the Siemens version and meet current emission standards. Most observers believe that solenoids will hold their market share among low-end car makers until 2010 when Euro 5 standards come into effect.

Figure 6-3 shows a piezo injector in cutaway view. Wafer expansion opens a servo valve that stands between the high-pressure fuel delivery circuit and the low-pressure return. With the valve open, pressure above the needle diminishes, while pressure in the nozzle cavity remains nearly constant. The differential unseats the needle to initiate injection.

Because wafer expansion is so miniscule, a hydraulic coupler is used to extend the stroke. The coupler also functions like a hydraulic valve lifter to compensate for lash caused by thermal expansion.

Service

While injectors are routinely replaced in the field, it should be noted that each injector is marked with a code, showing where it falls within production tolerances. For best performance, the code must be inputted into the EMS with a factory-supplied scan tool.

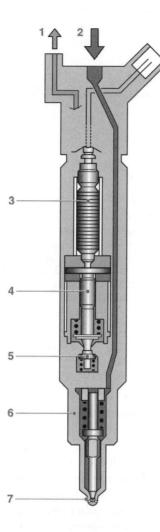

6-3 A piezo common-rail injector in cross-section, showing the fuel return (1), fuel inlet (2), piezo stack (3), hydraulic coupler (4), control valve (5), nozzle assembly (6), and injection orifices (7). Courtesy Bosch

Unit injection

While common rail is standard on most European auto and light-truck engines, Volkswagen remains committed to electronic unit injection. Figure 6-4 illustrates the Bosch unit injector of the type currently used by VW and its SEAT and Skoda affiliates. A solenoid-actuated valve controls injector timing and duration, which of course, must occur during periods when the UI is pressurized by the camshaft. Caterpillar and Detroit Diesel developed their own versions of electronic unit injectors (EUIs) described below.

In summary, computer-controlled c-r and unit-injector (UI) systems came about because of the need to reduce exhaust emissions by providing more precise control over fueling. Secondary effects are higher power outputs, more flexible torque curves, and better fuel economy.

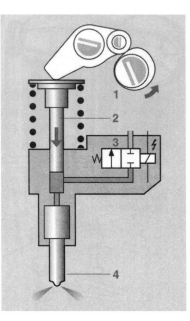

6-4 Cross-section of a solenoid-actuated EUI showing engine camshaft (1), plunger (2), solenoid-actuated spill valve (3), and nozzle (4). Piezo unit injectors are also available.
Courtesy Bosch

Analog versus digital

Engine management systems communicate through voltage changes that take an analog or digital form. Analog signals fluctuate over time, as shown by the waveform voltage in Fig. 6-5. The instantaneous value—the value of the signal at any moment in time—conveys data. Voltage spikes, bad grounds, and loose connections seriously compromise the ability of analog circuits to act as reliable messengers. Even so, most sensors are analog devices that respond to changes in pressure, temperature, or fluid level by varying their internal resistance. Supply voltage is constant (normally 5V +/− 0.5V), and output voltage changes as sensor resistance increases or diminishes. In order to process the analog data, the computer converts these voltages into digital form.

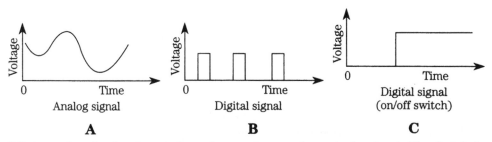

6-5 An analog signal varies continuously over time, as shown in drawing A. The digital signals in B and C have only two values—"on" (full voltage) and "off" (zero voltage).

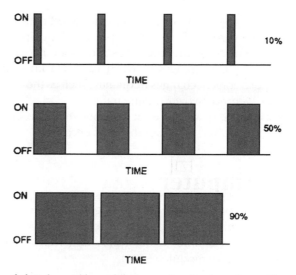

6-6 Pulse width modulation varies signal on-time with changes in the quality measured. Courtesy Caterpillar Inc.

Digital signals have only two states—"on" corresponding to 1, and "off," corresponding to 0. All the computer need do is to discriminate between binary on and off states, or bits. Depending upon the system, a string of 8, 16, or 32 bits make up a word, which conveys information. Circuit noise can destroy part of a word, making it unintelligible, but noise generally does not alter the meaning of complete words in the systematic fashion required to fool the computer.

A digital signal can also communicate data by means of pulse width modulation (PWM). As illustrated in Fig. 6-6, the width of the signal represents its duration expressed as the ratio of the percentage of "on time" divided by the percentage of "off-time." The most common application of PWM is for throttle position sensors that typically generate a pulse width of 15–20% at idle and 90% or more at full throttle.

Bosch CAN bus

A controller area network (CAN) employs a single line, or bus, to convey data to all components that make up the network. Individual components—computers, sensors, display panels, stepper motors, and other devices—respond only to those signals addressed to them. These components, or stations, still need power lines, but one signal line suffices for all. Multiplexing is not too different than old-fashioned telephone systems that rang every phone on the line. By counting the number of rings, you knew who should pick up the receiver.

To make production changes easier, message content, rather than the name of the intended receiving station, functions as the address. This enables new stations to be added without modifying the transmitters. An 11- or 29-bit word prefaced to each data packet identifies message content. Thus, engine-speed data will be received by all stations, but will be of interest only to the computer and the operator display panel.

Upon acceptance of a message, the receiving station scans it for errors and acknowledges receipt. Transmission speed ranges between 125 kb (1 kb = 1000 bits/ second) and 1Mb. Messages whose content changes frequently, such as the voltage drop that signals injector opening, have priority over routine traffic.

The CAN protocol has fairly wide use for automotive and truck applications, and will, experts say, become universal within the next decade.

On-board computer

The computer that oversees engine functions goes under various names— electronic or engine control unit (ECU), power train control module (PCM), or microcontroller. We'll follow the example of Caterpillar and Detroit Diesel and call the computer an electronic control module (ECM).

Central processing unit

The central processing unit or microprocessor (CPU) chip contains millions of tiny transistors that integrate sensor data and operator commands to generate the outputs necessary to control the engine (Fig. 6-7). The CPU performs arithmetic calculations, makes decisions on the basis of algebraic logic, and consults data held in memory. Data usually takes the form of maps, which plot changes in one parameter against two others. For example, injection timing is plotted against engine load and rpm. The CPU could calculate the required timing from sensor inputs, but it's faster to look up the answer on a map.

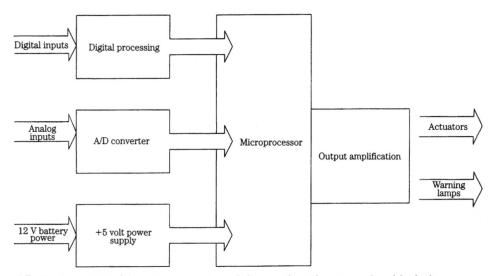

6-7 The basic ECM (electronic control module or onboard computer) in block diagram.

Memory

Transient data, such as ambient air temperature and fuel pressure that can be forgotten when the engine is shut down, are stored in random access memory (RAM represented diagrammatically in Fig. 6-8). Upon restarting, RAM collects new data, updated hundreds of times a second. Operating instructions, diagnostic routines, and other constants reside in read-only memory (ROM), where they remain as if chiseled in stone.

Flash memory is a semi-permanent repository that, like a Swiss bank account, requires a closely held password for access. Manufacturers use flash to adapt standard ECMs to various engine families, each of which has unique operating parameters. Factory technicians, armed with the correct password, can "reflash" this type of memory with updated programs and patches to correct the bugs that invariably appear in service. The Motorola 6800, a pioneer 8-bit microprocessor found in early Caterpillar systems and in millions of GMC automobiles, had 160 kb of flash memory and 32 kb of RAM. The 32-bit Adem 2000, now specified by Cat, has 256 kb of RAM and 1 Mb of flash. The Motorola MPC 500, currently specified by Magneti Marelli Power Trains for automotive applications, has similar capabilities.

Peripherals

In order to interact with the external world, the CPU requires peripherals, such as a converter to translate analog signals from temperature and pressure sensors into digital format. Driver circuits, often working through relays, provide the amperage for electronic injectors, fan motors, turbo wastegates, and other current-hungry actuators. One or more timers, controlled by oscillating crystals, give the CPU the precise time reference necessary to orchestrate injection timing and duration relative to inputs from the camshaft position sensor.

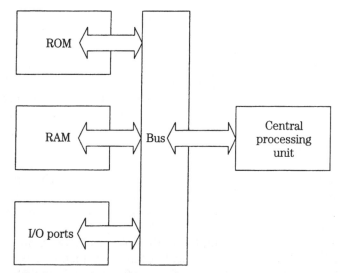

6-8 The central processing unit carries on an internal conversation with its several types of memory and communicates with the external world through input/output ports.

Computer capabilities evolve faster than the hardware: the MPC5500, released in 2003, can actuate electronic intake and exhaust valves, vary camshaft timing and control electro-mechanical braking, when these features become available.

Sensors

Advanced c-r systems for motor vehicles include the following sensors:

- Temperature—ambient, coolant, intake air, fuel, and exhaust (upstream and/or downstream of turbo). Most of these sensors are negative temperature coefficient types, which lose internal resistance as temperature increases. Coolant and exhaust temperature sensor elements collect deposits that should be periodically removed.
- Pressure—atmospheric, fuel, turbo boost, crankcase, exhaust, and compressed air (for heavy trucks). Exhaust-gas pressure sensors monitor pressure drops across particulate traps.
- Position—rack, idle mode, variable-geometry turbocharging (VGT) control, turbo wastegate, back-pressure, and exhaust gas recirculation (EGR) valves. These sensors generally take the form of a switch or variable resistor.
- Mass air flow entering the manifold.
- Oxygen content in exhaust gas.
- Velocity—crankshaft rpm, road speed and, for some applications, turbo rpm. A coil responds to shifts in the magnetic field by generating a small signal voltage.

One-off EM systems, custom-built for existing installations, often retain the mechanical governor as a way to reduce costs. Production systems incorporate electronic governing, whose operation is controlled by the computer in concert with a magnetic engine-speed sensor. The ECM also uses speed-sensor data to calculate injector pulse width. Figure 6-9 illustrates sensor operation, which uses gear teeth as markers. Some speed sensors pick up the rpm signal from a single gear tooth, which has a distinctive shape and magnetic signature.

In addition, a Hall-effect sensor, triggered by slit on the camshaft gear, generates a timing reference and, in some applications, reports engine speed. Failure of the camshaft position sensor shuts the engine down, often without warning. Bosch systems have a fail-safe feature that enables the computer to calculate timing from the crankshaft speed sensor when the primary reference is lost. The engine will continue to run, but starting may be more difficult. For some Cat models, either sensor can stand in for the other.

As mentioned earlier, accelerator position is reported as changes in the band width (duration of voltage) generated by the throttle position sensor (TPS). These sensors are not without problems, and Bosch again comes to the rescue by providing a second, backup throttle position sensor.

A mass air flow (MAF) sensor measures the volume of air entering the engine as a function of its cooling effect on a heated film or a platinum wire (Fig. 6-10). Supporting electronics measure how much current is needed to maintain the film at target temperature, which is about 75° F above ambient. As engine speed and air flow increase, correspondingly more current is needed. Current draw appears as an

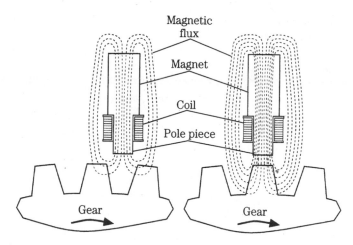

6-9 Engine speed sensors read crankshaft rpm as voltages magnetically induced in a sensor Gear teeth are convenient markers. These circuits may also respond to a magnetic disturbance caused by a tdc reference in the form of a modified gear tooth or a permanent magnet on the rim of the gear. Dynalco Controls, a unit of Main Controls Corp.

6-10 The hot-film air mass sensor measures the amount of current needed to maintain its sensing element, one side of which is cooled by incoming air, at a predetermined temperature. Courtesy Bosch

analog output voltage of between 0 and 5V. MAF sensors lose accuracy when contaminated and can be ruined by rough handling.

One would imagine that the computer, capable of several hundreds of thousands operations a second, would exert absolute control over the engine. But things get out of hand in ways the computer cannot predict. For example, fuel quality varies with each fill-up, injector nozzles wear unevenly, humid air exerts a greater cooling effect on MAF sensors than dry air. These and a hundred other variables affect combustion efficiency. The O_2, or lambda, sensor closes the control loop by monitoring the level of oxygen in the exhaust gases.

Unlike other sensors, the platinum-coated zirconium-oxide O_2 sensor generates its own voltage. Once it reaches operating temperature (approximately 700°C), the sensor develops nearly 0.7V in fuel-rich environments. As less fuel is burned and the oxygen content of the exhaust approaches that of the atmosphere, sensor voltage drops off to nearly zero. Signal voltage exhibits a sawtooth pattern. Normally, the computer uses this data, in conjunction with air-flow data, to deliver no more fuel than needed for combustion. Under load, the computer relaxes its standards a bit, richening the mixture for more torque. The O_2 sensor then functions as a smoke limiter. The sensor also influences the amount of exhaust gas recirculated into the combustion chambers.

Inline engines have an O_2 sensor threaded into the exhaust manifold at Y, where temperatures are high and exhaust gases are representative of all cylinders. V-type engines replicate the arrangement for each bank. An electrically heated four-wire (two wires to ground, one carrying signal voltage and the other battery voltage to the heater) O_2 sensor will also be installed downstream of the catalytic converter. This sensor monitors converter efficiency and has no effect upon mixture strength. OBD-2 sensors are good for 100,000 miles, although response slows with age and contamination.

In event of malfunction, the EMS can disregard the O_2 sensors and enter into open-loop operation. Working without the benefit of feedback, the computer adjusts fuel delivery to programmed values. Open-loop operation is, with good reason, called the "limp-home" mode.

Actuators

Sensors provide data to the ECM and actuators carry out the computer's commands. The primary actuators are the electronic injectors, part and parcel of any EMS. Another almost universal type of actuator is the electronic governor. Figure 6-11 sketches the "drive-by-wire" governor used on the Deere 7L 6076 H. A switch on the electronic control module enables the operator to select any of three speed-control programs—true all-speed governing (as used in vehicles), min-max governing (for generator applications), and full-power boost. The latter option has a timer associated with it to prevent engine damage.

Other actuators control fuel pressure, turbo boost and, in some applications, exhaust back pressure.

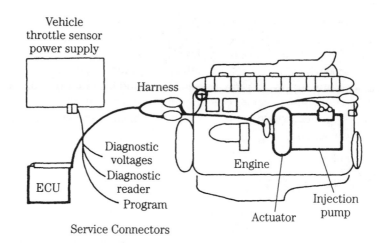

6-11 "Drive-by-wire" throttle control used on some Deere engines.

Tools

Negotiating one's way around computerized engines requires an investment in special tools. A basic troubleshooting technique is to listen in on the "conversation" between the ECM and its array of sensors and actuators. The ECM has some degree of diagnostic capability, in that it reacts to abnormal sensor and actuator behavior by generating trouble codes. These codes are the starting points for all diagnostic work.

Fault codes can be retrieved by counting the blinks made by the check engine lamp and looking up what the numbers mean. By law, motor vehicles sold in the United States must have their fault codes published on the Internet. (Try to find them.)

It is much simpler to use a scanner that reads off codes directly and, depending upon how much you are willing to pay, retrieves codes that have been erased, shows raw sensor data going into the computer, energizes actuators, and reports pulse-width percentages. Readers who simply want to stay current with the status of their vehicle can make do with an inexpensive scanner that reads active trouble codes and erases them with the push of a button.

While scanner-to-engine connections are standardized, the standards vary. OBD-2 (On-Board Diagnostics, 2nd generation) has been mandated for autos and light trucks sold in this country since the early 1990s. Protocols vary with the manufacturer. For example, we have PWM (Ford), VPW (General Motors), and ISO 9141 (Chrysler, European, Asian) and a scattering of Keyword 2000, aka ISO 14230. Diesel makers often specify SAE J 1587. The Bosch CAN protocol has become increasingly popular and, according to industry watchers, may become universal after 2008. The scanner must be compatible with the protocol of the engine under test.

Factory technicians have a decisive advantage in diagnostic tooling. Testers such as the Bosch KTS 650, Caterpillar Electronic Analyzer and Programmer, and Ford NGS (New Generation Star) were developed in tandem with the systems they monitor and often by the same personnel. These tools translate trouble codes, display raw sensor data, have limited reflash (reprogramming) capability, cycle injectors and other actu-ators, and display operating parameters a few seconds before and after a fault has been flagged. All report injector band width. The Ford NGS tool includes a buzz test for HEUI injectors, which while not definitive, is helpful. In addition, Hartridge and other European firms supply an array of bench test and calibration equipment.

You will also need an accurate digital multimeter, crimping tools for the various types of harness connectors, and, a breakout box, similar to the one described under "Throttle position sensor."

Troubleshooting

As far as they go, the troubleshooting procedures described in Chap. 4 apply to computerized engines. For example, blue- or blue-gray smoke means lube oil is being burned. Oil can enter the chambers by way of faulty valve or turbo seals, or past worn piston rings. EMC-equipped engines may have additional sources of oil intrusion, such as HEUI injector seals.

The first step in troubleshooting is to interview the operator to obtain as much information as possible about the complaint that might have nothing to do with the engine. For example, a truck driver's technique (or lack of it), vehicle effi-ciency, or operating conditions (such as weather, loads, and fuel) can have seri-ous effects upon engine performance and fuel economy. Custom-built fixed and marine installations require careful scrutiny to make sure they have been designed properly.

Once the problem has been narrowed down to the engine, walk around the unit looking for anything amiss, with emphasis on the cooling, air inlet, exhaust, and fuel systems. Table 6-1 is a generalized troubleshooting guide. More specific procedures are provided in the sections that deal with Caterpillar and Ford systems.

CAUTION: Do not disconnect or make up harness connectors or actuators while the engine is running or with the ignition key "on." The resulting voltage spikes can damage the computer.

Low-pressure fuel circuits should be tested for transfer-pump pressure, air leaks, and restrictions as described in Chap. 4 and in more detail in this chapter.

High-pressure fuel circuits take us into the realm of diagnostic trouble codes (DTCs) and highly specific repair procedures for which engine manufacturers issue multi-volume service manuals. But while no serious repair effort should be contem-plated without this documentation, certain generalizations can be made.

Broadly speaking, the most vulnerable components are:

- Harness connectors. Look for misaligned or loose connector pins, and for evidence of overheating. Wiggling the harness can sometimes reveal loose pin connections that generate transient faults in service. Arrange harness clamps to provide slack at the connectors.
- Sensors and sensor wiring. Most sensors have three leads—voltage input, voltage output, and ground. Determine that voltage is present (which means that the input circuit is probably okay) and that the output is neither shorted to ground nor open. Bad grounds, that is, loose tie-downs, rust, paint, or

Table 6-1. Basic troubleshooting—EMS-equipped engines

Item to check	Action
Trouble codes	Scan active and past trouble codes, and make the necessary repairs.
Fuel contamination	Look for evidence of water, macrobiotic growth, and solids.
Fuel specific gravity	Test with a hydrometer. No. 2 diesel should have a specific gravity (S.G.) of 0.840 or higher. Use of No. 1 diesel can cut power by as much as 7%. Mixing Nos. 1 and 2, a common practice in cold climates, costs a proportional amount of power.
Fuel shutoff valve—mechanical or solenoid	Make sure the valve opens fully.
Fuel temperature	Although fuel temperature varies with the application, 85°F is a reasonable figure. Every 10° increase in temperature above 100°F cuts power by 1%. Restrictions —crimped lines, clogged filters, sharp bends—increase pumping loads and heat. Verify that the fuel heater functions normally.
Restrictions in fuel filters and associated components	Change fuel filters. Look for restrictions at points where fuel flow is disturbed with particular attention to the water separator, check valves, and flow meter. Examine the lines for physical damage.
Tank vent restriction	A clogged vent causes the tank to develop vacuum as fuel is consumed.
Fuel leaks	Check for fuel leaks, which occur most often on the high-pressure side of the system.
Air intrusion	Air usually enters on the suction side of the transfer pump. Look for bubbles in the injector return line and in the filter-to-injector-pump line.
Turbo boost	A rubbing turbo wheel, bent blades, oil leaks, or failure of the VGT to function properly reduce turbo boost. Potential leak sites include the exhaust plumbing to the turbo and, on the outlet side, the intake manifold-gasket and aftercooler. Replace the air filter unless obviously new. Excessive turbo boost with normal fuel delivery indicates retarded injector timing, which works against the additional boost to reduce power, increase fuel consumption, and raise exhaust temperatures. These high temperatures accelerate exhaust-valve and turbocharger failure.
Air charge temperature	Consult the shop manual to determine the permissible charge-air temperature. In normal operation, an efficient aftercooler should hold the temperature rise to no more than 30°F over ambient. Maximum temperatures, as generated during full-throttle dyno tests of turbocharged engines, should be about 325°F without an aftercooler, 245°F with a jacket-water aftercooler, and 150°F with an ambient air cooler.
Injector-pump timing	Check injector-pump timing against manufacturer's specs with a dial indicator as described in Chap. 4.
Misfiring cylinders	Locate weak or misfiring cylinders by shutting down one injector at a time with the appropriate factory tool while the engine is running. Replace the associated injector(s). If that does not restore cylinder function, make a compression test as described in Chap. 4.

(*Continued*)

Table 6-1. Basic troubleshooting—EMS-equipped engines (*Continued*)

Item to check	Action
EGR function	Refer to manufacturer's manual for test procedures. Remove carbon deposits from valve and manifold inlet porting. Check turbo-pressure control and monitoring devices.
Exhaust restriction	Check exhaust piping for damage; verify back pressure against manufacturer's specs. Exhaust back pressure readings of more than 27 in./H_2O for turbocharged engines or 34 in./H_2O for naturally aspirated engines are cause for concern.
Parasitic loads	Check for high current draws, air leaks in compressed-systems, and other parasitic loads that reduce engine power output.
Governor seals	Look for damage to factory-sealed adjustments that, for example, can indicate that governed idle speed has been increased in an attempt to boost full-throttle power output. Using an accurate tachometer (not the control-panel unit), compare idle speed against the factory specification.

grease on the chassis or engine ground studs—can affect any circuit and generate noise that is fatal for band-width signals. If the circuitry is functional, the problem lies with the sensor that, in the absence of factory documentation, should be tested by substitution of a known good one. The camshaft position sensor has been a chronic problem on some engines, the throttle-position sensor on others.

- Actuators that live in bad environments. Exhaust recirculation valves (EGR) carbon over and stick, the exhaust back-pressure valve on Power Stroke exhaust systems rusts and binds, injectors malfunction, and glow-plug control modules sometimes melt.

  WARNING: Do not disconnect, test, or otherwise disturb injector wiring while the engine is running. Voltages are hazardous.

- Damage caused by contaminated fuel, off-road excursions, "high-performance" air filters and so on.

High-pressure components—computer-controlled injector pumps, fuel-pressure regulators and electronic injectors—do not lend themselves to field repair. Repairs must be accomplished by professionals equipped with the necessary test and calibration tooling. Denso has invested millions to upgrade its factory repair centers with Hartridge AVM2PC test benches.

Less expensive components—sensors, EGR valves, and control modules—are sacrificial items, discarded when they fail.

CAUTION: The speed of a diesel engine is entirely a function of fuel delivery. A mistake made when servicing the fuel system can result in a runaway engine. Caterpillar suggests that a helper stands by ready to block the air intake with a steel plate the first time the engine is started. The helper must be careful not to allow his fingers to be caught between the plate and the intake-manifold flange, especially if the engine is turbocharged.

Caterpillar EMS

Figure 6-12 illustrates the EMS used on Caterpillar 3116, 3176, 3406E, and 3500 engines. It is a relatively simple system, without the bells and whistles mandated by current emissions regulations. The ECM mounts on the engine, which reduces the electromagnetic radiation given off by the harness, simplifies packaging, and enables the computer to be cooled by fuel. A major engineering effort was required to isolate the electronics from the heat, vibration, solvents, steam, and water blasts that engines are exposed to.

All sensors, with the exception of the oil pressure sensor, input data for efficient fuel allocation. Abnormally low or high sensor readings cause the ECM to set one or

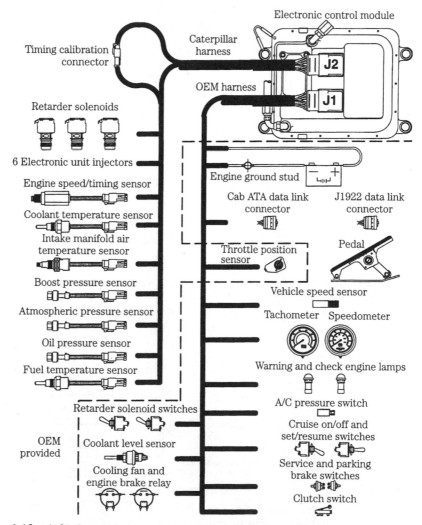

6-12 3406E Cat engine management system in block diagram.

more trouble codes, which can be retrieved by connecting a scan tool to the SAE J1922 data link connector. The only computer-controlled actuators in this particular system are the electronic unit injectors (EUIs), cooling fan, and cruise control.

A more sophisticated system used on current C-10, C-12, and C-15 truck engines is illustrated in Fig. 6-13. Sensor and actuator functions for this and the earlier Cat system

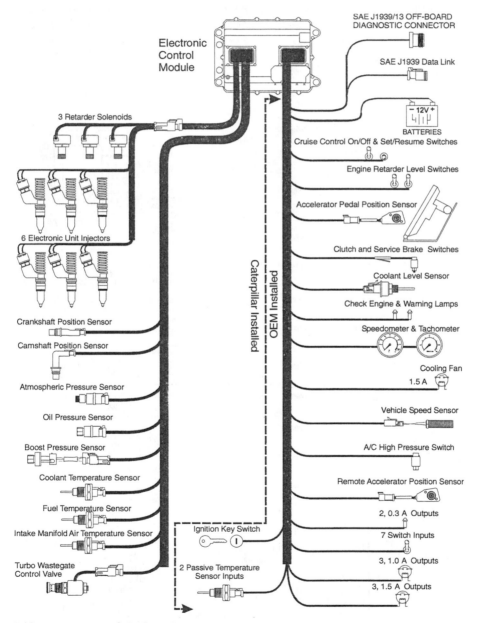

6-13 C-10, C-12, and C-15 engine management system. Courtesy Caterpillar Inc.

Table 6-2. Component overview—caterpillar EMS

Nomenclature	Description
Electronic unit injector (EUI)	Unit injector with injection timing and volume controlled by a solenoid slaved to the ECM.
Atmospheric pressure sensor	Absolute pressure sensor measuring pressure from 0–16.8 psi. Supplied with 5V.
Oil pressure sensor	Absolute pressure sensor measuring oil pressure from 0–165 psi. Data communicated to service tools is absolute pressure less atmospheric pressure.
Coolant temperature sensor	Thermistor not requiring an external voltage supply. Used to monitor engine operation during cold starts and to reduce power output when coolant temperature is excessive.
Inlet manifold air temperature sensor	Thermistor not requiring an external voltage source. Provides data for control of the inlet air heater, cold idle, and cooling fan.
Fuel temperature sensor	Used to adjust fuel delivery and to limit engine power when fuel temperatures exceed 86°F (30°C). Maximum power reduction occurs at 178°F (70°C). Higher fuel temperatures flag a trouble code.
Crankshaft and camshaft position sensors	Report engine rpm and timing data for fuel injection. Normally, the ECM monitors both the camshaft and crankshaft position sensors during cranking and the crankshaft sensor while running. If either sensor fails, the engine will start and run using data supplied by the other.
Turbo wastegate actuator	A solenoid that regulates turbo boost.

are outlined in Table 6-2. Figure 6-14 shows the location of EMS components on the C-10 and C-12 engines. Sensor location is similar for the C-15, except that boost and intake-manifold pressure sensors are on the left side forward rather than on the right.

Electronic unit injector

Both the 3100-3500 and C series engines employ EUIs that combine mechanical actuation with electronic control over fuel volume. Figure 6-15 illustrates an EUI in cross-section. A roller-tipped rocker arm, acting on in the injector plunger, provides the force necessary to generate the 28,000-psi injection pressure. The solenoid valve, shown on the left of the drawing, opens the spill/fill port to admit fuel into the injector barrel during the period of plunger retraction. A spring-loaded check valve, located near the injector tip and set to open at 5000 psi, remains closed during the fill process.

Injection can be initiated any time after the plunger starts its downward travel. But until the ECM signals the solenoid valve to close the spill/fill port, fuel merely cycles through the EUIs as a coolant and as a purge to remove any entrapped air.

Upon signal, the solenoid valve closes the port, trapping fuel in the injector barrel. Further downward movement of the plunger raises fuel pressure sufficiently to overcome spring tension acting on the check valve. The check valve opens and injection begins.

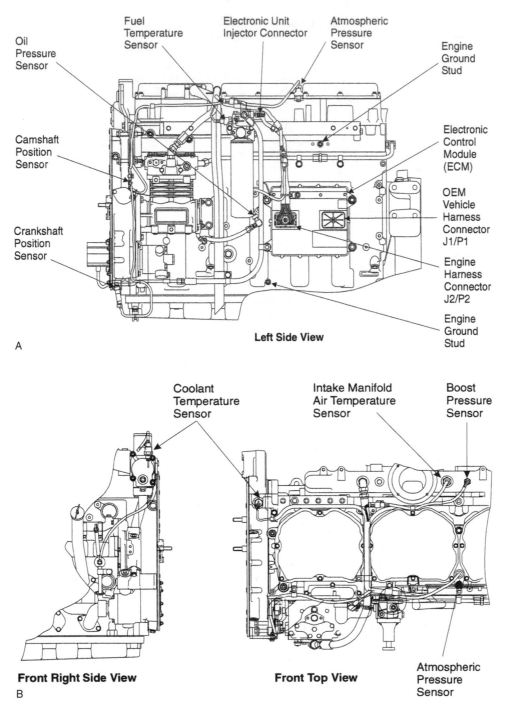

Oil
Pressure
Sensor

Fuel
Temperature
Sensor

Electronic Unit
Injector Connector

Atmospheric
Pressure
Sensor

Engine
Ground
Stud

Camshaft
Position
Sensor

Electronic
Control
Module
(ECM)

OEM
Vehicle
Harness
Connector
J1/P1

Crankshaft
Position
Sensor

Engine
Harness
Connector
J2/P2

Engine
Ground
Stud

Left Side View

A

Coolant
Temperature
Sensor

Intake Manifold
Air Temperature
Sensor

Boost
Pressure
Sensor

Front Right Side View

B

Front Top View

Atmospheric
Pressure
Sensor

6-14 EMS component location, C-10 and C-12 Caterpillar engines. Left-side view (A) and right-side and top view (B). Courtesy Caterpillar Inc.

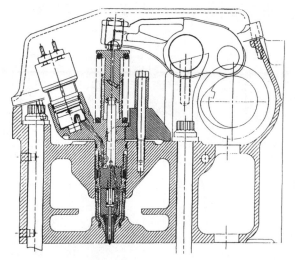

6-15 EUI in cross-section. The injector is a hybrid, with electronic control of injector timing and duration, and mechanical actuation.

Injection continues until the ECM signals the solenoid to open the spill/fill port. Responding to the sudden loss of pressure at the injector tip, the check valve snaps shut. Injection ceases. The plunger continues its downward stroke, displacing fuel through the open spill/fill port and into the fuel-return gallery.

Although the software has primary responsibility for injector timing, plunger movement is affected by rocker-arm lash. The onset of pressure rise depends upon the clearance between the rocker arm and the plunger. A Caterpillar PN J 35637 height gauge is used to establish this critical variable (Fig. 6-16).

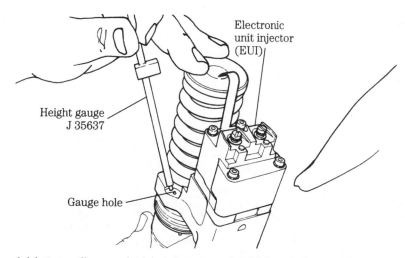

6-16 Caterpillar-supplied height gauge should be used to position EUI injectors relative to the camshaft.

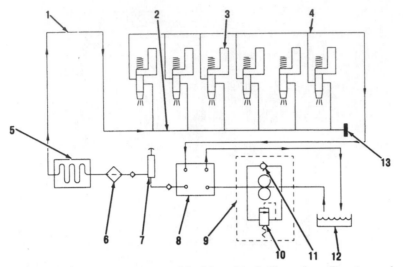

6-17 EUI fuel system consists of fuel line (1), fuel-supply gallery integral with cylinder head (2), EUIs (3), fuel-return gallery (4), fuel-cooled ECM (5), filter (6), manual priming pump (7), distribution block (8), transfer pump (9), pressure regulating valve (10), check valve (11), fuel tank (12), and drain plug (13). Courtesy Caterpillar Inc.

Low-pressure circuit

Figure 6-17 illustrates the low pressure circuit for EUI-equipped engines. A gear-type transfer pump (9) delivers fuel from the tank (12) to the injectors (3). The pump incorporates a check valve (11) that permits fuel to flow around it during manual priming and a pressure-relief valve (10) to protect the system from over-pressurization.

Cat supplies dealer mechanics with a special tool, consisting of a pressure gauge, a sight glass, and the necessary fittings for makeup to mechanical and electronic low-pressure systems. This tool is similar to the Ford tool described in Chap. 4.

Check fuel pressure by removing the fuel pressure sensor at the base of the fuel filter. Install a pressure gauge and run the engine at rated rpm and load. Pressure should be 90 psi. If pressure is 75 psi or lower, check the fuel level in the tank, verify that the fuel cap vent is open, and that supply and return lines are in good condition. Pay particular attention to the return line, which can collapse if exposed to excessive heat. Replace the fuel filter. If necessary, replace the lift pump.

If fuel pressure is high, that is, 100 psi or greater, remove the fuel-regulating valve (10) and make sure the orifices near the tip of the unit are open. Flush out any debris and attempt to find the source of the contamination. Check for a clogged fuel-return line. Verify that the transfer-pump relief valve functions.

High-pressure circuit

Deactivating one EUI at a time will identify misfiring or dead cylinders. As mentioned previously, current values are high, and for your own safety, do not tamper with EUI wiring while the engine is running. Professional-quality scanners can safely deny power to the injectors so that the effect, if any, on engine performance can be gauged. When a weak cylinder is located, replace the injector and retest. If a new injector does not solve the problem, check valve lash to make certain that the valves are seating and, if necessary, make a cylinder compression as described in Chap. 4. Caterpillar technicians use a boroscope, inserted through the injector mounting boss, to visually inspect cylinder internals.

Hydraulic/electronic unit injector

HEUI evolved from the EUI, and since mid-1994 has been used on Caterpillar, International T444E, DT466E, 1530E, certain Perkins models, and, most famously, on the Ford (International) Power Stroke. Unlike conventional camshaft-driven unit injectors, HEUIs are actuated by high-pressure crankcase oil. Since injection is no longer tied to camshaft motion, fuel can be injected at any crank angle. At higher-than-idle speeds, injector pressure is independent of engine rpm and can attain values of 30,000 psi.

Figure 6-18 illustrates the general layout of the Cat system that differs only in detail from Ford and other HEUI applications. Figure 6-19 shows the arrangement of the piping for the 3126B truck engine, which is the focus of this discussion. The high-pressure oil pump delivers crankcase oil to the injectors at cranking pressures of 500 psi that increase to as much as 3300 psi under speed and load. The computer-controlled injector actuation pressure control valve (IAPCV) regulates oil pressure by shunting pump output to the crankcase. Fuel pressure is regulated to 55 psi minimum.

HEUI consists of five major components—a solenoid, poppet valve, mushroom-shaped intensifier piston, barrel, and a seven-hole nozzle assembly (Fig. 6-20). Approximately 4A is required to energize the solenoid and 1.5A holds it open.

Upon command, the solenoid opens the poppet valve to admit high-pressure lube oil to the upper end of the intensifier piston (6 in Fig. 6-20). The oil forces the piston and plunger (7) down against fuel trapped in the injector body by a ball check valve. The fuel end of the plunger is smaller in diameter than the upper, or oil, end. The difference in cross-sectional area multiplies the hydraulic force acting on the fuel 7.5 times, to produce injection pressures of 3000–30,000 psi. Pressurized fuel lifts a second check valve, enters the cavity on the lower right of the drawing, and reacts against the needle shoulder to overcome spring tension. Nozzle-opening pressure (NOP) varies with the application. Some Cat injectors have an NOP of 4500 psi. Rebuilders of Power Stroke HEUIs look for 2750 +/− 75 psi.

Note that both band width (the time the injector is open) and actuator-oil pressure determine the amount of fuel delivery.

HEUI injectors have undergone constant refinement. The most significant change is the split-shot PRIME (preinjection metering) version that first appeared on 1994 California Ford F series trucks and has since become almost universal.

These injectors work like other HEUIs, except that the plunger incorporates a radial groove that receives fuel through six bleed ports drilled in the face of the plunger. As the piston moves downward, the groove aligns with a spill port in the barrel, shunting pressure. The output-side check ball seats and injection stops. Further piston movement masks the port and injection resumes.

Injector service

The three external o-rings are critical. The upper o-ring, now supplied with steel and elastomer backup rings, prevents lube oil from leaking up and out of the injector

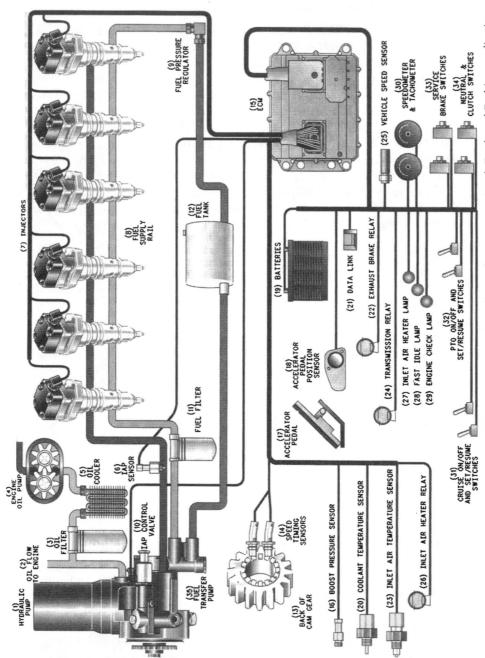

(1) HYDRAULIC PUMP
(2) OIL FLOW TO ENGINE
(3) OIL FILTER
(4) ENGINE OIL PUMP
(5) OIL COOLER
(6) IAP SENSOR
(7) INJECTORS
(8) FUEL SUPPLY RAIL
(9) FUEL PRESSURE REGULATOR
(10) IAP CONTROL VALVE
(11) FUEL FILTER
(12) FUEL TANK
(13) BACK OF CAM GEAR
(14) SPEED TIMING SENSORS
(15) ECM
(16) BOOST PRESSURE SENSOR
(17) ACCELERATOR PEDAL
(18) ACCELERATOR PEDAL POSITION SENSOR
(19) BATTERIES
(20) COOLANT TEMPERATURE SENSOR
(21) DATA LINK
(22) EXHAUST BRAKE RELAY
(23) INLET AIR TEMPERATURE SENSOR
(24) TRANSMISSION RELAY
(25) VEHICLE SPEED SENSOR
(26) INLET AIR HEATER RELAY
(27) INLET AIR HEATER LAMP
(28) FAST IDLE LAMP
(29) ENGINE CHECK LAMP
(30) SPEEDOMETER & TACHOMETER
(31) CRUISE ON/OFF AND SET/RESUME SWITCHES
(32) PTO ON/OFF AND SET/RESUME SWITCHES
(33) SERVICE BRAKE SWITCHES
(34) NEUTRAL & CLUTCH SWITCHES
(35) FUEL TRANSFER PUMP

6-18 HEUI schematic for the Caterpillar 3126B truck engine is broadly similar to International, Ford, and Perkins applications.

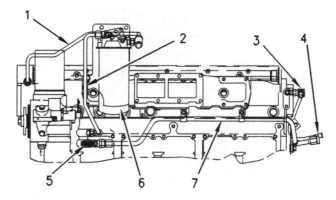

6-19 HEUI plumbing incorporates drilled galleries in the cylinder head for fuel delivery and return, tubing between engine components, and hoses. Major components as shown for the Cat 3126B engine are fuel filter to fuel-supply passage (1), lift pump to filter (2), pressure regulator (3), tank return line (4) fuel-filter inlet for the lift pump (5), filter (6) and fuel supply line (7).

body. The middle o-ring and its supporting rings prevents lube oil from mixing with the fuel. Sudden and dramatic increases in oil consumption can usually be attributed to failure of this middle ring assembly. The lower o-ring keeps fuel from puddling around the base of the injector.

Lubricate o-rings and wait ten minutes or so for the rings to assume their natural shape before installing injectors or other parts sealed by o-rings. Failure to do so can result in leaks. Seals on all plumbing, especially the fuel-return lines, are critical. Do not disturb connections without new seals in hand.

Injector failure is not always easy to diagnose, since rough idle, long cranking periods, and misfires at particular rpm's can have multiple causes, several associated with the high-pressure lube oil system. This subject is explored in more detail in the context of Power Stroke systems. It can be helpful to check individual exhaust runner temperatures with an infrared thermometer. All exhaust-manifold runners should be at or very close to the same temperature: a hotter-than-average runner indicates the associated cylinder is receiving too much fuel, and a cool runner means insufficient fuel. Both of these conditions point to injector failure.

HEUI injectors are not field repairable in the sense that the average mechanic can do the work. Rebuilt units are available from factory and independent suppliers such as Dipaco.

HEUI hydraulics

A seven-piston swash-plate pump, located forward on the left side of the 3126B engine, supplies oil pressure to the injectors through a passage in the engine cylinder head. Oil returns to the sump through ports under the valve cover. The pump, also used by Ford and International, is nonrebuildable.

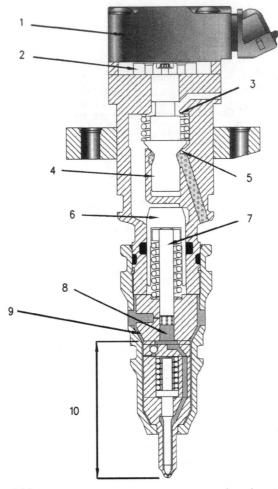

6-20 HEUI injector in cross-section, with solenoid (1), armature (2), upper poppet seat (3), poppet valve (4), lower poppet seat (5), intensifier piston (6), plunger (7), plunger cavity (8), barrel (9), and nozzle assembly (10). Force multiplication occurs between the oil and fuel ends of the intensifier piston. Courtesy Caterpillar Inc.

The injector actuation pressure control valve (IAPCV), mounted aft of the pump, regulates oil pressure (Fig. 6-21). The poppet valve floats in suspension between pump-discharge pressure that tends to move it to the left and force generated by the solenoid that pushes it to the right. When the oil pressure over-balances the solenoid, the poppet valve opens to shunt a portion of pump output back to the crankcase. Pressure falls.

How much force the solenoid generates depends upon the amount of current it receives. The ECM compares oil-pressure data, collected 67 times a second from a

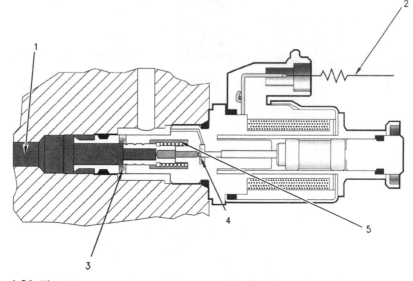

6-21 The injector actuator pressure control valve (IAPCV) regulates pressure by shunting oil from the pump discharge back to the sump. Outlet flow to injectors is shown at 1, return flow at 5. In this drawing the solenoid is energized, opening the poppet valve (4) to divert a portion of pump output back to the crankcase (5) through the drain port (3). Flow to the injectors exits on the left (1). Courtesy Caterpillar Inc.

sensor mounted in the cylinder-head gallery, with desired pressure and adjusts current intensity accordingly.

Note that HEUI injectors in this application require 870 psi to fire, a figure that varies with ambient temperature. The engine will not start unless firing pressure is reached. Pressure ranges between 580 psi during initial cranking to 3350 psi with 3500 psi occasionally seen.

When the IAPCV fails, replacing the o-rings can sometimes save the cost of a new part. One should be very careful when servicing these valves, since wrench torque can easily distort the housing. Caterpillar supplies a special tool for this purpose, which can be described as a flair-nut crow's-foot wrench. That is, the box end of the wrench has been partially cut away so that it will slip over the valve body. Torque to 37 ft-lb.

Fuel pressure

The piston-type transfer pump runs off an eccentric on the back of the injector oil pump. Nominal fuel pressure is 65 psi at the pump. When the measurement is made downstream of a new filter, expect to see a pressure drop of about 5 psi. If pressures are low, clean the inlet screen on the pump and replace the fuel pressure regulator with a known-good unit. If the problem persists, replace the transfer pump.

Boost pressure

The amount of boost relative to engine speed and load depends upon the application, with the specification given for 28.8 in./Hg dry barometric pressure, 77°F

(25°C) ambient temperature, and 35 API gravity fuel. If the air is denser, that is, if barometric pressure is higher and/or temperature lower, boost pressure will be higher than specified. Likewise, a heavier fuel increases turbo output. When documentation is unavailable, consult the factory for boost specifications. The wastegate is nonadjustable.

Low manifold pressure combined with high exhaust temperature suggests that the air filter is clogged or that the aftercooler, pressure-side turbo plumbing, or intake-manifold gasket leak.

Throttle position sensor

The TPS is located under the accelerator pedal and generates a pulse-width signal that can read with a Fluke Model 97 or equivalent multimeter and a breakout box. The breakout consists of three 12-in. long No.10 AWG wires that bridge the harness connectors and three 6-in. long No. 12 AWG wires that make up to the meter.

1. Using Caterpillar PN 1U5804, crimp Deutch DT 04-3P-E008 male connectors to one end of each of the three long wires and DT 04-3P-E003 female connectors to the other ends. See Fig. 6-22.
2. Make up DT 04-3P-E003 female connectors to one end of each of three short wires. These connectors plug into the meter receptacle.
3. Solder the three short wires to the centers of the longer wires.

Set the meter to measure pulse-width percentage and, with the ignition switch "on," depress the accelerator pedal. Pulse width varies with make and model, but should be in the neighborhood of 80–90% at wide-open throttle and drop to 10 or 20% with the pedal at rest. A spare TPS is good insurance.

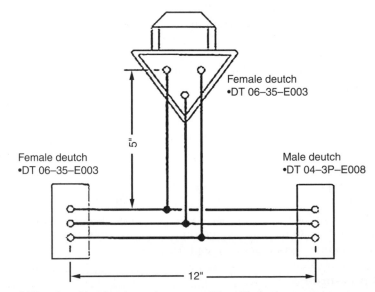

6-22 A breakout box must be compatible with the harness connector used. <small>Courtesy Caterpillar Inc.</small>

Table 6-3. Partial listing of caterpillar HEUI trouble codes

PID and FMI	Flash code	Description
01–11, 2–11	72	Cylinder 1 or 2 fault
3–11, 4–11	73	Cylinder 3 or 4 fault
5–11, 6–11	74	Cylinder 5 or 6 fault
22–13	42	Check timing sensor calibration
41–02, 41–03	21	8V supply above or below normal
42–11	18	IACPV fault
64–02, 64–11	34	Loss of engine rpm signal, or loss of signal pattern—No. 2 speed and timing sensor
91–08	32	Invalid throttle signal
91–13	28	Check throttle sensor calibration
102–03, 102–04	25	Boost pressure sensor circuit open or shorted
105–03, 105–04	38	Intake-manifold air temp. sensor circuit open or shorted
108–03, 108–04	26	Atmospheric pressure sensor circuit open or shorted
110–10	61	High coolant temp. warning
110–03, 110–04	27	Coolant temp. sensor circuit open or shorted
111–01	61	High coolant temperature
111–02	12	Coolant level sensor fault
164–00	17	Injector-actuation pressure out of range
164–02	15	Injector actuator pressure sensor erratic
164–03, 164–02	15	Injector actuator pressure circuit open or shorted
164–11	39	Injector actuation pressure system fault
168–02	51	Intermittent battery power to ECM
190–02, 190–11	34	Loss of engine rpm signal or no pattern to signal—No. 1 speed and timing sensor

Trouble codes

Table 6-3 lists the most critical DTCs for this system. PID (parameter identifier) is a two- or three-digit SAE numerical code assigned to each component. For example, 91 refers to the throttle position sensor. FMI is a failure mode identifier used to describe the kind of failure detected, and flash refers to the number of blinks the code triggers on the "check engine" lamp. Some codes are merely informational and do not affect engine performance. Active codes, that is, those that turn on the diagnostic lamp and keep it on, mean that the problem requires immediate attention. The lamp will go out when the malfunction is repaired. Intermittent malfunctions, often caused by loose harness connectors or bad grounds, cause the lamp to blink and go out. These codes are then logged in computer memory, where they can be retrieved and erased with the appropriate scanner. Intermittent codes that are logged repeatedly need investigation.

Ford (International) 7.3L DI power stroke

Four engines, three of them built by International, go by the name "Power Stroke." The first was the direct injection turbocharged 7.3L V-8. This engine, also known as the International T444E, powered F series trucks and E series vans between 1994 and 2003. The 32-valve 6.0L Power Stroke, or International VT 365, has been used on pickups and vans since 2004. Production of the 6.0 will end in 2007, when it will be replaced by a 6.4L V-8. The Ford V-6, aka DT17, has been used on low cab forward (LCF) trucks since 2005.

In many ways, the original 7.3L Power Stroke is the most interesting of the group. It carried Ford to first place in light diesel truck sales and generated a cult following not seen since the days of the Mustang. People have taken out second mortgages on their homes to buy these trucks.

The material dealing with the operation of HEUI injectors, the high-pressure oil pump, and the IAPCV also applies to the Power Stroke, although Caterpillar and Ford parts rarely interchange and pressure specifications vary. Table 6-4 translates Ford acronyms and describes sensor function. "Supply" is the voltage going to the sensor; "signal" is the sensor output to the computer (ECM).

Slow or no-start

Begin with a walk-around the vehicle to determine that the tank is at least a third full, batteries are charged, battery terminals are clean and secure, and the radiator is topped off. Check the crankcase and high-pressure pump reservoir oil levels. Ford requires a 15W40 oil that meets SH, CG-4, and Mack E-0L specifications, and recommends that Motorcraft 15W40 "black-lid" be used because of its resistance to aeration. The oil level should come to within one inch of the top of the pump reservoir.

Question the operator to learn as much as possible about the problem and conditions that led up to it. Did it appear suddenly? After repairs to the engine? What parts were replaced? Does the malfunction occur at all engine temperatures or only during cold starts?

Mechanics differ about what checks to make first. But, however you sequence them, the main bases to touch with 7.3L are:

1. Replace the air filter and examine the exhaust system for restrictions. The exhaust butterfly valve, which should cycle closed on cold startups, may stick shut. The back-pressure regulator that controls this valve can also fail.
2. Retrieve active (current) and historical trouble codes. P1111 means that no codes are set and, as far as the ECM knows, things are normal.
3. Drain a sample of fuel from the filter while cranking. If water is present, drain the tank(s) and refill with clean fuel before proceeding. Inspect the water separator for oxidation damage and, if necessary, replace it.
4. Check the glow-plug relay, which is a high-mortality item and especially so in its original oval form. The newer round-case relay interchanges with the older unit. The red cable, connected via fusible links to the starter relay, should have battery voltage at all times. If voltage is not present, check the fusible links and connections for opens. The heavy 10-gauge brown wire

Table 6-4. Power stroke sensor nomenclature and function

AP	Accelerator position sensor
	Along with other variables, the ECM uses the AP signal to determine injector oil-pressure, pulse width and timing. Supply 5.0V +/– 0.5V, signal 0.5–0.7V at idle, 4.5V at wide open throttle.
BARO	Barometric pressure sensor
	The ECM uses the BARO signal to adjust fuel timing, fuel quantity, and glow-plug on-time during high altitude operation. Supply 5.0V +/– 0.5V, signal 4.6V at sea level, decreasing at higher altitudes.
CMP	Camshaft position sensor
	The ECM uses the CMP signal to monitor engine rpm and tdc for Nos. 1 and 4 cylinders. This Hall-effect sensor generates a digital voltage signal of 12.0V high, 1.5V low.
DTC	Diagnostic trouble code
EBP	Exhaust back pressure sensor
	The ECM uses the EBP signal to control the exhaust pressure regulator (EPR). Supply 5.0V +/– 0.5V, signal 0.8–1.0V @ 14.7 psi at idle, increasing with engine speed and load, decreasing with altitude.
EOT	Engine oil temperature sensor
	The ECM uses the EOT signal to control glow-plug on time, EPR, idle rpm and fuel delivery and timing. Supply 5.0V +/– 0.5V, signal 4.37V @ 32°F, 1.37V @ 176°F, 0.96V @ 205°F.
EPR	Exhaust back pressure regulator
	The EPR operates hydraulically from oil taken off the turbocharger pedestal mount. When intake air temperature is less than 37°F (50°F on some models) and the engine oil temperature is less than 140°F (168°F on some models) the ECM energizes a solenoid valve that causes oil pressure to close a butterfly at the turbo exhaust outlet. The valve opens under load and as the engine warms.
GPC	Glow plug control
	The ECM energizes the GPC relay for 10–120 seconds depending on engine oil temperature and barometric pressure.
GPL	Glow plug light
	The ECM turns the "wait to start" lamp "on" for 1–10 seconds, depending on engine oil temperature and barometric pressure.
GPM	Glow plug monitor
	Used on 1997 and later California vehicles to report if glow plugs malfunction.
IAT	Intake air temperature sensor
	The ECM uses the IAT signal to regulate exhaust backpressure. Supply 5 V+/– 0.5V, signal 3.90V @ 32°F, 3.09V @ 68°F, 1.72V @ 122°F.
IPC	Injection pressure control sensor
	The ECM uses the IPC signal to match fuel delivery with load and to stabilize idle rpm. Signal 1.00V @ 580 psi, 3.22V @ 2520 psi.
IDM	Injector driver module
	The IDM receives cylinder-identification and fuel-demand signals from the ECM, and generates a 115 VDC, 10A signal for the appropriate injector, varying pulse width as required.
IPR	Injection pressure regulator
	The ECM varies the duty cycle of the IPR to control oil pressure and the volume of fuel delivered. 0% = full return to sump (open valve), 100% = full flow to injectors (closed valve). Functioning is monitored by the Injector pressure control sensor.

(Continued)

Table 6-4. Power stroke sensor nomenclature and function (*Continued*)

IVS	Idle validation switch
	The IVS is an on-off switch that signals the ECM when the engine is idling. Signal 0V at idle, 12V off-idle.
MAP	Manifold absolute pressure sensor
	The MAP measures manifold pressure to limit turbo boost, optimize timing, and reduce over-fueling and smoke. Signal frequency: 111 Hz @ 14.7 psi, 130 Hz @ 20 psi, 167 Hz @ 30 psi.
MIL	Malfunction indicator lamp
	"Check engine" or "service engine" lamp that the ECM illuminates when certain system faults are present. Can also be used to retrieve trouble codes.
PCM	Powertrain control module
	PCM is the onboard computer that receives sensor inputs, calculates output signals to actuators, and generates diagnostic codes. The computer also controls transmission shift points, anti-skid braking, and other powertrain functions. Referred to in this text as the ECM.
PID	Parameter identification
	PID, or the data stream, is the sensor data read by scan tools.

going to glow plugs is hot when the relay is energized. One of the 18-gauge wires (usually red with a light green tracer) is the signal wire, energized by the lube-oil temperature sensor. This wire must have battery voltage for the relay to function. Depending upon the circuit, the "wait-to-start" lamp may signal when the glow plugs energize or merely count off seconds. Individual glow-plugs should have a resistance of about 0.1Ω to ground when cold and 2Ω or more after cylinder temperatures normalize.

5. Check fuel pressure while cranking for 15 seconds. If pressure is less than 50 psi, replace the fuel-filter element, remove any debris on the fuel screen and on the screen protecting the IPR deceleration orifice. See the "IPR" section for more information.

6. Check injector-oil pressure during cranking. A pressure of 500 psi must be present to enable the injectors. Expect to see 960–1180 psi at high idle and 2500+ psi during snap acceleration. If the computer senses that the IPR has malfunctioned, it holds idle oil pressure at a constant 725 psi. Trouble code 1280 means low IPR signal voltage, 1281 high signal voltage, and 1212 abnormally high (at least 1160 psi) oil pressure with engine off. See "Low oil pressure" below.

7. While cranking, check for the presence of an rpm signal. A failed CMP can hold oil pressure below the 500-psi injector-enabling threshold. Refer to "Camshaft position sensor" section below.

8. Check power at the injector solenoids with an appropriate scan tool. Zero pulse width on all injectors means a bad CMP sensor or IDM. Failure may trip trouble code 1298. The high power levels—10A at 115 VDC—put severe demands on the injector drive module. In addition, modules on Econoline 7.3L vans built before 4–11–96 have problems with water intrusion.

WARNING: Injector voltage can be lethal. Do not disconnect or otherwise tamper with the injector wiring while the engine is running or the ignition key is "on."

9. Finally, check for air in the circuit by inserting a length of transparent tubing in the return fuel line.

Erratic idle

A frequent complaint is a rough idle, which can originate from the fuel system or from one or more weak cylinders. Follow this procedure:

1. Verify that the disturbance at idle is not caused by exhaust pipe contact with the vehicle frame, a loose air-conditioner compressor, or other mechanical sources.
2. Scan for trouble codes.
3. Perform steps 1 through 9 in the preceding "Slow or no-start" section.
4. If you have access to a Ford factory tester, make a "buzz" test on the injectors, listening for differences in the sound. A bad injector often buzzes at a lower frequency than the others.
5. Remove the valve covers and, with the engine idling, watch the oil flow out the injector spill ports. Each time a HEUI closes, the oil charge above the plunger drains out the spill port. Thus, amount of oil spilling out of an injector relates to fuel delivery. Replace any injector that passes noticeably less oil than the others.
6. Test engine compression as described in Chap. 4.
7. As a last resort, replace the injectors.

Low oil pressure

If you have not already done so, check the oil level in the pump reservoir, which should come to within an inch of the top of the unit. If the engine has been idle for a long period, oil drains back out of the reservoir, making for hard starting.

Verify that there are no leaks at the line connections, around HEUI bodies and into the fuel system. Remove the valve covers and look for cracks in the casting, caused by mis-drilled oil galleries. Small leaks can be detected by pressurizing each gallery with an injector tester or a grease gun and a 3000-psi pressure gauge. Charge the gallery with oil at 2000 psi and wait several hours for any leaks to register on the gauge.

The next step in the search for causes of low oil pressure is to check the IPR (functionally similar to the Caterpillar IAPCV). If the 5V +/− 0.5V reference signal is present and the output circuit is not shorted or open, replace the regulator o-rings with parts available from Ford International or aftermarket sources. The DF6TZ-9C977-AN Dipaco repair kit services IPRs used before and after engine SN 187099. If o-rings do not solve the problem, the regulator should be tested by substitution of a known good unit and, if necessary, replaced. Torque to 35 ft/lb and do not use sealant that could clog the orifice on the threaded section.

Air in lube oil

Aeration often results in hard or no-starts, erratic idle, shutdowns on deceleration, and loss of rpm. CG4/SH oil helps control the problem. As originally specified, 1994-model engines held 12 quarts of crankcase oil. This specification has been revised to 14 quarts. The dipstick should be recalibrated or replaced with Ford PN F4TZ-6750-E.

Leaks at pump pickup tube o-ring can be detected by jacking the rear wheels 10 in. off the ground and overfilling the crankcase with three quarts of oil. If the idle smoothes out, replace the o-ring seal.

In addition to its effect on performance, air in the oiling system causes rapid injector wear. Whenever an injector is changed, the associated cylinder bank should be purged:

- Disconnect the CMP to prevent the engine from starting.
- Mechanical fuel pumps (pre-1998 models)—disconnect the return line at the fuel pressure regulator on the fuel filter. Crank the engine in 15-second bursts, allowing ample time for the starter to cool and until a steady stream comes out of the line.
- Electric fuel pumps—crack a vent port on each cylinder head and run the pump until air is purged.

If head galleries have been drained, prefill the galleries with pressurized oil as described under "Low oil pressure."

Injector drive module

The IDM delivers a low-side signal to control the timing, duration, and sequence of injector opening. The high-side delivers 115 VDC at 10A to the HEUI solenoids.

WARNING: Do not disconnect, attempt to measure or otherwise disturb IDM circuits while the engine is running. Voltages are lethal.

Air in fuel circuit

Air leaks can cause rough idle, shutdowns on deceleration, and no-starts. The time required for air intrusion to make itself known depends upon the proximity of the leak to the transfer pump. For example, the engine might start and run several minutes before air entering from a fuel tank connection reaches the pump. Air entry closer to the pump has more immediate effects.

Multiple leak points—Schrader valves, bleed screws, fuel-line connections—make the filter/water separator a prime suspect. Use OEM clamps on hose connections, new seals, and dope screw threads with sealant. Surplus fuel from the injectors recycles back to the filter. The connection at the filter incorporates a check valve that closes when the engine stops. Should the check valve leak, air from the return line will be drawn into the filter as fuel in the canister cools and shrinks.

Camshaft position sensor

Sensors for mid-1994 to 1996 models are marked C96 or C 97; C-92 sensors were used from late-1997 to the end of production. These OEM units can cause no-start, hard-start, erratic idle, and shutdown on deceleration problems unless shimmed 0.010 in. off their mounting surfaces.

Replacement sensors have been shortened and do not require shimming. The CMP for mid-1994 to 96 models carries International part number C92 F7TZ-12K03-A, and can be recognized by the tin-coated connectors. The latest C98-F7TZ-12K073-A version features gold-plated connectors, which gives one an idea of the problems associated with these units. And while the newer CMP has better reliability than the previous type, its gold connectors present an electrolysis problem when mated with the tin connectors used on early 7.3L production. Replacement sensors cost about $100 from International. It's good insurance to carry a spare.

The CMP mounts on the front of the engine, adjacent to the camshaft, and is secured by two 10-mm bolts. Trouble code 0344 will be flagged if sensor response is intermittent; 0341 means that enough electrical noise has been detected to affect engine operation. Check connectors for loose, bent, or spread pins and clean grounds. Code 0340 indicates the absence of a sensor signal while cranking. In this case, the engine will not start.

Detroit Diesel

Now in its sixth generation, DDEC (Detroit Diesel Electronic Control or "dee-deck") was the first EMS designed for heavy-duty diesels. Later versions follow European practice by incorporating VGT, large amounts of EGR, and exhaust aftertreatment.

As described in more detail in Chap. 9, VGT generates boost across the whole rpm band. It also enables large amounts of exhaust gas to be recirculated under load, when EGR is needed most. Exhaust aftertreatment traps particulates and converts oxides of nitrogen, which are smog precursors, into nitrogen and water.

The complexity of later DDEC systems and a design philosophy that ties the product closely to service facilities severely limits what can be done without access to factory documentation and a DDR/DDL (Diagnostic Data Reader/Diagnostic Data Link). What follows pretty well sums up what a nonfactory technician, armed with a volt-ohmmeter and a generic J 1587 data-link scanner can accomplish. Table 6-5 lists Detroit Diesel nomenclature.

Figure 6-23 illustrates Series 60 DDEC V component locations and Figs. 6-24A and B illustrate schematics for injector and VPOD wiring. In addition to those called out in the schematics, DDEC V and VI systems incorporate sensors that monitor:
- Ambient air temperature and pressure
- EGR Delta-P and flow rate
- Turbo boost and rpm
- VGT vane position

Flash and PID codes are listed in Table 6-6.

Electronic unit injectors

Series 50 and 60 solenoid-actuated unit injectors are replaced as assemblies, using new seals, washers, and hold-down bolts. Disconnect the battery to protect the computer from voltage spikes, remove the hold-down bolt, and lift the injector free. Gentle taps with a rubber hammer should be enough to separate it from its sleeve. Kent-Moore catalogs an extraction tool (PN J47372) for stubborn cases.

If the injector is to be reused, carbon accumulations can be removed from the nozzle body with a wire brush, emory paper, or Scot-Bright. But keep abrasives clear

Table 6-5. Detroit diesel electronic control nomenclature

Acronym	Term	Description and notes
DDR	Diagnostic Data Reader	Scanner
DPS	Delta-P Sensor	Measures pressure drop across a component
EFPA	Electronic Foot Pedal Assy.	Includes the throttle position sensor
EOP	Engine Over-Temperature Protection	Monitors coolant temperature
ESH	Engine Sensor Harness	
FPS	Fuel Pressure Sensor	
FRS	Fuel Restriction Sensor	
LSG	Limiting Speed Governor	
OLS	Oil Level Sensor	
OPS	Oil Pressure Sensor	
OTS	Oil Temperature Sensor	
PWM	Pulse Width Modulated	
RHS	Relative Humidity Sensor	
SRS	Synchronous Reference Sensor	Identifies specific cylinder in the firing order
TBS	Turbocharger Boost Sensor	
TCI	Turbo Compressor Inlet	Used as a modifier, e.g., "TCI temperature"
TCO	Turbo Compressor Outlet	Same as previous note
TPS	Throttle Position Sensor	
TRS	Timing Reference Sensor	Provides timing data
VGT	Variable Geometry Turbocharger	
VNT	Variable Nozzle Turbine	Same as VGT
VPOD	Variable Output Pressure Device	VGT and ERG valve actuator.
VSG	Variable Speed Governor, also referred to as power takeoff, PTO	
VSS	Vehicle Speed Sensor	

of the nozzle orifices. Using a hand-held brush—not a power tool—clean carbon from the sleeve and vacuum up the particles. This procedure reduces contamination of the drilled fuel passages.

Lubricate components with clean diesel fuel, and install the injector with new seals, washers, and clamp bolt, as supplied under PN 2353711 (Fig. 6-25). Note that the flat side of the copper washer goes down, toward the cylinder head. Fit the clamp over its locating pin, run the bolt down finger-tight, and

- torque the bolt to 50 N-m (37 lb-ft),
- back off the bolt 60° (one bolt flat),
- torque to 35 N-m (26 lb-ft),
- tighten the bolt 90° (one-quarter turn).

Internal fuel galleries must be flushed before starting the engine. Prime the fuel system and, with the key "off," disconnect the ECM at the fuse box or harness

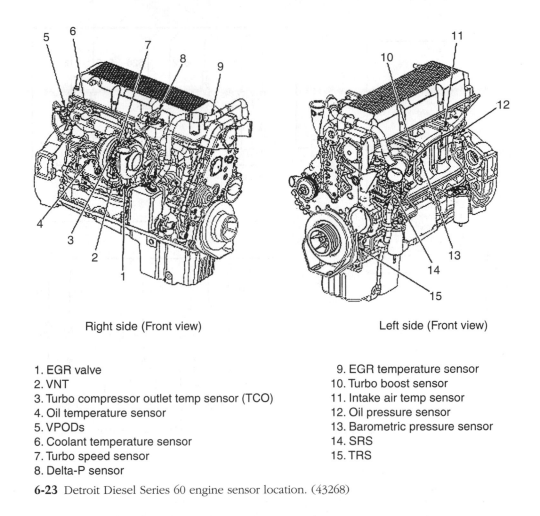

Right side (Front view) Left side (Front view)

1. EGR valve
2. VNT
3. Turbo compressor outlet temp sensor (TCO)
4. Oil temperature sensor
5. VPODs
6. Coolant temperature sensor
7. Turbo speed sensor
8. Delta-P sensor

9. EGR temperature sensor
10. Turbo boost sensor
11. Intake air temp sensor
12. Oil pressure sensor
13. Barometric pressure sensor
14. SRS
15. TRS

6-23 Detroit Diesel Series 60 engine sensor location. (43268)

connector. Remove the combination check valve and pressure regulator, which are located at the rear of the cylinder head at the return-line elbow. Connect a hose to the gallery outlet, and crank the engine in three 15-second bursts, allowing ample time for the starter motor to cool between engagements. Once the galleries are flushed, replace the regulator, make up the fuel-return line and connect the wiring harness to the computer. Run the engine up to operating temperature and check for fuel leaks.

Detroit Diesel supplies an upgrade kit (PN 23528939) for Series 50 and 60 injectors that consists of a spring 11-mm longer than the original, a cam-follower retainer and new hold-down screws. Apparently, the original springs allowed injector plungers to "float" at high rpm. Remove the rocker-arm assembly, place it on a clean surface, and, working with one injector at a time to avoid mixing parts, remove the two 5-mm Allen screws that secure the follower retainer. Lift the retainer and the follower free (Fig. 6-26). Clean the parts in diesel fuel, and install the new spring, follower and retainer, using the screws provided in the kit. Torque to 25–28 N-m (22–25 lb-in.).

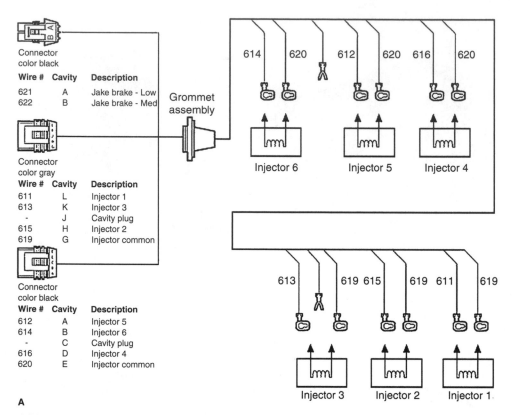

Connector
color black

Wire #	Cavity	Description
621	A	Jake brake - Low
622	B	Jake brake - Med

Grommet assembly

Connector
color gray

Wire #	Cavity	Description
611	L	Injector 1
613	K	Injector 3
-	J	Cavity plug
615	H	Injector 2
619	G	Injector common

Connector
color black

Wire #	Cavity	Description
612	A	Injector 5
614	B	Injector 6
-	C	Cavity plug
616	D	Injector 4
620	E	Injector common

A

6-24 Wiring diagram for 2002 model Series 60 engines with EGR: injector harness (A) and engine sensor harness (B). Courtesy Detroit Diesel Corp. (43290 and 43292)

No start

Table 6-1 outlines basic troubleshooting procedures for this and other EMS-equipped engines. Trouble codes, the most critical of which are listed in Table 6-6, should help pinpoint starting problems. As far as the high-pressure fuel system is concerned, failure to start can result from:

- No voltage to the injectors. With the ignition "off," remove the rocker covers and disconnect the ground-return wire 619 or 620 (shown back at Fig. 6-24B) from one injector. Make up a 6V test lamp between the ground wire and an engine ground. Crank the engine to verify that the injector receives voltage. Reconnect the lead and repeat the test for each of the injectors.
- Malfunctioning synchronous reference sensor/timing reference sensor. The SRS/TRS provides the timing reference necessary to initiate and time injection (Fig. 6-27). Flash code 41 indicates that the TRS signal is missing or that the SRS generates phantom pulses. Check the security of the sensor mounting and the pulse wheel for chipped teeth or other damage. The air gap between the TRS and any pulse-wheel tooth should be .020–.040 in. Verify with a depth

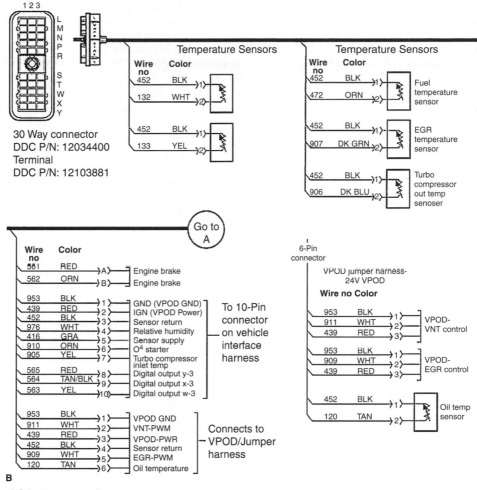

6-24 (*Continued*)

micrometer and adjust sensor position as necessary. Each sensor has two leads, which should be checked for opens or shorts.

VGT/EGR

The variable geometry turbocharger and the exhaust gas recirculation system are intimately related, with overlapping trouble codes. Depending upon its J 1587 addendum, flash code 48 translates as low air, fuel, or EGR pressure, or as low EGR, turbocharger inlet (TCI), or turbocharger outlet (TCO) temperature. A very low level of fuel in the tank or restrictions in the fuel supply lines generally account for loss of fuel pressure. EGR pressure is sensed upstream of the turbocharger. Consequently, low EGR pressure

Table 6-6. Major DDEC flash and PID codes

Flash	PID	Malfunction
11	187	Low voltage—Throttle Position Sensor (TPS)
	187	Not responding—TPS
12	187	High voltage—TPS
13	111	Input voltage low—Add Coolant Level Sensor
14	52	Input voltage high—Intercooler Coolant Temp. Sensor
	110	Input voltage high—Coolant Temperature Sensor
	175	Input voltage high—Oil Temperature Sensor
15	52	Input voltage low—Intercooler Coolant Temp.Sensor
	110	Input voltage low—Coolant Temperature Sensor
	175	Input voltage low—Oil Temp. Sensor
16	111	Input voltage high—Coolant Level Sensor
	111	Input voltage high—Add Coolant Level Sensor
17	111	Input voltage high—Throttle Plate Position Sensor
	72	Input voltage high—Blower Bypass Position Sensor
	354	Failure—Humidity Sensor (DDEC release 33.0 and later)
18	51	Input voltage low—Throttle Plate Position Sensor
	72	Input voltage low—Blower Bypass Position Sensor
	354	Failure, low—Relative Humidity Sensor (DDEC release 33.0 and later)
21	91	Input voltage high—TPS
22	91	Input voltage low—TPS
23	174	Input voltage high—Fuel Temperature Sensor
	—	Input voltage high—Oxygen Content Circuit
25	—	No codes
26	—	Aux. Shutdown No.1 active
	—	Aux. Shutdown No. 2 active
27	105	Input voltage high—Intake Manifold Temp. Sensor
	171	Input voltage high—Ambient Air Temp. Sensor
	172	Input voltage low—Ambient Air Temp. Sensor
28	105	Input voltage low—Intake Manifold Temp. Sensor
	171	Input voltage low—Ambient Air Temp. Sensor
	172	Input voltage low—Air Temperature Sensor
29	351	Failure—Turbo Compressor Inlet Temp. Circuit
	404	Input voltage low—Turbo Compressor Temp. Sensor
33	102	Input voltage high—Turbo Boost Pressure Sensor
34	102	Input voltage low—Turbo Boost Pressure Sensor
35, 36	19	Input voltage high—High Range Oil Pressure Sensor
	100	Input voltage low—Oil Pressure Sensor
37	18	Input voltage high—High Range Fuel Pressure Sensor
	94	Input voltage high—Fuel Pressure Sensor
	95	Input voltage low—Fuel Restrictor Sensor
38	18	Input voltage low—High Range Fuel Pressure Sensor
	94	Input voltage low—Fuel Pressure Sensor
	95	Input voltage low—Fuel Restrictor Sensor

Table 6-6. Major DDEC flash and PID codes (*Continued*)

Flash	PID	Malfunction
39	146	EGR leak—boost power
	146	EGR leak—boost jake
	146	No response—EGR valve
	147	No response—Variable Nozzle Turbo (VNT) vanes (DDEC release 33.0 and later)
	147	No response—VNT vanes, boost jake (DDEC release 33.0 and later)
	—	No response—EGR valve (DDEC release 29.0 and later)
	—	No response—VNT vanes (DDEC release 29.0 and later)
41	—	Too many synchronous cylinder references (SRS-related malfunction) or no timing signal (TRS-related malfunction)
42	—	Too few synchronous cylinder references (SRS-related malfunction)
43	111	Low—coolant level
44	52	High—intercooler coolant temperature
	105	High—intake-manifold temperature
	110	High—coolant temperature
	172	High—air intake temperature
	175	High—oil temperature
45	19	Low—high range oil pressure
	100	Low—oil pressure
46	168	Voltage low—ECM battery
	—	Voltage low—sensor supply
47	18	High—high range fuel pressure
	94	High—fuel pressure
	102	High—turbo boost pressure
	106	High—air inlet pressure
	164	High—injector control pressure
48	18	Low—high range fuel pressure
	94	Low—fuel pressure
	106	Low—air inlet pressure
	164	Low—injector control pressure
	351	Below range—turbo compressor inlet temperature (DDEC release 33.0 and later)
	404	Low—turbo compressor outlet temperature (DDEC release 33.0 and later)
	411	Low—EGR OPU (DDEC release 33.0 and later)
	412	Low—EGR temp. (DDEC release 33.0 and later)
	—	Low—EGR Delta-P (DDEC release 33.0 and later)
49	351	High—turbo compressor inlet temp. (DDEC release 33.0 and later)
	404	High—turbo compressor outlet temp. (DDEC release 33.0 and later)
51	351	Failure—turbo compressor inlet temp. circuit (DDEC 33.0 and later)

(Continued)

Table 6-6. Major DDEC flash and PID codes (*Continued*)

Flash	PID	Malfunction
51	409	Input voltage low—Turbo Compressor Outlet Sensor
52	—	Failure—A/D conversion
55, 56, 57	—	Fault—J 1939 data link
61	—	Slow injector response time
77	810	High—exhaust backpressure
81	—	Voltage high—individual Exhaust Port Temp. Sensors
82	—	Voltage low—individual Exhaust Port Temp. Sensors
84	98	Low—crankcase oil level
	101	Low—extended crankcase oil level
86	73	Supply voltage high—fuel pump pressure sensor
	108	Supply voltage high—barometric pressure sensor
87	73	Supply voltage low—fuel pump pressure sensor
	108	Supply voltage high—barometric pressure sensor
88	20	Low—high range coolant pressure
	109	Low—coolant pressure
89	95	High—fuel restriction across filter
	111	Maintenance alert—coolant level

comes about because of leaks in exhaust tubing going to the turbocharger or because the EGR valve sticks open to pass more than the normal volume of exhaust gases.

EGR, TCI, or TCO temperatures are influenced by the heat content of the exhaust gases mixing with incoming air. Abnormally low temperatures result from exhaust leaks in the EGR cooler or inlet piping, or from failure of the EGR valve to open sufficiently.

Code 49 signals high TCI or TCO temperatures, which can be traced to a clogged air filter, excessive exhaust backpressure, or to the EGR valve. If the EGR valve were to stick open, the higher-than-normal input of exhaust gas would be registered as an increase in turbo temperature.

Fig. 6-28 illustrates the VGT turbo, EGR valve (7) and the tube-and-shell EGR cooler (1), which is distinguished from the earlier tube-and-fin type by its rounded

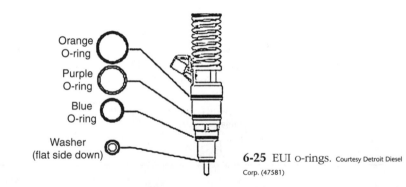

6-25 EUI o-rings. Courtesy Detroit Diesel Corp. (47581)

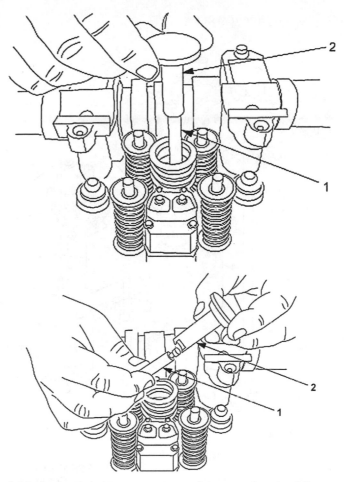

6-26 Detroit Diesel injector-spring replacement, showing follower
(1) and plunger (2). (40839)

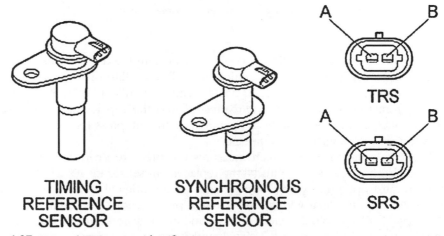

TIMING
REFERENCE
SENSOR

SYNCHRONOUS
REFERENCE
SENSOR

TRS

SRS

6-27 SRS and TRS sensor identification. Courtesy Detroit Diesel Corp. (31245)

137

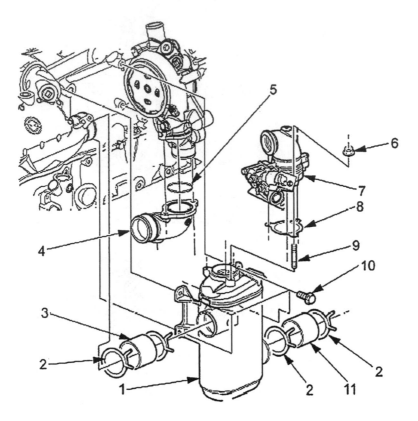

1. EGR Cooler
2. Spring Wire Clamp (Re-use from engine)
3. EGR Cooler-to-Oil Cooler Hose (Re-use from engine)
4. Water Pump Elbow (Re-use from engine)
5. Elbow-to-Water Pump Seal Ring
6. M8 Nut (3)

7. EGR Valve (Re-use from engine)
8. EGR Cooler Gasket
9. M8 Stud (3)
10. M10 Bolt (Re-use from engine)
11. EGR Cooler-to-Water Pump Hose
     (Re-use from engine)

6-28 New-style EGR cooler and related parts. Courtesy Detroit Diesel Corp. (45473)

contours. The newer, more efficient cooler can be retrofitted to earlier Series 60 engines with kit PN 23533985, an operation that calls for reflashing the computer on DDEC IV engines built before 06R0755298. The hydraulic EGR unit must be extracted from the original cooler and pressed into the replacement part. Kent-Moore supplies special tools for these operations, although press fixtures and a suitable extractor are not difficult to fabricate.

The EGR Delta-P sensor, which measures the pressure drop across the EGR valve, has also been revised. To install the replacement sensor kit PN 23532364, the technician must open the wiring harness, snip off the original Delta-P sensor connector, splice in additional wiring, and reassign pin functions. As originally wired, pin 1 supplied voltage, pin 2 went to ground, and pin 3 carried signal voltage. Under the new scheme, pin 1 is ground, 2 signal, and 3 supply.

7
CHAPTER

Cylinder heads and valves

The cylinder head acts as the backing plate for the head gasket and must, above all, be rigid. Iron is the preferred material, although automotive and other light-duty engines often are fitted with aluminum heads. Because of its good conductivity, aluminum assists cooling and eliminates the steep thermal gradients that iron is heir to. In other words, aluminum heads are less likely to develop local hot spots.

But, in common with most other structural materials, the weight savings of aluminum come at the cost of rigidity. Aluminum has a third the density of iron and a third the rigidity. To achieve the same level of stiffness, an aluminum head would have to weigh as much as the iron head it replaces. Good design practice and closely spaced head bolts ameliorate the problem, but aluminum heads are always less rigid than their cast-iron equivalents.

And while aluminum can be heat-treated to T6 or T7 hardness, few manufacturers bother, since the benefits of heat treating disappear at high temperatures. The light-metal alloys in general use exhibit a rapid loss of strength at temperatures above 200°C (392°F). Overheating an engine is almost guaranteed to warp the head. To further complicate matters, cast-iron has a lower rate of thermal expansion than aluminum. When bolted to an iron block, an aluminum head tends to bow upward, reducing the clamping force on the gasket. This phenomenon is discussed in the section that deals with head straightening.

The cylinder head forms an important part of the cooling system. The water jacket should be cast with large passages that resist clogging. Some designs employ diverters, or baffles, to direct the coolant stream to valve seats and other critical areas. Standard practice is to integrate block and head cooling, but several industrial engines separate the two so that a blown head gasket does not contaminate the crankcase oil with antifreeze.

Fasteners—capscrews or studs—should be arranged symmetrically insofar as rocker pedestals, valve ports, and injector and glow-plug bosses permit.

Combustion chamber types

A vast amount of development work was done in the 1920s and 30s on optimum combustion chamber shapes. The effort continues—with emphasis on control of emissions—but the main outlines of the problem and the alternatives appear to be firmly established.

The central problem is to provide a mechanism for mixing fuel with the air. The fuel spray must be dispersed throughout the cylinder so that all available oxygen takes part in the combustion process. As fuel droplets move away from the nozzle, the chances of complete combustion are progressively lessened. There is less oxygen available because some of the fuel has already ignited. In addition, the droplets shrink as they travel and successive layers of light hydrocarbons boil off.

Direct injection

Direct Injection (DI) chambers, also called open or undivided chambers, resemble those used in carbureted engines (Fig. 7-1). These symmetrical chambers have small surface areas and, to reduce heat losses further, generally take the form of a cavity in the piston crown.

Until the advent of multiple-orifice, staged injectors and the high-pressure fuel systems necessary to support them, DI chambers compensated for poor fuel penetration by imparting energy to the air charge. Two mechanisms were involved: swirl and squish. The exit angle of the intake-valve seat imparted a spinning motion, or swirl, to incoming air stream. Squish was achieved by making the edges of the piston

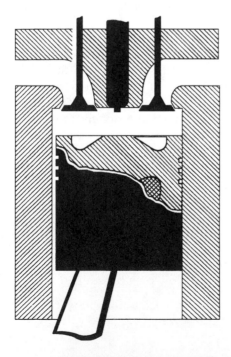

7-1 DI or open chamber.

crown parallel to the chamber roof. As the piston neared top dead center (tdc), air trapped between these two faces "squished" inward, toward the piston cavity.

Mixing was less than perfect, which resulted in ignition delay and rapid rises in cylinder pressure as the accumulated fuel charge exploded. As a result, DI was confined to stationary, heavy trucks and marine engines where noise, vibration, and exhaust smoke counted for less than fuel economy.

The breakthrough came in the late 1980s in the form of electronic injectors that sequenced fuel delivery to "soften" combustion and reduce ignition delay. Ultrahigh fuel pressures, coupled with orifice diameters as small as 0.12 mm, or twice the thickness of a human hair, atomize the fuel for better mixing. The fuel charge became the primary vehicle for mixing that, in some cases, enabled designers to eliminate the pumping losses associated with generating air turbulence. DI also opened the way for massive power increases. By substituting DI for the Ricardo Comet V cylinder head originally fitted, enlarging the bore, and adding a turbocharger, Volkswagen increased the power output of its four-cylinder engine by a factor of 3.4.

Indirect injection

Indirect injection (IDI) uses energy released by combustion to drive the fuel charge deeply into the air mass. The combustion chamber is divided into two sections with the larger chamber formed by the piston top and cylinder-head roof. Combustion begins in a smaller chamber, usually located over the piston. A passage, often in the form of a venturi, connects the two chambers.

Fuel mixing, at least with low-pressure injectors, is enhanced since the fuel droplets exit the smaller chamber at very high velocities. Because peak pressures occur in the smaller chamber and do not act on the piston, IDI reduces noise and combustion roughness. The first diesel passenger cars would not have been commercially acceptable without IDI, and many light-duty engines continue to use this technology.

On the debit side, indirect injection imposes pumping losses since the piston must work to pressurize the small antechamber. In addition, this chamber acts as a heat sink, bleeding thermal energy that could be better employed in turning the crankshaft. For cold starting, the air charge must be heated with glow plugs. All of these chambers expel a jet of burning fuel that rebounds off the piston. A supercharger or turbocharger increases the temperature of the air charge and the vulnerability of the piston to meltdown.

Precombustion chamber

The precombustion chamber was first used on the Hornsby-Ackroyd low-compression oil engine and subsequently by Caterpillar, Deutz, and Mercedes-Benz, and other diesel manufacturers (Fig. 7-2). The precombustion chamber, also known as a hot bulb because of its shape and absence of any provision for cooling, occupies 25–40% of the total swept volume.

As the piston approaches tdc, the injector opens to send a solid stream of fuel into the hot bulb. The charge ignites, bulb pressure rises, and a stream of burning fuel jets through the connecting channel and into the main chamber, where there is sufficient air to complete combustion. The smaller the channel, the greater the acceleration and, all things equal, the more complete the fuel-air mixing. These chambers employ pintle-type injectors.

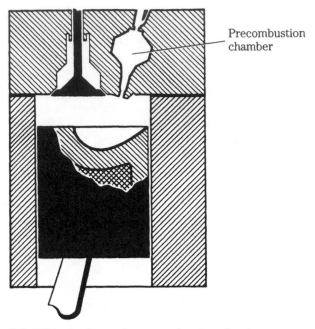

Precombustion chamber

7-2 IDI in the form of a precombustion chamber.

Swirl chamber

The swirl, or turbulence, chamber is similar in appearance to the precombustion chamber, but functions differently (Fig. 7-3). During compression, the disc-shaped or spherical antechamber imparts a circular motion to the air, which accelerates as the piston approaches tdc. The injector is timed to open at the peak of vortex speed. As the piston rounds tdc, combustion-induced pressure in the antechamber reverses the flow. A turbulent stream of burning fuel and superheated air exits the antechamber and rebounds off the piston to saturate the main chamber.

The swirl chamber was invented by Sir Harry Ricardo during the late 1920s and underwent numerous alterations during its long career. Except for Mercedes-Benz, most diesel cars and light commercial vehicles of the postwar era and for many years after employed the Ricardo Comet V chamber. These chambers are more economical than hot-bulb chambers, but are noisier.

The 6.25L engine used by General Motors for pickup trucks during the 1980s and early 90s demonstrates the tradeoffs implicit in combustion-chamber design. While the Ricardo chamber depends upon swirl for mixing, velocity is also important. Initially, these GM engines were set up with a small-diameter port between the main and swirl chambers. The pressure drop across the port generated velocity that, in conjunction with swirl, resulted in good air-fuel mixing and fuel economy. Near the end of the production run, GM acquiesced to customer demands for more power by enlarging the connecting port. More fuel could be passed, but thermal efficiency suffered. Fuel economy, which had approached 20 miles/gal, dropped to 14 or 15 mpg.

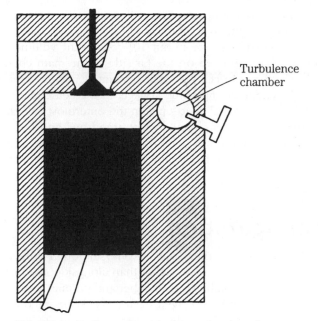

7-3 IDI in the form of a turbulence chamber, the most common of which were developed by Ricardo Ltd.

Energy cell

Examples of the energy cell, or Lanova divided chamber, can still be encountered in vintage Caterpillar tractor engines (Fig. 7-4). The cell consists of a kidney-shaped main chamber, located over the piston, and a secondary chamber, or energy

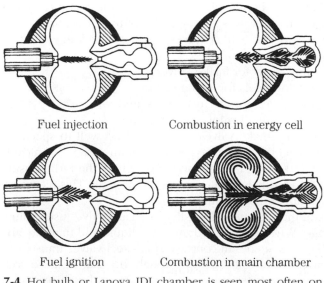

Fuel injection   Combustion in energy cell

Fuel ignition   Combustion in main chamber

7-4 Hot bulb or Lanova IDI chamber is seen most often on older Cat engines.

cell, which is divided into two parts. The cell opens to a narrow throat, situated between the two lobes of the main chamber.

During the compression stroke about 10% of the air volume passes into the energy cell. The injector, mounted on the far side of the main chamber, delivers a solid jet of fuel aimed at the cell. A small percentage of the fuel shears off and remains in the main chamber and some collects in the cell chamber closest to the piston bore. But most of the fuel dead-ends in the outermost cell chamber. .

Ignition occurs in this outer chamber. Unburned fuel and hot gases accelerate as they pass through the venturi between the two cell chambers and enter the main chamber. The rounded walls of the main chamber impart a swirl to the charge to promote better mixing.

Valve configuration

Modern engines employ overhead valves (ohv), although side-valve engines still find buyers. Mounting the valves over, rather than alongside, the piston makes for a cleaner, less heat-absorbent combustion chamber and permits larger valve diameters.

Overhead valves are generally driven from a block-mounted camshaft through pushrods and rocker arms (Fig. 7-5). High-performance auto engines and commercial engines with unit injectors employ a single overhead camshaft (ohc) that acts on the valves through rockers. At present, diesel engines do not turn fast enough to justify the reduction in reciprocating weight that double overhead camshafts (dohc) would provide.

For good volumetric efficiency, valves should be as large as possible. Some American trucks and many European passenger cars have four valves per cylinder to improve breathing and power output. With a pent-roof chamber, four valves can provide a flow area of about a third of the area of the piston crown. In contrast, the flow area of two parallel valves is only a fifth of the crown area.

When purchasing aftermarket valves, it is useful to know that the SAE classifies valve steels into four groups:

- NV low-alloy steels for intake valves,
- HNV high-alloy steels for intake valves,
- EV austenitic steels for exhaust valves; and
- HEV high-strength alloy for exhaust valves.

When upgrading intake valves, one would do well to specify SAE-rated EV8 21-4N austenitic steel that contains 21% chromium and 3.75% nickel, and tolerates temperatures of 1600°F (871°C). Inconel 751, which the SAE classifies as HEV3, is preferred for exhaust valves because of its hot strength. This nickel-based alloy contains 16% chromium and 3% titanium.

Regardless of the steel used, long-lived exhaust valves use sodium as a medium of heat exchange. When heated, sodium turns liquid and transfers about 40% of the heat load to the valve guide. In contrast, solid-steel valves transfer only 25% of the heat to the guide, the rest of it going through the seat. Because sodium combines explosively with water, these valves should be treated with respect. Any discarded valves should be clearly marked as containing sodium.

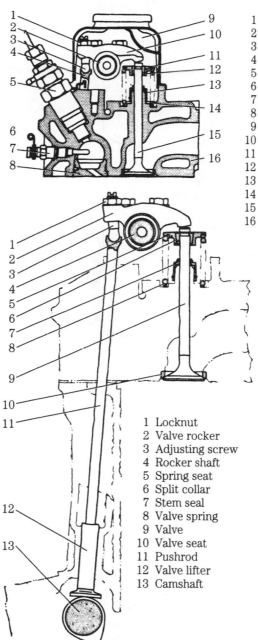

1 Locknut
2 Rocker shaft
3 Adjusting screw
4 Pushrod
5 Nozzle assembly
6 Glow plug cable
7 Glow plug
8 Combustion chamber
9 Rocker cover
10 Valve rocker
11 Spring seat
12 Split collar
13 Stem seal
14 Valve spring
15 Valve
16 Valve seat

1 Locknut
2 Valve rocker
3 Adjusting screw
4 Rocker shaft
5 Spring seat
6 Split collar
7 Stem seal
8 Valve spring
9 Valve
10 Valve seat
11 Pushrod
12 Valve lifter
13 Camshaft

7-5 Typical ohv valve train, driven from the camshaft via pushrods and rocker arms. Marine Engine Div., Chrysler Corp.

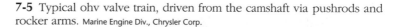

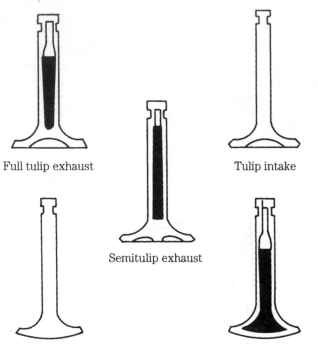

Full tulip exhaust

Semitulip exhaust

Tulip intake

Mushroom intake

Mushroom exhaust

7-6 Tulip and mushroom valves can still be encountered on older engines. But contemporary practice is to flatten the valve crown to reduce heating.

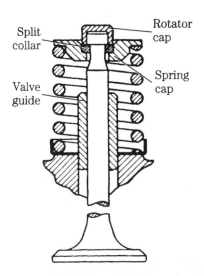

Split collar

Rotator cap

Valve guide

Spring cap

7-7 The GM Bedford valve assembly includes rotator cap (Rotcap), split collar keepers, spring cap, and spring.

Valve inserts, or seats, are an interference fit in the cylinder head. In addition to forming a gas seal with the valve face, the seats provide a major path for heat transfer from the valve face. Consequently, the seats are adjacent to fins on air-cooled engines or surrounded by water on liquid-cooled types. Sintered-iron alloys are often used for intake seats and nickel-steel for the exhausts.

While valves are forced open mechanically by the camshaft, spring pressure seats them. As the springs weaken in service, it is possible for the valve train to "lead" the cam. The acceleration imposed during lift times the reciprocating weight of the train (cam followers, pushrods, half of the mass of the rocker arms, and valves) overcomes spring tension and the valves float open. A similar problem occurs during seating: insufficient spring tension permits valves to bounce several times before coming to rest. These effects become more severe as engine speeds increase.

The angle of the faces varies with many manufacturers using a 30° angle on the intakes and 45° on the exhausts. The steeper angle imposes a flow restriction, but generates the higher seating forces needed for exhaust valves. Valve guide material ranges from chilled iron to manganese or silicon bronze.

Before you begin

Some repairs, such as head-gasket replacement, are normally done on site. Other operations require the assistance of a machinist. But the mechanic needs to understand enough about the compromises involved in returning worn parts to service to be able to evaluate the machinist's work. And, as far as possible, the mechanic should keep current with the technology, such as diamond-like carbon (DLC), titanium nitride (TnN), and chromium nitride (CrN) anti-friction coatings, and isotropic polishing. The latter involves the use of irregularly shaped ceramic pellets in a vibratory machine. The process, which takes about ten hours to complete, extends valve and valve-train life by removing stress risers and microscopic peaks.

Replacement parts can be purchased from the original equipment manufacturer (OEM) or, for the more popular engines, from the aftermarket. Some aftermarket parts are equal to those supplied to the OEM and may, in fact, be the same parts. Others are not so good in ways that might not be apparent until the engine runs a few hundred hours.

Counterfeit parts, disguised as OEM parts, are widely marketed and sometimes by authorized factory dealers. Be suspicious of a part that costs considerably less than it should or that differs from the original in finish, code number, or packaging.

Diagnosis

Head-related problems usually involve loss of compression in one or two cylinders, a condition that is signaled by a ragged idle, by increased fuel consumption, and in some instances, by exhaust smoke. An exhaust temperature gauge, with switch-controlled thermocouples on each header, will give early warning. Exhaust from weak cylinders will be cooler than the norm. You can cross-check by disabling one injector at a time while the engine ticks over at idle. A cylinder that causes less of an rpm drop than the others does not carry its share of the load. (See chapter 4 for additional information about this procedure.)

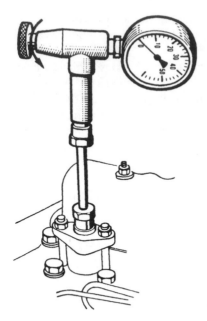

7-8 Cylinder compressor gauge and adapter. Peugeot.

At this point, you can check the injectors (the usual suspects) or else go to the heart of the matter with a cranking compression test (Fig. 7-8). As detailed in the diagnostics chapter, we are looking for cylinders with dramatically (at least 20%) lower compression than the average of the others. If the weak cylinder is flanked by healthy cylinders, the problem is either valve- or head-gasket related; or very low compression in an adjacent cylinder points to gasket failure. Abnormally high readings on all cylinders indicate heavy carbon accumulations, a condition that might be accompanied by high pressures and noise. The next step is to make a cylinder leak-down test, which will distinguish between valve and gasket failure.

While most compression leaks bleed into adjacent cylinders or across the fire deck to the atmosphere, it is possible for a leak path to open into water jacket. The engine might seem healthy enough, but overheat within a few minutes of start-up. Coolant in the header tank might appear agitated and might spew violently with the cap removed. A cooling system pressure test will verify the existence of a leak, which can be localized with a cylinder leak-down test. However, the leak-down test cannot distinguish between cracks in the casting and a blown gasket.

Fortunately, it is rare for an engine that has not suffered catastrophic overheating to leak coolant into the oil sump, where it can be detected visually or, in lesser amounts, by a spectrographic analysis. Likely sources are casting cracks, cracked (wet-type) cylinder liners, and liner-base gasket leaks.

The cylinder head casting, like the fluid end of a high-pressure pump, will eventually fail. After a large, but finite, number of pressure cycles, the metal crystallizes and breaks. Owners of obsolete engines for which parts are no longer available would do well to keep a spare head casting on hand. Even so, most cylinder heads fail early, long before design life has been realized, because of abnormally high combustion pressure and temperature.

Combustion pressure and heat can be controlled by routine injector service (dribbling injectors load the cylinders with fuel), attention to timing, and conservative pump settings. Cooling system maintenance usually stops when the temperature gauge needle remains on the right of center. But local overheating is as critical as radiator temperature and is rarely addressed. According to engineers at Detroit Diesel, ¼ in. of scale in the water jacket is the thermal equivalent of 4 in. of cast iron. Eroded coolant deflectors and rounded-off water pump impeller blades can also produce local overheating, which will not register as a rise in header-tank temperature.

Local overheating on air-cooled engines can usually be traced to dirty, grease-clogged fins or to loose shrouding.

Gasket life can be extended by doing what can be reasonably done to minimize potential leak paths. As explained below, some imperfection of the head and fire deck seating surfaces must, as a practical matter, be tolerated. Fasteners should be torqued down to specifications and in the suggested torque sequence, which varies between engine makes and models. Asbestos gaskets, without wire or elastomer reinforcements, "take a set" and must be periodically retightened. Newer, reinforced gaskets hold initial torque, but proper diesel maintenance entails retightening head bolts periodically.

Disassembly

The drill varies between makes and models, and is described in the manufacturer's manual. Here, I merely wish to add some general information, which might not be included in the factory literature.

It is good and sometimes necessary practice to align the timing marks before the head is dismantled. Bar crankshaft over—in its normal direction of travel—to tdc on No. 1 cylinder compression stroke. Tdc will be referenced on the harmonic balancer or flywheel; the compression stroke will be signaled by closed intake and exhaust valves on No. 1 cylinder. In most cases, the logic of timing marks on overhead cam and unit injector engines will be obvious; when it is not, as for example, when camshaft timing indexes a particular link of the drive chain, make careful notes. A timing error on assembly can cost a set of valves.

Experienced mechanics do not disassemble more than is necessary. Normally you will remove both manifolds, unit injector rocker mechanisms, and whatever hardware blocks access to the head bolts. Try to remove components in large bites, as assemblies, by lifting the intake manifold with turbocharger intact, removing the shaft-type rocker arms at the shaft hold-down bolts, and so on.

Miscellaneous hardware should remain attached, unless the head will be sent out for machine work. In this case, it should be stripped down to the valves and injector tubes. Otherwise, the head might be returned with parts and fasteners missing.

Note: Injector tubes on some DI heads extend beyond the head parting surface and, unless removed, might be damaged in handling.

Examine each part and fastener as it comes off. If disassembly is extensive, time will be saved by storing the components and associated fasteners in an orderly fashion. You might wish to use plastic baggies, labeled with a Sharpie pen or Marks-A-Lot for this purpose. Keep the old gaskets for comparison with the replacements.

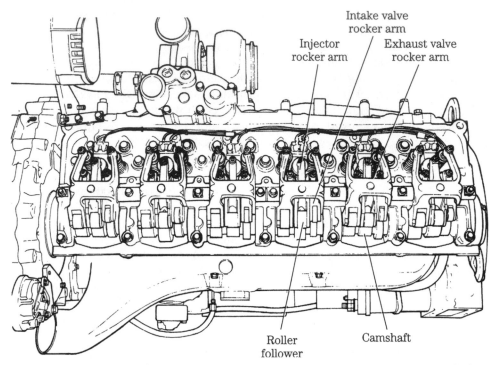

7-9 Detroit Diesel Series 60 valves and injectors actuate from a single overhead camshaft.

Rocker assemblies

Two rocker arm configurations are used on OHV engines. Commercial and automotive engines developed from industrial engines generally pivot the rockers on single or double shafts, as illustrated several times in this chapter, including Fig. 7-6 and, as applied to ohc engines, in Figs. 7-9 and 7-10. The earlier drawing is the most typical, because a single rocker drives each intake and exhaust valve. Figures 7-9 and 7-10 illustrate Detroit Diesel solutions to the problem of driving two valves from the same rocker arm. Regardless of the rocker configuration, the hollow pivot

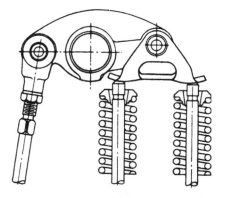

7-10 Detroit Diesel dual valve actuating mechanism, used on the firm's two-cycle engines.

shaft doubles as an oil gallery, distributing oil to the rocker bushings through radially drilled ports. Rockers are steel forgings, case-hardened on the valve end and generally include provision for valve lash adjustment.

Shaft-type rockers are detached from the head as an assembly with the shaft while the pushrods are still engaged. Note the lay of the shaft and rockers, and if no identification marks are present, tag the forward end of the shaft as an assembly reference. Loosen the hold-down bolts slowly, a half-turn or so at a time following the factory-recommended breakout sequence. If no information is provided on this point, work from the center bolt outward. This procedure will distribute valve spring forces over the length of the shaft.

Set the rocker arm assembly aside and remove the pushrods, racking them in the order of removal with the cam-ends down. A length of four-by-four, drilled to accept the pushrods and with the front of the engine clearly marked, makes an inexpensive rack.

Although the task is formidable on a large engine, the rocker arms, together with spacers, locating springs, and wave washers, should be completely disassembled for cleaning and inspection. Critical areas are

- *Adjusting screws*—check the thread fit and screw tips. Screws are case-hardened: once the carburized "skin" is penetrated, it will be impossible to keep the valves adjusted.
- *Rocker tips*—with proper equipment, worn tips can usually be recontoured, quieting the engine.
- *Rocker flanks*—check for cracks radiating out from the fulcrum. Cracks tend to develop on the undersides of the rockers at the fillets.
- *Bushings*—wear concentrates on the engine side of the bushing and, when severe, is accompanied with severe scoring, which almost always involves the shaft. Clearance between the bushing and an unworn part of the shaft should be on the order of 0.002 in. Replacement bushings are generally available from the original equipment manufacture (OEM) or aftermarket. The old bushing is driven out with a suitable punch, and the replacement pressed into place and reamed to finish size. Aligning the bushing oil port with the rocker arm port is, of course, critical.
- *Shaft*—inspect the surface finish, mike bearing diameters, and carefully clean the shaft inner diameter (ID), clearing the oil ports with a drill bit. Do the same for the oil supply circuit. Rocker arms are remote from the pump, and lubrication is problematic.

Small engines in general and automotive plants derived from SI engines use pedestal-type rockers, of the type illustrated in Fig. 7-11. This technology, pioneered by Chevrolet in 1955, represents a considerable cost saving because the pivots compensate for dimensional inaccuracies and the rockers are steel stampings. Most examples lubricate through hollow pushrods. If rockers are removed, it is vital that they be assembled as originally found, together with fulcrum pieces and hold-down hardware. Rack the pushrods as described in the previous paragraph.

The locknuts that secure the rockers to their studs should be renewed whenever the head is serviced. Other critical items are

- *Studs*—check for thread wear, nicks, distortion, and separation from the head. As far as I am aware, all diesel pedestal-type rocker studs thread into the cylinder head.

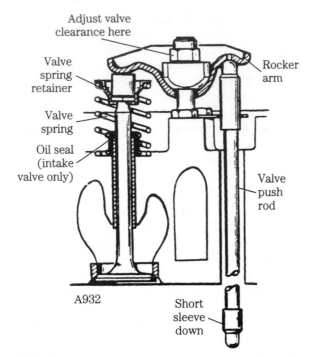

Adjust valve
clearance here

Valve
spring
retainer

Rocker
arm

Valve
spring

Oil seal
(intake
valve only)

Valve
push
rod

A932

Short
sleeve
down

7-11 Pedestal-type rocker arm, pivots on an adjustable
fulcrum for lash adjustment. Onan

- *Rocker pivots*—the rocker pivots on a ball or a cylindrical bearing, secured by
  the stud nut, and known as the fulcrum seat (Fig. 7-12A). Reject the rocker, if
  either part is discolored, scored, or heat checked. How much wear is permissible on the rocker pivot is a judgment call.
- *Fasteners*—replace locknuts if nut threads show low resistance to turning, or
  for the considerable insurance value. Replace stud nuts if faces exhibit fractures (Fig. 7-12B).
- *Rocker tips*—look for evidence of impact damage that could point to a failed
  hydraulic lifter and possible valve tip, valve guide, or pushrod damage. For
  want of a better rule, replace the rocker when tip wear is severe enough to hang
  a fingernail.

Pushrods

Inspect the push rod for wear on the tips and for bends. The best way to determine
trueness is to roll the rods on a machined surface or a piece of optically flat plate glass.

Valve lifters

Valve lifters, or tappets, are serviced at this time when the lifters are driven from
an overhead camshaft or when battered rocker arm tips indicate the need. It should
be mentioned that GM 350 hydraulic lifters must be bled down before the cylinder
head is reinstalled. Factory manuals recommend that lifters be collapsed twice, the
second time 45 minutes after the first. According to technicians familiar with these
engines, the factory is not kidding.

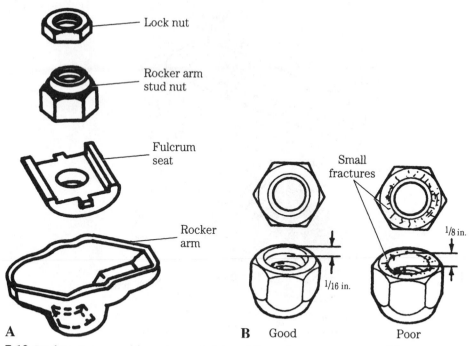

Lock nut

Rocker arm
stud nut

Fulcrum
seat

Rocker
arm

Small
fractures

1/8 in.

1/16 in.

A   **B**   Good   Poor

7-12 Rocker arm assembly and nomenclature (A); stud nut wear pattern (B). Ford Motor Co.

Figure 7-13 illustrates the roller tappet used on GM two-cycle engines. This part is subject to severe forces and, in the typical applications, experiences fairly high wear rates. Inspect the roller for scuffing, flat spots, and ease of rotation. Damage to the roller OD almost always is mirrored on the camshaft.

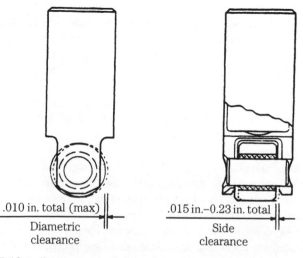

.010 in. total (max)
Diametric
clearance

.015 in.–0.23 in. total
Side
clearance

7-13 Roller tappet, with wear limits. General Motors Corp.

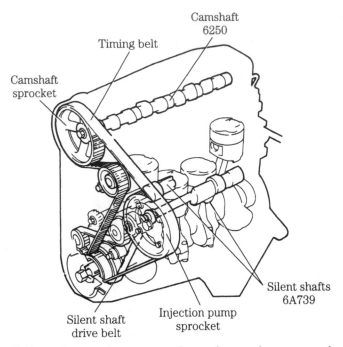

7-14 Ford 2.3L turbo is a new design that employs a cogged (or Gilmer) belt for camshaft drive. Gilmer belts might fail without obvious warning and should be replaced on a rigid engine-hours or mileage schedule. Belts must be tensioned as described by the engine maker and should not be subjected to reverse rotation. Baring the engine backwards can shear or severely damage the teeth.

Roller and pin replacement offers no challenge, but lubrication is critical. During the first few seconds of operation, the only lubrications the follower receives is what you provide. Engine-supplied lube oil is slow to find its way between the roller and pin.

Ensure proper lubrication by removing the preservative from new parts with Cindol 1705; clean used parts with the same product. Just before installation, soak the followers in a bath of warm (100°–125°F) Cindol. Turn the rollers to release trapped air.

Install with the oil port at the bottom of the follower pointed away from the valves. There should be 0.005-in. clearance between the follower legs and guide. The easiest way to make this adjustment is to loosen the bolts slightly and tap the ends of the guide with a brass drift. Bolts should be torqued to 12–15 ft-lb.

Overhead camshafts

Disengage the drive chain or belt (Fig. 7-14). Camshaft mounting provisions vary; some ride in split bearings and are lifted vertically, others slip into full-circle bearings and might require a special tool to open the valves temporarily for lobe clearance during camshaft withdrawal. Split bearing journals must be assembled exactly as originally found. Make certain that bearing caps are clearly marked for number and orientation. Loosen the caps one at a time, working progressively from the center

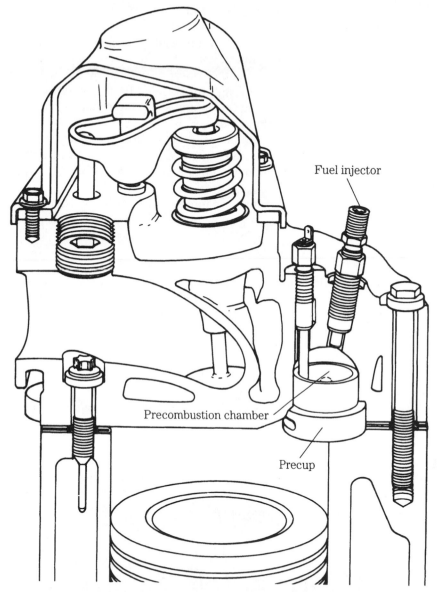

Fuel injector

Precombustion chamber

Precup

7-15 Peek-a-boo head bolts hide under pipe plugs on GM 350 engines. Upon assembly, plug threads should be coated with sealant. Fel-Pro, Inc.

cap out to the ends of the shaft. (Center, first cap right of center, first cap left of center, second cap right of center, and so on.)

Head bolts

Head bolts should now be accessible, but not always visible. Olds 350 engines hide three of the bolts under pipe plugs (Fig. 7-15); some Japanese engines secure the timing cover to the head with small-diameter bolts that, more often than not, are submerged in a pool of oil.

The practice of using an impact wrench on head bolts should be discouraged. A far better procedure and one that must be used on aluminum engines is to loosen the bolts by hand in three stages and in the pattern suggested by the manufacturer. Make careful note of variations in bolt length and be alert for the presence of sealant on the threads. Sealant means that the bolt bottoms into the water jacket, a weight-saving technique inherited from SI engines.

Note: To prevent warpage, the head must be cold before the bolts are loosened.

Clean the bolts and examine carefully for pulled threads, cracks (usually under the heads), bends, and signs of bottoming. Engines have come off the line with short bolt holes.

GM and a few other manufacturers use torque-to-yield bolts, most of which are throwaway items. When this is the case, new bolts should be included as part of the gasket set.

Lifting

Large cylinder heads require a lifting tackle and proper attachment hardware (Fig. 7-16); heads small enough to be manhandled might need a sharp blow with a rubber mallet to break the gasket seal. Lift vertically to clear the alignment pins, which are almost always present.

Mount the head in a holding fixture (Fig. 7-17), or lacking that, support it on wood blocks.

Valves

Valves are removed by compressing the springs just far enough to disengage the keepers, or valve locks (Fig. 7-18). It is good practice to replace valve rotators,

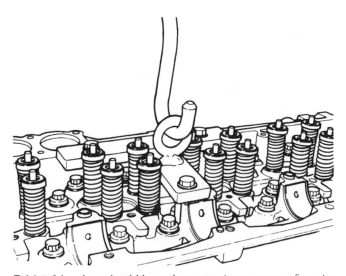

7-16 Lift brackets should be at the approximate center of gravity of the casting, so that the head lifts vertically off its alignment pins. Detroit Diesel

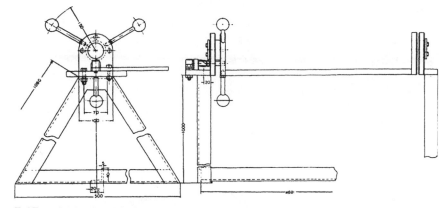

7-17 Cylinder head holding fixture. Marine Engine Div., Chrysler Corp.

shown in Fig. 7-7. Seals (Fig. 7-19) should also be renewed. Specify high temperature Viton for the seal material.

If all is right, the valves will drop out of their guides from their own weight. The usual cause of valve bind is a mushroomed tip, which can be dressed smooth with a stone and which means that the engine has been operating with excessive lash. The associated camshaft lobe might be damaged. A bent valve suggests guide seizure or piston collision.

Cleaning

Cleaning techniques depend on the available facilities. In the field, cleanup usually consists of washing the parts in kerosene or diesel fuel. Gasket fragments can be scraped off with a dull knife (a linoleum knife with the blade ground square to the handle is an ideal tool). The work can be speeded up by using one of the aerosol preparations that promise to dissolve gaskets. In general, it is not a good practice to use a wire brush on head and cooling system gaskets that, until recently, contained asbestos. Sealant, sometimes used in lieu of conventional valve cover and cooling system gaskets, can be removed with 4 in. 3M Scotch-Brite Surface Conditioning Discs, mounted on a high-speed die grinder. Use the coarse pad (3M 07450) on steel surfaces, the medium (3M 07451) on aluminum.

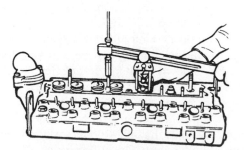

7-18 Lever-type valve spring compressor.

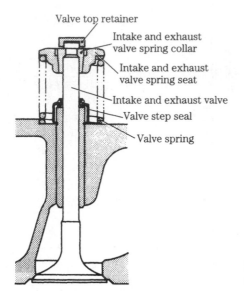

Valve top retainer

Intake and exhaust
valve spring collar

Intake and exhaust
valve spring seat

Intake and exhaust valve

Valve step seal

Valve spring

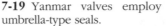

7-19 Yanmar valves employ
umbrella-type seals.

Carbon responds to a dull knife and an end-cutting wire wheel. Clean the piston tops, rotating the crank as necessary. Overhead camshaft drive chains will foul if the crankshaft is turned, and one cannot pretend to do much by way of piston cleaning on these engines.

Machine shops and large-scale repair depots employ less labor-intensive methods. Some shops still use chlorinated hydrocarbons (such as perchloroethylene and trichloroethylene) for degreasing, although the toxicity of these products has limited their application. A peculiar side effect of trichloroethylene (TCE) exposure is "degreaser's flush." After several weeks of contact with the solvent, consumption of alcohol will raise large red welts on the hapless degreaser's face.

Once the head (and other major castings) are degreased, ferrous parts are traditionally "hot-tanked" in a caustic solution, heated nearly to the boiling point. Caustic will remove most carbon, paint, and water-jacket scale. Parts are then flushed with fresh water and dried. In the past, some field mechanics soaked iron heads in a mild solution of oxalic acid to remove scale and corrosion from the coolant passages.

Caustic and other chemical cleaners pose environmental hazards and generate an open-ended liability problem. The Environmental Protection Agency holds the producer of waste responsible for its ultimate disposition. This responsibility cannot be circumvented by contract; if a shop contracts for caustic to be transported to a hazardous waste site, and the material ends up on a country road somewhere, the shop is liable.

Consequently, other cleaning technologies have been developed. Pollution Control Products is perhaps the best-known manufacturer of cleaning furnaces and claims to have more than 1500 units in service. These devices burn natural gas, propane, or No. 2 fuel oil at rates of up to 300,000 Btu/hour to produce temperatures of up to 800°F. (Somewhat lower temperatures are recommended for cylinder heads and blocks.) An afterburner consumes the smoke effectively enough to meet EPA emission, OSHA workplace safety, and most local fire codes.

7-20 Walker Peenmatic shot blaster adds an "as-new" finish used casting.

Such furnaces significantly reduce the liability associated with handling and disposal of hazardous materials, but are not, in themselves, the complete answer to parts stripping. Most scale flakes off, but some carbon, calcified gasket material, and paint might remain after cleaning.

Final cleanup requires a shot blaster such as the Walker Peenimpac machine illustrated in Fig, 7-20. Parts to be cleaned are placed on a turntable inside the machine and bombarded with high-velocity shot. Shot size and composition determine the surface finish; small diameter steel shot gives aluminum castings a mat finish, larger diameter steel shot dresses iron castings to an as-poured finish. Delicate parts are cleaned with glass beads.

Head casting

Make a careful examination of the parting surfaces on both the head and block, looking for fret marks, highly polished areas, erosion around water-jacket ports, and scores that would compromise the gasket seal.

The next step is to determine the degree of head distortion, using a machinist's straightedge and feeler gauges, as illustrated in Fig. 7-21. Distortion limits vary with the application and range from as little as 0.003–0.008 in. or so. In theory, the block deck has the same importance as the head deck; in practice, a warped block would not be welcome news when head work was all that was contemplated, and most mechanics let sleeping dogs lie.

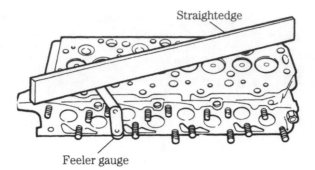

Straightedge

Feeler gauge

7-21 Head warp should be checked in three planes: diagonally as shown, longitudinally, and transversely with particular attention to areas between combustion chambers. Ford Motor Co.

However head bolt holes deserve special attention. Be alert for

- *Stripped threads*—strip-outs can be repaired with Heli-Coil inserts. It is rare to find more than one bolt hole stripped; if the problem is endemic, check with the manufacturer's technical representative on the advisability of using multiple inserts. At least one manufacturer restricts the number of Heli-Coils per engine and per cylinder.
- *Pulled threads*—this condition, characterized by slight eruption in the block metal adjacent to the bolt holes, can be cured by chamfering the uppermost threads with an oversized drill bit or countersink. Limit the depth of cut to one or two threads.
- *Dirty threads*—chase head bolt and other critical threads with the appropriate tap. In theory, one should use a bottoming tap, recognized by its straight profile and squared tip. In practice, one will be lucky to find any tap for certain metric head bolt threads, which are pitched differently than the run of International Standards Organization (ISO) fasteners. Also realize that "bargain" taps can, because of dimensional inaccuracies, do more harm than good. Blow out any coolant that has spilled into the bolt holes with compressed air, protecting your eyes from the debris.

Head resurfacing

Minor surface flaws and moderate distortion can usually be corrected by resurfacing, or "milling." However, there are limits to how much metal can be safely removed from either the head or the block. These limits are imposed by the need to maintain piston-to-valve clearance and, on overhead cam engines, restraints imposed by the valve actuating gear. The last point needs some amplification. Reducing the thickness of the head retards valve timing when the camshaft receives power through a chain or belt. Retarded valve timing shifts the torque curve higher on the rpm band. The power will still be there, but it will be later in coming. The effect of lowering a gear-driven camshaft is less ambiguous; the gears converge and ultimately jam.

Fire deck spacer plates are available for some engines to minimize these effects, and ferrous heads have been salvaged by metal spraying. These options are worth investigating, but one is usually better off following factory recommendations for head resurfacing, as in other matters.

The minimum head thickness specification, expressed either as a direct measurement between the fire deck and some prominent feature on the top of the casting or as the amount of material that can be safely removed from the head and block. Detroit Diesel allows 0.020 in. on four-cycle cylinder heads and a total of 0.030 in. on both the heads and block. Other manufacturers are not so generous, especially on light- and medium-duty engines. For example, the head thickness on Navistar 6.9L engines (measured between the fire deck and valve cover rail) must be maintained at between 4.795 and 4.805 in., a figure that, when manufacturing tolerances are factored in, practically eliminates the possibility of resurfacing. As delivered, the exhaust valve might come within 0.009 in. of the piston crown, and the piston crown clears the roof of the combustion chamber by 0.025 in. Admittedly, these are minimum specifications, but it is difficult to believe that any 6.9L can afford to lose much fire deck metal. Navistar's 9.0L engine, the Ford 2.2L, and the Volkswagen 1.6L engines simply cannot be resurfaced and remain within factory guidelines. Thicker-than-stock head gaskets are available for the VW, but are intended merely to compensate for variations in piston protrusion above the fire deck.

The cardinal rule of this and other machining operations is to remove as little metal as possible, while staying within factory limits. Minor imperfections (such as gasket frets and corrosion on the edges of water jacket holes,) should be brazed slightly overflush before the head is milled.

Precombustion chambers, or precups, are pressed into the head or Caterpillar engines, threaded in. In some cases, precups can remain installed during resurfacing; others must be removed and (usually) machined in a separate operation (Fig. 7-22) When it is necessary to replace injector tubes, the work is done after the head has been resurfaced and is always followed by a pressure test.

Note: Ceramic precups are integral components, serviced by replacing the head.

Most shops rely on a blanchard grinder, such as the one shown in Fig. 7-23 for routine resurfacing. Heavier cuts of 0.015 in. or more call for a rotary broach, known in the trade as a "mill." When set up correctly, a mill will give better accuracy than is obtainable with a grinder. Some shops, especially those in production work, use a movable belt grinder. These machines are relatively inexpensive, require zero set-up time, and produce an unsurpassed finish. However, current belt grinders, which support the workpiece on a rubber platen, are less accurate than blanchard grinders.

Typically, iron heads like a dead smooth surface finish in the range of 60 and 75 rms (root mean square). Some machinists believe that a rougher finish provides the requisite "tooth" for gasket purchase, although there is little evidence to support the contention.

Crack detection

Cylinder heads should be crack tested before and after resurfacing. The apparatus used for ferrous parts generates a powerful magnetic field that passes through the part under test. Cracks and other discontinuities at right angles to the field become polarized and reveal their presence by attracting iron filings. The magnetic

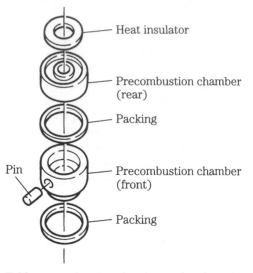

Heat insulator

Precombustion chamber (rear)

Packing

Pin

Precombustion chamber (front)

Packing

7-22 Precombustion chamber and gaskets. An installed precup is shown in Fig. 7-15. Yanmar Diesel Engine Co. Ltd.

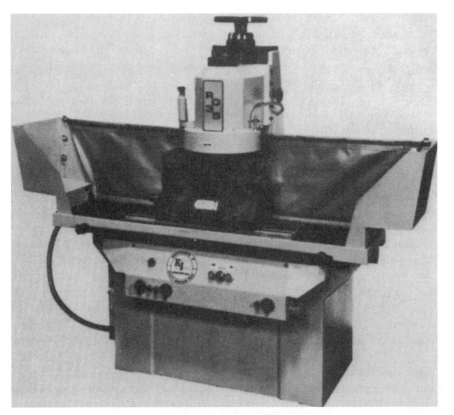

7-23 A blanchard grinder is used for light head milling, manifold milling, and other jobs where some loss of precision can be tolerated.

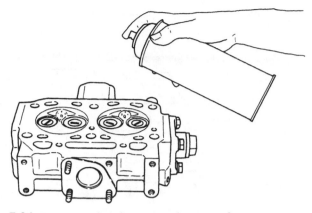

7-24 Penetrant dye detects cracks in nonferrous parts, a category that includes Valve heads. Yanmar Diesel Engine Co. Ltd.

particle test, known generically by the trade name Magnaflux, is useful within its limits. It cannot detect subsurface flaws, nor does it work on nonferrous metals. But fatigue and thermal cracks always start at the surface, and will be seen.

Nonferrous parts are tested with a penetrant dye (Fig. 7-24). A special dye is sprayed on the part, the excess is wiped off, and the part is treated with a developer that draws the dye to the surface, outlining the cracks. In general, penetrant dye is considered less accurate than Magnaflux, but, short of x-ray, it remains the best method available for detecting flaws in aluminum and other nonmagnetic parts.

Neither of these detection methods discriminates between critical and superficial cracks. The cylinder head might be fractured in a dozen places and still be serviceable. But cracks that extend across a pressure regime—that is, from the combustion chamber, cylinder bore, water jacket, or oil circuits—require attention.

Mack and a handful of other manufacturers build surface discontinuities into the roofs of the combustion chambers, which appear as cracks under Magnafluxing, but which are intended to stop crack propagation. These built-in "flaws" run true and straight, in contrast to the meandering paths followed by the genuine article. In general, cracks that are less than a $\frac{1}{2}$ in. long and do not extend into the valve seat area might be less serious than they appear. Short cracks radiating out from precombustion chamber orifices can also be disregarded.

Crack repairs

Assuming that both ends of the crack are visible, it is normally possible to salvage an iron head by gas welding. For best results, the casting should be preheated to 1200°F. Skilled TIG practitioners can do the same for aluminum heads.

Some shops prefer to use one of several patented "cold stitching" processes on iron castings. The technician drills holes at each end of the crack (to block further propagation) and hammers soft iron plugs into the void, which are then ground flush. The plugs must not be allowed to obstruct coolant passages.

Metal spraying is demonstrated to have utility, but rarely practiced.

Test strips

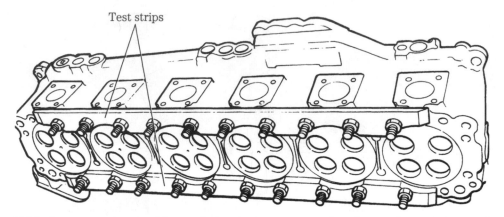

7-25 The water jacket should be sealed as shown, pressurized to about 30 psi, and immersed in hot water. Bubbles indicate the presence of cracks.

Pressure testing

Figure 7-25 illustrates a Detroit Diesel cylinder head, partially dressed out for pressure testing. At this point, most shops would introduce high-pressure water into the jacket and look for leaks, which might take the form of barely perceptible seepage. Detroit Diesel suggests that air be used as the working fluid. Their approach calls for pressurizing the head to 30 psi and immersing the casting in hot (200°F) water for 30 minutes. Leaks register as bubbles.

Straightening

Aluminum heads usually carry the camshaft and are often mounted to a cast-iron block. The marriage is barely compatible. Aluminum has a thermal coefficient of expansion four times greater than that of iron. Even at normal temperatures, the head casting "creeps," with most of the movement occurring in the long axis (Fig. 7-26). Pinned at its ends by bolts, the head bows upward—sometimes as much as 0.080 in. The camshaft cannot tolerate misalignments of such magnitude and, if it does not bend, binds in the center bearings.

Such heads are best straightened by stress relieving. The process can be summarized as follows: the head is bolted to a heavy steel plate, which has been drilled and tapped to accommodate the center head bolts. Shims, approximately half the thickness of the bow, are placed under the ends of the head, and the center bolts are lightly run down. Four or five hours of heat soak, followed by a slow cool down, usually restores the deck to within 0.010 in. of true. Camshaft bearings are less amenable to this treatment, and will require line boring or honing.

Corrosion can be a serious problem for aluminum heads, transforming the water jacket into something resembling papier-mâché. Upon investigation, one often finds that the grounding strap—the pleated ribbon cable connecting the head to the firewall—was not installed.

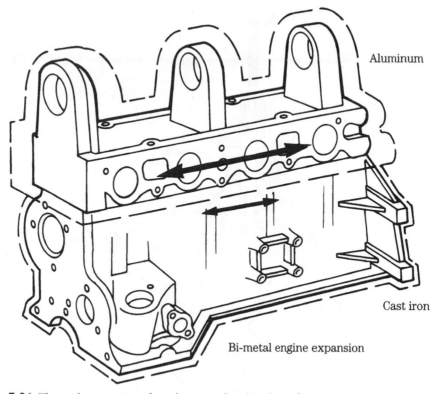

Aluminum

Cast iron

Bi-metal engine expansion

7-26 Thermal expansion of an aluminum head is about four times greater than for a cast iron block. Something must give. Fel-Pro, Inc.

Valve guides

Rocker-arm geometry generates a side force, tilting the valves outward and wearing away the upper and lower ends of the guides. The loss of a sharp edge at the lower end of the guides encourages carbon buildup and accelerates stem wear; the bellmouth at the upper end catches oil, which then enters the cylinder. Figure 7-27 illustrates a split ball gauge used to determine guide ID.

Nearly all engines employ replaceable guides or, if lacking that, have enough "meat" in the casting to accept replaceable guides (Fig. 7-28). A Pep replacement guide for 0.375-in. valve stem measures 0.502 in. on the OD. The BMW 2.4L engine is one of the few for which replacement guides are not available. However, the integral (i.e., block metal) guides can be reamed to accept valves with oversized stems. The GM 350 offers the option of replaceable guides, plus oversized stems The latter are apparently a manufacturing convenience and such parts are difficult to come by.

Old guides drive out with a punch, and new guides install with a driver sized to pilot on the guide ID (Fig. 7-28). Cast-iron heads can be worked cold, but a careful technician will heat aluminum heads so that the guide bores do not gall. Engine manufacturers seem to prefer perlitic cast-iron or iron-alloy guides; many machinists claim phosphor bronze has better wearing qualities. Whatever the material, replacement guides rarely are concentric with the valve seat, and some corrective machine work

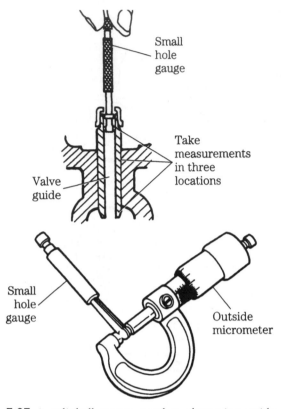

Small
hole
gauge

Take
measurements
in three
locations

Valve
guide

Small
hole
gauge

Outside
micrometer

7-27 A split-ball gauge, used to determine guide
wear. Ford Motor Co.

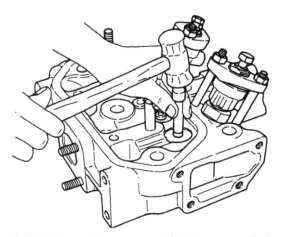

7-28 Valve guides are installed to a specified
depth by means of a factory-supplied driver. Yanmar
Diesel Engine Co. Ltd.

is almost always in order. Stem-to-guide clearances vary with engine type and service; light- and medium-duty engines will remain oil-tight longer with a 0.0015-in. clearance. Heavy-duty engines, which run for long periods at hall rated power, need to be set up looser—as much as 0.005 in. when sodium-cooled valves are fitted.

Valve service

Stem and guide problems can be the result of wear or carbon and gum accumulations that hold the valve open against spring pressure. These deposits are caused by the wrong type of lubricant, ethylene glycol leakage into the sump, and low coolant temperatures (below 160°F). Low temperatures are usually the result of long periods at idle and aggravate any mismatch between fuel characteristics and engine demands. Fuel that burns cleanly at normal temperatures can gum the guide and carbon over the valve heads when the engine runs consistently cool. Sticking can also be caused by bent stems.

Premature valve burning, which in extreme cases torches out segments of valve face and seat, has one cause: excessive heat. This can be the result of abnormal combustion chamber temperatures or of a failure of the cooling system. Faulty injector timing and failure of the exhaust gas recirculation (EGR) and sustained high speed/high-load operation system will create high cylinder temperatures. The thermal path from the valve face to the water jacket can be blocked by insufficient lash (which holds the valve off its relatively cool seat), local water-jacket corrosion, or a malfunction associated with flow directors. Also known as diverters, flow directors are inserted into the water jacket to channel coolant to the valve seats and injector sleeves. These parts can corrode or vibrate loose. Of course, a generalized failure of the cooling system, affecting all cylinders, is possible.

Valve breakage usually takes the form of fatigue failure from repeated shock loads or bending forces. Fatigue failure leaves a series of rings, not unlike tree growth rings, on the parting surfaces. Shock loads are caused by excessive valve lash, which slams the valve face against the seat, or weak valve springs. Bending forces are generated if the seat and/or guide is not concentric with the valve face.

The most dramatic form of valve failure comes about because of collision with the piston. As mentioned previously, Navistar allows as little as 0.009 in. between the exhaust valve and the piston crown at convergence, which occurs $4\frac{1}{4}°$ btdc. Admittedly, the 6.9 is a "tight" engine, but no diesel will tolerate excessive valve lash, weak valve springs, or overspeeding.

Valve service involves two operations, both of which are done by machine. The face is reground to the specified angle, which is usually (but not always) 30° on the intake and 45° on the exhaust, less a small angle, usually about 1.5°. Thus, the 30° intake seats specified for Bedford engines (Fig. 7-29) would be cut to 28.5°. A valve face grinder adjusts to any angle. Seats are cut true to specification, which creates a slight mismatch, known as the interference angle. Once the engine starts, valve faces pound into conformity with the seats. The interference angle ensures a gastight seal on initial start-up and eliminates the need for lapping.

Seats are typically ground in three angles: *entry* (which ranges from about 60°–70°), seat, and exit (which is usually 15°). Thus seat width can be controlled by undercutting either of the flanking angles (Fig. 7-30). The seat should center on the valve face.

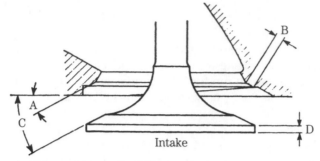

Intake

A. Valve seating angle: 30°
B. Valve seating width: 0.055–0.069 in.
C. Valve seat angle: 29°
D. Valve head minimum thickness: 0.035 in. valve head depth in relation to cylinder head face (minimum permissisble): 0.023 in.

A

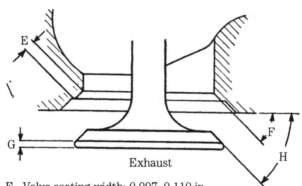

Exhaust

E. Valve seating width: 0.097–0.110 in.
F. Valve seating angle: 45°
G. Valve head minimum thickness: 0.035 in.
   Valve head depth in relation to cylinder head face (minimum permissible): 0.041

B

7-29 Typical valve specifications, Bedford engines.

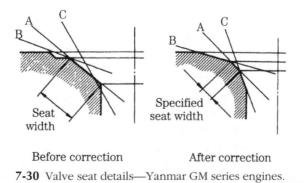

Before correction   After correction

7-30 Valve seat details—Yanmar GM series engines.

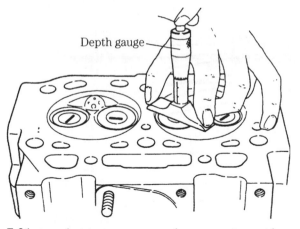

Depth gauge

7-31 A technician measures valve protrusion with a depth micrometer. Yanmar Diesel Engine Co., Ltd.

Valve protrusion or, as the case might be, recession, must be held to tight limits on these engines. A valve that extends too far into the chamber might be struck by the piston; one that is sunk into its seat hinders the combustion process, sometimes critically. Figure 7-31 illustrates how the measurement is made. Recession can be cured by replacing the valve and/or seat. Protrusion is a more difficult problem, encountered when the head has been resurfaced; a deep valve grind might help, provided one has the requisite seat thickness and can tolerate the increase in installed valve height (see immediately below). A handful of OEMs supply thinner-than-stock seat inserts. These inserts require corresponding thicker valve spring seats or shims, in order to maintain valve spring tension. Ultimately, one might be forced to replace the cylinder head.

Installed valve height, measured from the valve seat to the underside of the spring retainer, or cap, has become increasingly significant (Fig. 7-32). Springs on newer engines come perilously close to coil bind with as little as 0.005 in. separating the coils in the full open position.

Finally, the machinist dresses the valve stem tip square to restore rocker-tip to increase available contact area and to center hydraulic-lifter pistons. The average lifter has a stroke of 0.150 in. If the plunger is centered, available travel is 0.075 in., before it bottoms and holds the valve off its seat. But it is rare to find new engines with precisely centered lifter plungers, and available stroke might be considerably less than indicated. Deep valve grinding, head and block resurfacing, and camshaft grinding will, unless compensated for, collapse the lifters. The machinist can remove about 0.030 in. from the tip without penetrating the case hardening. Another limiting factor is rocker/spring-cap contact.

Seat replacement

Valve seat inserts are pressed into recesses machined into the head. (Some very early engines used spigoted inserts with mixed results.) Seats must be replaced when severely burned, cracked, loose, or as a means of obtaining the correct valve protrusion. Many shops routinely replace seats during an overhaul for the insurance value.

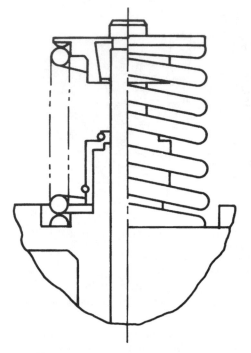

7–32 Installed valve height is the distance between the spring seat and the underside of the spring retainer. Often the factory neglects to provide the specification, and it must be determined by direct measurement during disassembly.

Seats in iron heads are customarily driven out with a punch inserted through the ports, although more elegant tools are available (Fig. 7-33); seats in aluminum heads should be cut out to prevent damage to the recess (Fig. 7-34).

While new seats can be installed in the original counterbores, it is good practice to machine the bores to the next oversize. Replacement seats for most engines are available in 0.010-, 0.015-, 0.020-, and 0.030-in. oversizes. Material determines the fit: iron seats in iron heads require about 0.005-in. interference fit; Stellite expands less with heat, and seats made of this material should be set up a little tighter; seats in aluminum require something on the order of 0.008-in. interference.

Seat concentricity should be checked with a dial indicator mounted in a fixture that pilots on the valve guide. Often the technician finds it necessary to restore concentricity by lightly grinding the seat.

Springs

Valve spring tension is all that keeps the valves from hitting the pistons. A "swallowed" valve is the mechanic's equivalent of the great Lisbon earthquake or the gas blowout at King Christian Island, which illuminated the Arctic night for eight months and could be seen from the moon. Thus, I suggest that valve springs be replaced (regardless of apparent condition) during upper engine overhauls.

If springs are to be used, inspect as follows:

- Carefully examine the springs for pitting, flaking, and flattened ends.
- Measure spring freestanding height and compare with the factory wear limit.

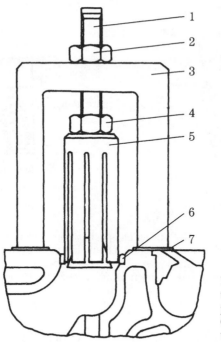

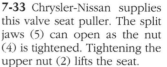

7-33 Chrysler-Nissan supplies this valve seat puller. The split jaws (5) can open as the nut (4) is tightened. Tightening the upper nut (2) lifts the seat.

- Stand the springs on their ends and, using a feeler gauge and machinist's square, determine the offset of the uppermost coil (Fig. 7-35). Compare with the factory specification (in angular terms, maximum allowable tilt rarely exceeds 2°). Most keeper failures arise from unequal loading.
- Verify that spring tension falls within factory-recommended norms. Figure 7-36 illustrates the tool generally used to make this determination.

Valve spring shims have appropriate uses, chiefly to restore the spring preload lost when heads and valve seats are refurbished. But shims should not be used as a tonic for tired springs, because the fix is temporary and can result in coil bind.

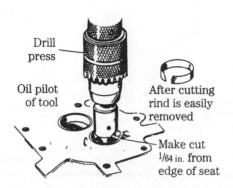

Drill press

Oil pilot of tool

After cutting rind is easily removed

Make cut 1/64 in. from edge of seat

7-34 Seats in aluminum heads should be machined out, rather than forced out. Onan

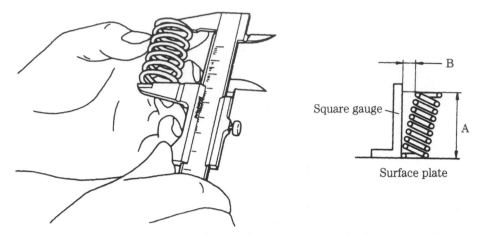

7-35 Most engine makers provide a linear valve spring tilt specification; Yanmar is more sophisticated. First spring free length is determined as shown in the left-hand drawing. The amount of offset is measured with a machinist's square. Offset (B) divided by free length (A) gives the specification, which for one engine series must not exceed 0.0035 in.

Top clearance

Top clearance, or the piston-to-head clearance at tdc, is critical. Unfortunately, the position of piston crown varies somewhat between cylinders because of the stacked tolerances at the crankshaft, rod, piston pin bosses, and deck (which might be tilted relative to the crankshaft centerline). No two pistons have the same spatial relationship to the upper deck. Resurfacing, or decking, the block lowers the fire deck 0.010 in. or more with no better accuracy than obtained by the factory.

Most manufacturers arrive at top clearance indirectly by means of a piston deck height specification. Either of these measurements must be made when

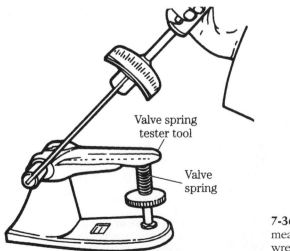

7-36 Most valve spring testers measure force with a torque wrench. Ford Motor Co.

- the block is resurfaced.
- the manufacturer supplies replacement head gaskets in varying thickness to compensate for production variations.

The geometry of some engines (flat pistons and access to the piston top with head in place) invites direct measurement of top clearance. The cylinder head is installed and torqued, using a new gasket of indeterminate thickness. The mechanic then removes No. 1 cylinder glow plug and inserts the end of a piece of soft wire, known as fuse wire, into the chamber. Turning the flywheel through tdc flattens the wire between the piston crown and cylinder head. Wire thickness equals top clearance, or piston deck height plus compressed gasket thickness (Fig. 7-37).

Scrupulous engine builders sometimes make the same determination using modeling clay as the medium. Upon disassembly the clay is removed and carefully miked. This method applies equally well to flat and domed pistons.

The more usual approach is to measure piston deck height with a dial indicator. The procedure involves three measurements, detailed as follows:

1. Zero the dial indicator on the fire deck, with the piston down (left-hand portion of Fig. 7-38).
2. Position the indicator over a designated part of the piston crown, shown as A in Fig. 7-38.
3. Turn the crankshaft in the normal direction of rotation through tdc. Note the highest indicator reading.
4. Repeat Step 3, taking the measurement at B.

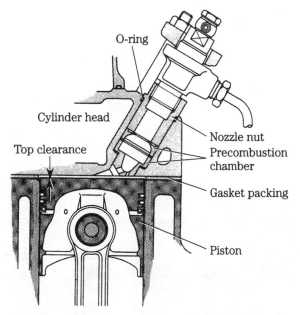

7-37 Top clearance equals piston deck height plus the thickness of the compressed head gasket. Yanmar Diesel Engine Co. Ltd.

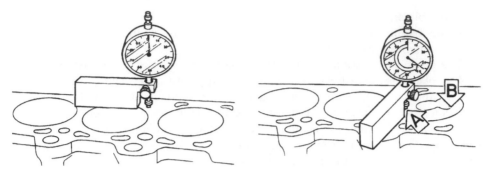

7-38 The first step when measuring piston deck height is to zero the dial indicator on the fire deck. The second step in the measurement process is to determine piston protrusion at two points, the average of which gives piston protrusion. Ford Motor Co.

5. Average measurements A and B.
6. Repeat the process for each piston. Use the piston with the highest average deck height to determine the thickness of the replacement head gasket.

Assembly

The manufacturer's manual provides detailed assembly instructions but includes little about the things that can go wrong. Most assembly errors can be categorized as follows:

- *Insufficient lubrication.* Heavily oil sliding and reciprocating parts, lightly oil head bolts and other fasteners, except those that penetrate into the water jacket. These fasteners should be sealed with Permatex No. 2 or the high-tech equivalent.
- *Reversed orientation.* Most head gaskets, many head bolt washers, and all thermostats are asymmetrical.
- *Mechanical damage.* Run fasteners down in approved torque sequences and in three steps—1/2, 2/3, and 1/1 torque (Fig. 7-39). Exceptions are torque-to-yield

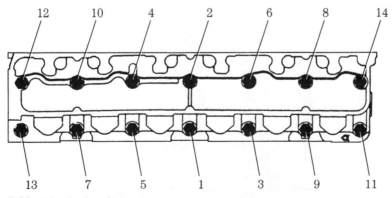

7-39 Cylinder head torque sequence. Gm Bedford Diesel

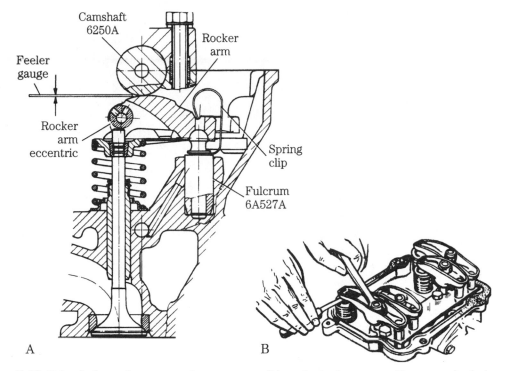

7-40 Valve lash can be measured at any accessible point in the system. For example, lash for the Ford-supplied 2.4L ohc is read as cam-to-rocker clearance (A); Onan measures the clearance between rocker arm tip and valve stem (B).

head bolts and rocker arm shaft fasteners. The former are torqued as indicated by the manufacturer, whose instructions will be quite explicit. The latter—rocker shaft fasteners—should be brought down in very small increments, working from the center bolts out.

Gaskets, especially head gaskets, might also be damaged during assembly. Lower the head on a pair of guide pins lightly threaded into the block. Pins can be fashioned from discarded head bolts by cutting the heads off. If pins are too short to extend through the head casting, slot the ends for screwdriver purchase.

Set initial valve lash adjustments, bleed the fuel system, start the engine. Final lash adjustments are usually made hot, after the engine has run for 20 minutes or so on the initial settings (Fig. 7-40).

8
CHAPTER

Engine mechanics

A mechanic needs to be a part-time electrician, semipro fuel system specialist, self-taught millwright, amateur machinist, and back-bench welder. But what he or she is supposed to do is to rebuild engines, the subject of this chapter.

Scope of work

Block assemblies can be *repaired, overhauled,* or *rebuilt* (Fig. 8-1). Spot repairs are either triggered by local failure (e.g., a sticking oil pressure relief valve or a noisy valve train bearing) or by a need to extract a few more hours from a worn-out engine. Many a poor mechanic has replaced an oil pump more out of hope than conviction.

While an overhaul is also an exercise in parts replacement, the scope is wider and usually occasioned by moderate cylinder and crankshaft-bearing wear. At the minimum, an overhaul entails grinding the valves and replacing piston rings and bearing inserts and whatever gaskets have been disturbed. The effort might extend to a new oil pump, timing and accessory drive parts, oil seals, cylinder liners (when easily accessible), together with new piston, ring, and wrist pin sets. Because the block and crankshaft remain in place, machine work is necessarily limited to the cylinder head.

In the classic sense, rebuilding an engine means the restoration of every frictional surface to its original dimension, alignment, and finish. The engine should theoretically be as good as new, or even better than new in the sense that used castings tend to hold dimension better than "green" parts. (Repeated heating and cooling cycles relieve stresses introduced during the casting process.) In addition, an older engine might benefit from late-production parts.

Although some mechanics would disagree, the rebuilding process cannot repeal the law of entropy. A competently rebuilt engine will be durable over the long run and will be reasonably reliable in the short term, but it will not quite match the factory norms. Subsurface flaws will not be detected. Metal lost to water jacket corrosion is irretrievably lost. Nor can original deck height, timing gear mesh, main bearing cap height, and camshaft geometry be achieved in any commercially practical sense. And

1 Cam follower
2 Camshaft
3 Cylinder block
4 Crankshaft
5 Crankshaft
   counterweight
6 Oil pan
7 Connecting rod
8 Liner packing rings
9 Cylinder liner
10 Piston
11 Piston pin
12 Piston rings
13 Oil pump

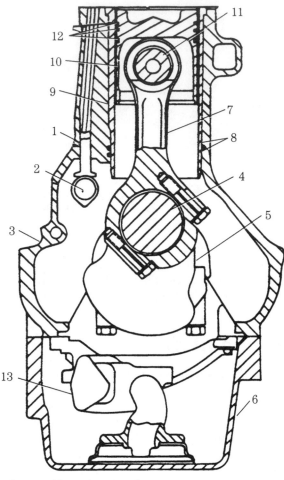

8-1 Sectional view of Deere 6076 block assembly and nomenclature.

the potential for error, on the part of both the machinist and the assembler, affects reliability. More often than not, a freshly rebuilt engine will experience "teething" difficulties.

On the other hand, the cost should not exceed half of the replacement cost, and engine life will be nearly doubled.

Traditionally, the work is divided among operator mechanics, who remove the engine from, its mounts, dismantle it, and consign the components to a machinist for inspection and refurbishing. The machinist might supply some or all of the replacement parts, which, together with the reworked parts, are returned to the mechanics for final assembly.

This approach organizes the work around specialists' skills, keeps the critical business of assembly in-house (where it probably belongs), and minimizes out-of-pocket expenses. One working mechanic—not the shop foreman—should have undiluted responsibility for the job, a responsibility that includes new and refurbished parts quality control (QC), assembly, installation, and start-up.

Engine machine work is an art like gunsmithing or watch repair in the sense that proficiency comes slowly, through years of patient application. In my anachronistic opinion, the best work comes out of small shops, where Model T crankshafts stand in racks, waiting for customers that never come, and the coffee pot hasn't been cleaned since 1940. These shops, in short, are places where a mill means a Bridgeport, a grinder is a Landis, and the lathes were made in South Bend.

Diagnosis

Before you begin you should have a good idea—or at least a plausible theory—about the nature of the problem. The diagnostic techniques described in Chap. 4 indicate whether or not major work is in order and, when supplemented by oil analysis, will suggest which class of parts—rings, gears, soft metal bearings, and so on—are wearing rapidly.

Test/analysis data, combined with a detailed operating history, should fairly well pinpoint the failure site (cylinder bore, crankshaft bearings, accessory drive, and the like). But analysis should not stop with merely verbal formulations. For example, it is hardly meaningful to say that a bearing or a piston ring set has "worn out" or "overheated." One should try to identify what associated failure or special operating condition selected those parts to fail. City-bus wheel bearings are a good example of the selection process; experience shows that the right front bearings tend to fail more often than those on the left. Traces of red oxide in the lubricant suggest that failure comes about because of moisture contamination, (i.e., water splash), which is more likely to occur on the curb side of the vehicle.

Once the mechanic understands the failure mechanism, it might be possible to correct matters by performing more frequent maintenance, upgrading parts quality, or modifying operating conditions.

Rigging

Figure 8-2 illustrates the proper lifting tackle for a large engine. Note how the spreader bar and adjustable crossheads keep the chains vertical to hook loads.

The engine shown incorporates lift brackets; less serious engines do not, and the mechanic is left to his own devising. In general, attachment points should straddle the center of gravity of the engine in two planes, so that it lifts horizontally without tilting. Chains must clear vulnerable parts, such as rocker covers and fuel lines. Lift brackets made of ⅜ in. flame-cut steel plate are the ideal, but *forged* eyebolts (available from fastener supply houses) are often more practical. A spreader bar will eliminate most bending loads, but there are times when chain angles of less than 90° cannot be entirely avoided. Reduce the bending load seen by the eyebolt with a short length of pipe and heavy washer, as shown in Fig. 8-3.

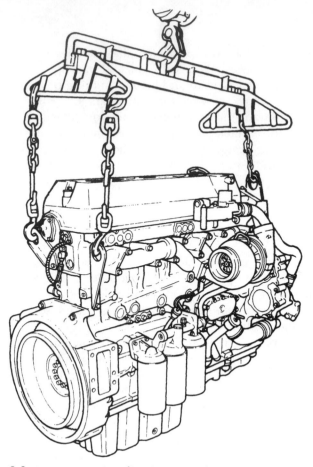

8-2 A proper engine sling is a necessity. Detroit Diesel

1/8 in. (3 mm) thick flat washer

Pipe

8-3 Hardened eyebolts occasionally must be used in lieu of lift brackets. Bolts should thread into a minimum depth of three times the diameter and should be reinforced as shown to limit bending forces.

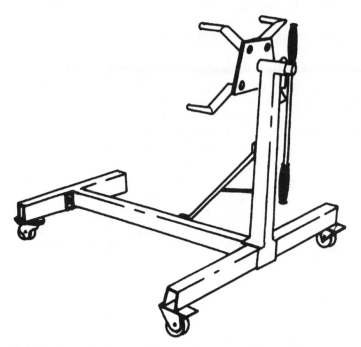

8-4 Purchase the best engine stand you can afford, with a weight capacity that provides a comfortable safety margin.

In no case should chains be bolted directly to the block without the intermediary of a bracket or eyebolt. Nor is multistrand steel cable appropriate for this kind of knock-about service.

Figure 8-4 illustrates a minimal engine stand, suitable for engines in the 600-lb. class. Better stands usually attach at the side of the block (as opposed to the flywheel flange) and can be raised and lowered.

Note: A mechanic can get into trouble with one of these revolving-head stands. Inverting an assembled engine with the turbocharger intact can dump oil from the turbocharger sump into one or more cylinders. Subsequent attempts to start the flooded engine might result in bent connecting rods or worse.

Special considerations

Mechanics who are knowledgeable about gasoline engine repair will find themselves doing familiar things, but to more demanding standards. Diesel engines are characterized by

Close tolerances Close tolerances impose severe requirements in terms of inspection, cleanliness, and torque limits. Tolerance stack—unacceptable variations in dimension of assemblies made up of components that fall on the high or low ends of the tolerance range—becomes a factor to contend with. As delivered from the factory, some engines employ a selective assembly of bearing inserts and pistons.

High levels of stress Hard-used industrial engines are subject to structural failures, a fact that underscores the need for careful inspection of crankshafts, main-bearing webs and caps, connecting rods, pistons, cylinder bore flanges, harmonic balancers, and all critical fasteners.

Inflexible refinishing and assembly norms Stress and close tolerances give little latitude for quick fixes and, unless contradicted by hard experience, factory recommendations should be followed to the letter.

Special features While nothing in four-cycle diesel crankcases can be considered uniquely diesel, some features of these engines might be unfamiliar to most gasoline-engine mechanics. Nearly all engines employ oil-cooled pistons. This is accomplished with rifle-drilled connecting rods or, as is more often the case today, by means of oil-spray tubes, or jets, which direct a stream of oil to the piston undersides (Fig. 8-5). Jets that bolt or press into place must be removed for cleaning and usually require alignment.

Diesel engines often employ removable cylinder sleeves, pressed or slipped into counterbores and standing proud of the fire deck. Installation is quite critical and special honing techniques are usually recommended.

Another difference is the apparent complexity of power transmission, which on some engines can have an almost baroque ornamentation. Figure 8-6 illustrates the bull gear/idler gear and accessory drive gear constellation on the DDA Series 60 engine. Belt drives may be hardly less imposing, as the drawing in Fig. 8-7 indicates.

In fact, this complexity is more illusory than real. One merely comes to terms with timing marks, deals with one component at a time, and builds the power train

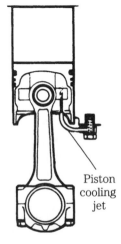

Piston
cooling
jet

8-5 Spray jet aiming is critical.
Ford Motor Co.

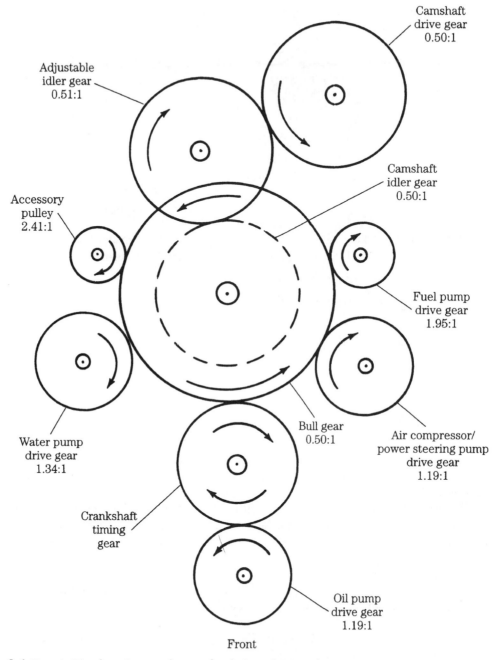

Camshaft
drive gear
0.50:1

Adjustable
idler gear
0.51:1

Camshaft
idler gear
0.50:1

Accessory
pulley
2.41:1

Fuel pump
drive gear
1.95:1

Water pump
drive gear
1.34:1

Bull gear
0.50:1

Air compressor/
power steering pump
drive gear
1.19:1

Crankshaft
timing
gear

Oil pump
drive gear
1.19:1

Front

8-6 Detroit Diesel engines are known for their sophisticated gear trains.

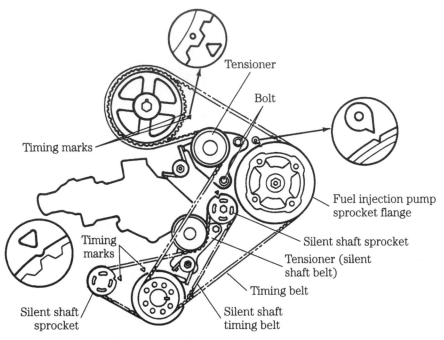

Tensioner

Bolt

Timing marks

Fuel injection pump
sprocket flange

Silent shaft sprocket

Timing
marks

Tensioner (silent
shaft belt)

Timing belt

Silent shaft
sprocket

Silent shaft
timing belt

8-7 Ford 2.3L turbocharged diesel makes extensive use of toothed belts.

brick by brick. Of course, the work goes more slowly than it would on a simpler engine, and the parts cost can be daunting, especially when gears need to be replaced.

The power train can include one or more balance shafts, a technology that is rarely seen on spark ignition engines (the Mitsubishi/Chrysler 2.6L is one of the few exceptions).

Figure 8-8 illustrates the balance shaft configuration for a four-cylinder in-line engine. Two contra-rotating shafts, labeled *Silent shafts* in the drawing, run at twice crankshaft speed to generate forces that counter the "natural" vibration of the engine.

Engines of this type employ single-plane, two-throw crankshafts. Pistons 1 and 4 move in concert, as do pistons 2 and 3. When pistons 1 and 4 are down, 2 and 3 are up. Consequently, vertical forces generated by pistons 1 and 4 (indicated by the dotted line in Fig. 8-9) oppose the forces generated by the center pair of pistons (represented by the fine line). These vertical forces, known as primary shaking forces, almost cancel and can be ignored.

Secondary forces pose a more serious problem. Created by connecting-rod angularity and by piston acceleration during the expansion stroke, these forces tend both to rotate the engine around the crankshaft centerline and to shake the engine vertically. Magnitude increases geometrically with rpm, to produce the dull rattle characteristic of in-line four-cylinder engines at speed.

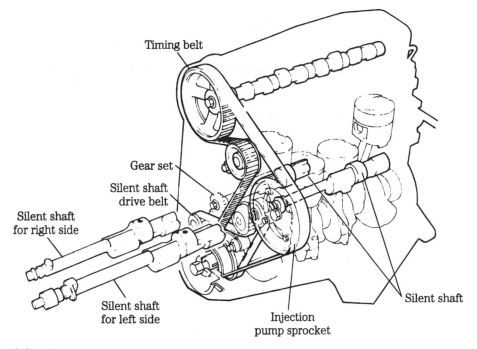

8-8 Balancing secondary forces requires two counterweighted shafts running at twice crankshaft speed. Ford Motor Co.

Secondary vertical and rolling forces are neutralized by deliberately induced imbalances in the balance shafts. Figure 8-10 diagrams the sequence of countervailing forces through full crankshaft revolution (two balance shaft revolutions).

Detroit Diesel approaches the question of balance differently on its two-cycle engines. Here the concern is to balance the rocking couple created by crankpin offset. Such couples exist unless all pistons share the same crankpin, as for example, in a radial engine with one connecting rod articulated from a central master rod. DDA practice is to use a counterweight can and balance shafts driven at crankshaft speed through counterweighted gears. In other words, each shaft has two sources of imbalance, one integral with the drive gear and the other in the form of a bob weight on

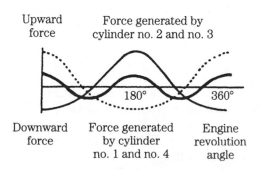

8-9 Primary forces in in-line four-cylinder engine very nearly balance. In an opposed four, with two crankshaft throws in the same plane 180° apart, balance is nearly perfect. A slight rocking couple, imposed by the offset between paired connecting rods on each crankshaft throw, does, however, exist. Ford Motor co.

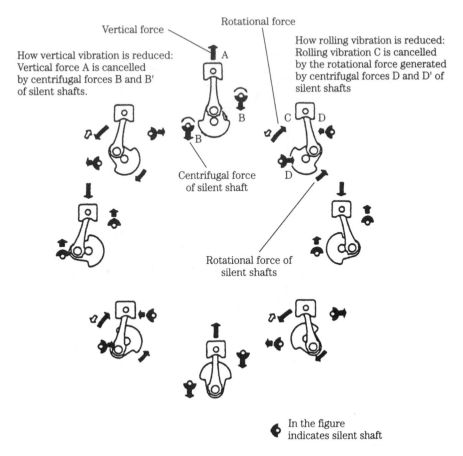

Vertical force

Rotational force

How vertical vibration is reduced:
Vertical force A is cancelled
by centrifugal forces B and B'
of silent shafts.

How rolling vibration is reduced:
Rolling vibration C is cancelled
by the rotational force generated
by centrifugal forces D and D' of
silent shafts

Centrifugal force
of silent shaft

Rotational force of
silent shafts

In the figure
indicates silent shaft

8-10 Balance shaft/crankshaft forces during a complete engine revolution. Ford Motor co.

the free end of the shaft. Shaft counterweights and shaft gear weights are disposed radially to create a countervailing couple, which acts in opposition to the crankshaft-induced couple. No attempt is made to balance secondary forces.

From a mechanic's point of view, the critical aspects of this technology are the shaft bearings, which are subject to severe radial loads and catastrophic failure. In some cases, bearing bosses must be sleeved before new bearings can be installed. Give the oiling circuit close scrutiny; endemic bearing can justify modifications to increase the rate of oil flow. And of course it is necessary to time the shafts relative to each other and to the crankshaft.

Fasteners

Contemporary foreign and, to a great extent, American engines are built to the metric ISO (International Standards Organization) standards, developed from the European DIN. For most practical purposes DIN and ISO fasteners interchange. A JIS standard also exists, but most Japanese fasteners made since the early 1970s follow the ISO pattern. Some JIS bolts interchange (although head dimension can differ) with those built to the current standard; others make up just enough to strip out.

Few American manufacturers remain wedded to the inch standard, although leaving is hard to do. Engines come off the line with both ISO and fractional fasteners, inch-standard pipe fittings and metric fuel systems.

ISO fasteners are classed by nominal bolt diameter in millimeters and thread pitch measured as millimeters between adjacent thread crests. Thus, a specification might call for an M8 × 1.0 cap screw or stud. Wrench size markings reflect bolt diameter, not the flat-to-flat distance across the screw head. Yield strength is indicated by a numeric code embossed on the screw head. The higher the number, the stronger the cap screw. Metric hex nuts often carry the same numerical code, but this practice is not universal.

Figures 8-11 and 8-12 supply identification data and suggested torque limits for U.S. and metric cap screws. Torque limits were calculated from bolt yield strength ratings, and do take into account the effects of clamping forces on vulnerable parts or gaskets. Consequently, these values should not be used when the engine manufacturer provides a different torque limit or tightening procedure for a specific application. Tighten plastic insert- or crimped steel-type locknuts to about half the amount shown in the charts; toothed- or serrated-type locknuts receive full torque. Replace fasteners with the same or higher grade, except in the case of shear bolts, which are grade specific. When substituting a better-grade fastener, torque it to the value of the original.

SAE Grade	Head Markings	SAE Grade	Nut Markings	SAE Grade	Head Markings	SAE Grade	SAE Grade	Nut Markings	SAE Grade
SAE GRADE 1 — SAE GRADE 2	No Mark	2	No Mark	SAE GRADE 5 SAE GRADE 5.1 SAE GRADE 5.2		5 Nut Markings	SAE GRADE 8 SAE GRADE 8.2	Nut Markings	8

DIA.	WRENCH SIZE	SAE GRADE 1		*SAE GRADE 2		SAE GRADE 5		SAE GRADE 8	
		OIL	DRY	OIL	DRY	OIL	DRY	OIL	DRY
		N·m(lb-in)	N·m(lb-in)	N·m(lb-in)	N·m(lb-in)	N·m(lb-in)	N·m(lb-in)	N·m(lb-in)	N·m(lb-in)
#6		0.5(4.5)	0.7(6)			1.4(12)	1.7(15)		
#8		0.9(8)	1.2(11)			2.4(21)	3.2(28)		
#10		1.4(12)	1.8(16)			3.4(30)	4.6(41)		
#12		2(19)	2.8(25)			5.4(48)	7.3(65)		
		N·m(lb-ft)	N·m(lb-ft)	N·m(lb-ft)	N·m(lb-ft)	N·m(lb-ft)	N·m(lb-ft)	N·m(lb-ft)	N·m(lb-ft)
1/4	7/16	3.5(2.5)	4(3)	5(4)	7(5)	8(6)	11(8)	12(8.5)	16(12)
5/16	1/2	7(5)	9(6.5)	10(7.5)	14(10)	16(12)	23(17)	24(18)	33(24)
3/8	9/16	12(8.5)	16(12)	19(14)	24(18)	30(22)	41(30)	41(30)	54(40)
7/16	5/8	19(14)	26(19)	30(22)	41(30)	47(35)	68(50)	68(50)	95(70)
1/2	3/4	24(21)	41(30)	47(35)	61(45)	75(55)	102(75)	102(75)	142(105)
9/16	13/16	41(30)	54(40)	68(50)	88(65)	108(80)	142(105)	149(110)	203(150)
5/8	15/16	54(40)	75(55)	88(65)	122(90)	149(110)	197(145)	203(150)	278(205)
3/4	1-1/8	102(75)	136(100)	163(120)	217(160)	258(190)	353(260)	366(270)	495(365)
7/8	1-5/16	163(120)	244(165)	163(120)	224(165)	414(305)	563(415)	590(435)	800(590)
1	1-1/2	244(180)	332(245)	244(180)	332(245)	624(460)	848(625)	881(650)	1193(880)
1-1/8	1-11/16	346(255)	468(345)	346(255)	468(345)	780(575)	1058(780)	1248(920)	1695(1250)
1-1/4	1-7/8	488(360)	664(490)	488(360)	665(490)	1098(810)	1492(1100)	1763(1300)	2393(1765)
1-3/8	2-1/16	637(470)	868(640)	637(470)	868(640)	1438(1061)	1953(1440)	2312(1705)	3140(2315)
1-1/2	2-1/4	848(625)	1153(850)	848(625)	1153(850)	1912(1410)	2590(1910)	3065(2260)	4163(3070)

8-11 Inch cap screw torque values. 1990.

Property Class and Head Markings	4.6		4.8		8.8	9.8	10.9		12.9	
Property Class and Nut Markings	5		5		10		10		12	

| DIA. | WRENCH SIZE | 4.6 | | 4.8 | | 8.8 or 9.8 | | 10.9 | | 12.9 | |
		OIL	DRY	OIL	DRY	OIL	DRY	OIL	DRY	OIL	DRY
		N•m(lb-ft)	N•m(lb-ft)	N•m(lb-ft)	N•m(lb-ft)	N•m(lb-ft)	N•m(lb-ft)	N•m(lb-ft)	N•m(lb-ft)	N•m(lb-ft)	N•m(lb-ft)
M3	5.5mm	0.4(0.2)	0.5(0.3)	0.5(0.4)	0.7(0.5)	1(0.8)	1.3(1)	1.5(1)	2(1.5)	1.5(1)	2(1.5)
M4	7mm	0.9(0.6)	1.1(0.8)	1(0.9)	1.5(1)	2.5(1.5)	3(2)	3.5(2.5)	4.5(3)	4(3)	5(4)
M5	8mm	1.5(1)	2.5(1.5)	2.5(1.5)	3(2)	4.5(3.5)	6(4.5)	6.5(4.5)	9(6.5)	7.5(5.5)	10(7.5)
M6	10mm	3(2)	4(3)	4(3)	5.5(4)	7.5(5.5)	10(7.5)	11(8)	15(11)	13(9.5)	18(13)
M8	13mm	7(5)	9.5(7)	10(7.5)	13(10)	18(13)	25(18)	25(18)	35(26)	30(22)	45(33)
M10	16mm	14(10)	19(14)	20(15)	25(18)	35(26)	50(37)	55(41)	75(55)	65(48)	85(63)
M12	18mm	25(18)	35(26)	35(26)	45(33)	65(48)	85(63)	95(70)	130(97)	110(81)	150(111)
M14	21mm	40(30)	50(37)	55(41)	75(55)	100(74)	140(103)	150(111)	205(151)	175(129)	240(177)
M16	24mm	60(44)	80(59)	85(63)	115(85)	160(118)	215(159)	235(173)	315(232)	275(203)	370(273)
M18	27mm	80(59)	110(81)	115(85)	160(118)	225(166)	305(225)	320(236)	435(321)	375(277)	510(376)
M20	30mm	115(85)	160(118)	165(122)	225(166)	320(236)	435(321)	455(356)	620(457)	535(395)	725(535)
M22	33mm	160(118)	215(159)	225(167)	305(225)	435(321)	590(435)	620(457)	840(620)	725(535)	985(726)
M24	36mm	200(148)	275(203)	285(210)	390(288)	555(409)	750(553)	790(583)	1070(789)	925(682)	1255(926)
M27	41mm	295(218)	400(295)	415(306)	565(417)	810(597)	1100(811)	1155(852)	1565(1154)	1350(996)	1835(1353)
M30	46mm	400(295)	545(402)	565(417)	770(568)	1100(811)	1495(1103)	1570(1158)	2130(1571)	1835(1353)	2490(1837)
M33	51mm	545(402)	740(546)	770(568)	1050(774)	1500(1106)	2035(1500)	2135(1575)	2900(2139)	2500(1844)	3390(2500)
M36	55mm	700(516)	950(700)	990(730)	1345(992)	1925(1420)	2610(1925)	2740(2021)	3720(2744)	3205(2364)	4355(3212)

8-12 Metric cap screw torque values. 1990. Dears & Co.

Cleaning

Ford Motor and other manufacturers say that dirt is the chief cause of callbacks after major work. The direct effect is to contaminate the oil supply; the indirect effect is to create an environment that makes craftsmanship difficult or impossible.

The need for almost septic standards of cleanliness argues against the practice of opening the engine for less than comprehensive repairs. Of course, it happens that such repairs must be made, regardless of the long-term consequences. Nor is it possible to maintain reasonable standards of cleanliness during in-frame overhauls, although the damage can be minimized by moving in quickly, cleaning only those friction surfaces that are opened for inspection, and getting out. Dirt accumulations on internal parts of the engine cannot be removed from below, while parts are still assembled, and attempts to do so will only release more solids into the oil stream.

When, on the other hand, the engine is rebuilt, the block, cylinder head, pan, and other steel stampings are sent out for thermal or chemical cleaning (see Chap. 7). These processes also remove the paint, which is all to the good. Crankshaft, piston assemblies, and other major internal parts receive a preliminary wash-down for inspection by the mechanic in charge of the job. These parts are then forwarded to the machine shop for evaluation. When pistons are reused, the machinist will chemically clean the grooves and piston undersides—chores that save hours of labor. About all that remains for the shop mechanic is to degrease fasteners and accessories that have been detached from the block and remove the rust preventative from new parts.

Teardown

Drain the oil and coolant and degrease the outer surfaces of the engine. Disconnect the battery, wiring harness (make a sketch of the connections if the harness is not keyed for proper assembly), and exhaust system. Attach the sling and undo the drive line connection and the motor mounts. With the engine secured in a stand, detach the manifolds, cylinder heads, and oil sump. The block should be stripped if you contemplate machine work or chemical cleaning of the jackets.

Lubrication system

The first order of business should be the lubrication system. To check it out you must have a reasonably good notion of the oiling circuits. Figure 8-13 is a

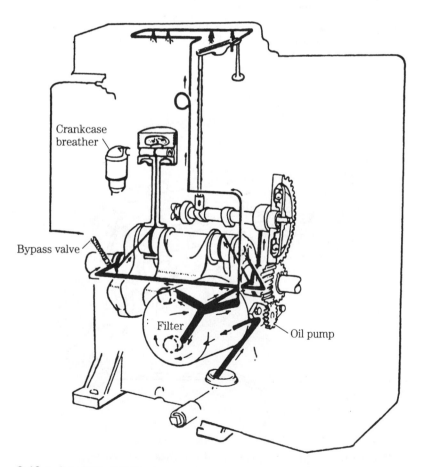

8-13 Lubrication system. Onan

drawing of the Onan DJ series lubrication system. The crankcase breather is included because it has much to do with oil control. Should it clog, the engine will leak at every pore. Oil passes from the screened pickup tube (suspended in the pan) to the pump, which sends it through the filter. From there the filtered oil is distributed to the camshaft, the main bearings (and through rifle-drilled holes in the crankshaft to the rod bearings and wrist pin), and to the valve gear. Valve and rocker arm lubrication is done through a typically Onan "showerhead" tube. Tiny holes are drilled in the line and deliver an oil spray at about 25 psi. On its return to the sump, the oil dribbles down the push rods to lubricate the cam lobes and tappets.

The system in Fig. 8-14 is employed in six-cylinder engines. From the bottom of the drawing, oil enters the pickup tube, then goes to the Gerotor pump. Unlike most oil pumps, this one is mounted on the end of the crankshaft and turns at engine speed. The front engine cover incorporates inlet and pump discharge ports. Oil is sent through a remote cooler (6), then directed back to the block, where it exits again to the filter bank (11). Normally oil passes through these filters. However, if the filters clog or if the oil is thick from cold, a pressure differential type of bypass valve (12) opens and allows unfiltered flow.

The main oil gallery (3) distributes the flow throughout the engine. Some goes to the main bearings and through the drilled crankshaft to the rods. The camshaft

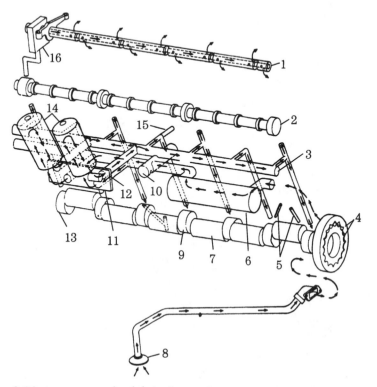

8-14 A more complex lubrication system. International Hamster

bearings receive oil from the same passages that feed the mains. The rearmost cam journal is grooved. Oil passes along this groove and up to the rocker arms through the hollow rocker shaft. On its return this oil lubricates the valve stems, pushrod ball sockets, tappets, and cam lobes. Other makes employ similar circuits, often with a geared pump.

Figure 8-15 describes the Ford 2.3L lubrication system, which is surprisingly sophisticated for a small and, by diesel standards, inexpensive product. Pressurized oil goes first to the filter, described in the following section.

Depending on oil temperature, output from the filter goes either directly to the main distribution gallery or to the gallery by way of an oil cooler (Fig. 8-16). A thermostatic valve, located in the filter housing, directs the flow (Fig. 8-17). Cooling the oil during cold starts is counterproductive, and the valve remains closed; as oil temperature increases, the valve extends to split the flow between the gallery and cooler. At approximately 94°C, all flow is diverted to the cooler. As a safety measure, the bypass valve opens if cooler pressure drop exceeds 14 psi. Thus, a stoppage will not shut down oil flow, although long-term survivability is compromised.

Camshaft lubrication, a weak point on many ohc engines, shows evidence of careful attention. A large-diameter riser, feeding from the main gallery, supplies oil to the center camshaft bearing. At that point the flow splits, part of it entering the hollow camshaft and part of it going to an overhead spray bar. The camshaft acts as an oil gallery for the remaining two shaft bearings; the spray bar provides lubrication for camshaft lobes, rockers, and valve tips.

The main bearings are lubricated through rifle-drilled ports connecting the webs and main gallery. As a common practice, diagonal ports drilled in the crankshaft convey oil from the webs to adjacent crankpins. Spray jets, again fed by the main gallery, cool the undersides of the pistons and provide oil for the piston pins. A gear and crescent pump—the lightest, most compact type available—supplies the necessary pressure.

This cursory examination of the oil circuitry on three very different engines should underscore the need to come to terms with these systems. No other system has such immediate implications for the integrity of the mechanic's work.

Clogging is the most frequent complaint, caused by failure to change oil and filters at proscribed intervals, and abetted by design flaws. Expect to find total or partial blockage wherever the oil flow abruptly changes direction or loses velocity. Prime candidates are the cylinder head/block interface, spray jets, and the junctions between cross-drilled passages. Neither immersion-type chemicals, heat, nor compressed air can be depended on to remove metal chips and sludge. Drilled passages must be cleared by hand, using riflebore brushes and solvent.

Thorough cleaning is never more important than after catastrophic failure, which releases a flood of metallic debris into the oiling system. In this case, the oil cooler can present a special problem. Modern, high-density coolers do not respond well to chemical cleaners and frustrate even the most flexible swabs. A new or used part—whose history is known—is sometimes the only solution.

Leaks can develop at the welch (expansion) plugs or, less often, at the pipe plugs that blank off cross-drilled holes and core cavities. It is also possible for cracks to open around lifter bores and other thin-section areas. A careful mechanic

A

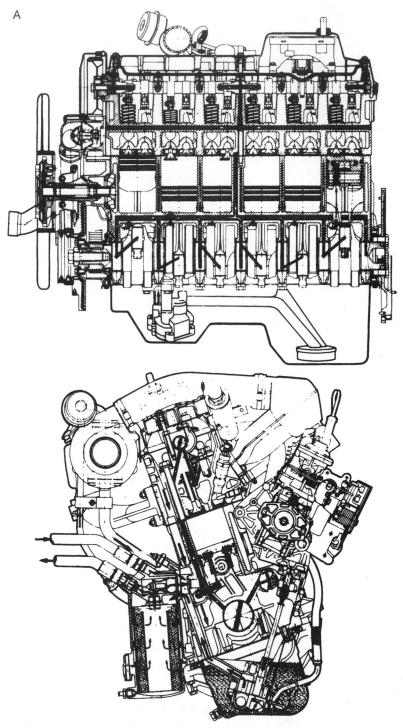

8-15 Ford 2.4L lubrication system.

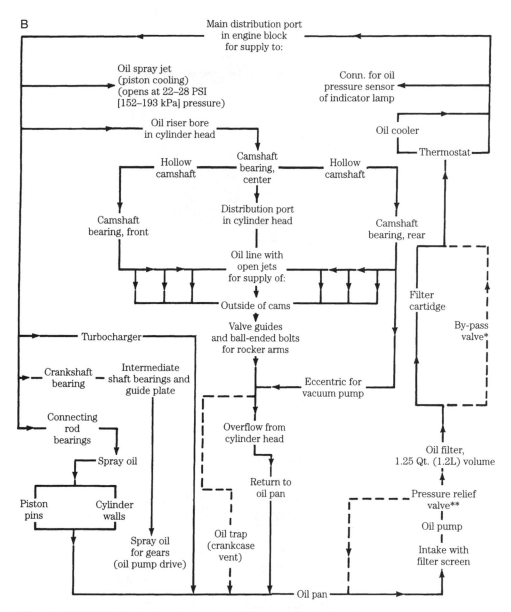

B

Main distribution port
in engine block
for supply to:

Oil spray jet
(piston cooling)
(opens at 22–28 PSI
[152–193 kPa] pressure)

Conn. for oil
pressure sensor
of indicator lamp

Oil riser bore
in cylinder head

Oil cooler

Thermostat

Camshaft
bearing,
center

Hollow
camshaft

Hollow
camshaft

Camshaft
bearing, front

Distribution port
in cylinder head

Camshaft
bearing, rear

Oil line with
open jets
for supply of:

Filter
cartridge

By-pass
valve*

Outside of cams

Valve guides
and ball-ended bolts
for rocker arms

Turbocharger

Crankshaft
bearing

Intermediate
shaft bearings and
guide plate

Eccentric for
vacuum pump

Oil filter,
1.25 Qt. (1.2L) volume

Connecting
rod
bearings

Spray oil

Overflow from
cylinder head

Pressure relief
valve**

Piston
pins

Cylinder
walls

Return to
oil pan

Oil pump

Spray oil
for gears
(oil pump drive)

Oil trap
(crankcase
vent)

Intake with
filter screen

Oil pan

*(Opens at 36 PSI [241 kPa]) oil supply guaranteed if filter cartridge is plugged

**(Oil pump) with cold oil (opens with 79–92 PSI [544–648 kPa] pressure)

8-15 (*Continued*)

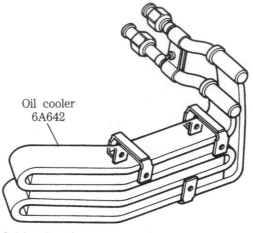

Oil cooler
6A642

8-16 Oil cooler.

will discard welch plugs (after determining that spares are available!) and apply fresh sealant to pipe plug threads during a rebuild. Most oil-wetted cracks can be detected visually.

Internal leaks that have been missed will show when the system is filled with pressurized oil before start-up.

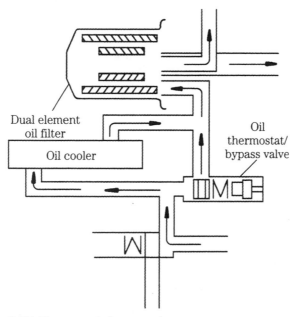

Dual element
oil filter

Oil cooler

Oil
thermostat/
bypass valve

8-17 Thermostatic bypass valve.

Filters

All contemporary engines employ full-flow filtration, using single or tandem paper-element filters in series with the pump outlet. Pleated paper filters can trap particles as small as 1 μm but are handicapped by limited holding capacity. Consequently, such filters incorporate a bypass valve that opens to shunt the element when the pressure drop reaches about 15 psi (Fig. 8-18), an action that should trigger a warning lamp on the instrument panel (Fig. 8-19). Otherwise the operator should occasionally feel the filter canister to verify that warm oil is circulating.

The Ford 2.3L filter consists of two concentric elements, one in series with the oiling circuit, the other in parallel (Fig. 8-20). As long as the bypass valve, shown at the top of the drawing, remains seated, pump output passes through the outer, series-connected element and into the lubrication circuit. Pump delivery rates are greater than the system requires and the surplus migrates through the inner filter and returns to the sump through the 9-mm orifice at the base of the assembly. I have been unable to obtain precise data for this engine, which was designed in Japan. However, standard design practice is to calculate diesel lube oil requirements as 0.003 times the ratio of oil to piston displacement times rpm. Pumps are then sized to meet twice these requirements. If this holds, the parallel filter makes a major contribution.

The most important single maintenance activity is to change the oil and filter on the engine-maker's schedule, which is more demanding than suggested for equivalent spark ignition (SI) engines. A turbocharger increases bearing loads and blowby, further shortening oil and filter life.

The final steps in an overhaul or rebuild are to prime the oil pump by rotating the shaft with the impeller submerged in a container of clean motor oil, fill the turbocharger cavity and feed line, and pressurize the oiling circuit. The latter is accomplished by connecting a source of oil pressure to the main distribution cavity, usually by way of the oil-pressure sender tap. This step is sometimes omitted and the

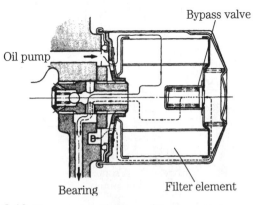

8-18 Yanmar cartridge-type filter mounts downstream of the pressure-regulating valve and includes a bypass valve.

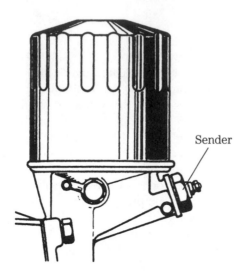

Sender

8-19 Peugeot automotive diesels alerted the driver when the filter clogged.

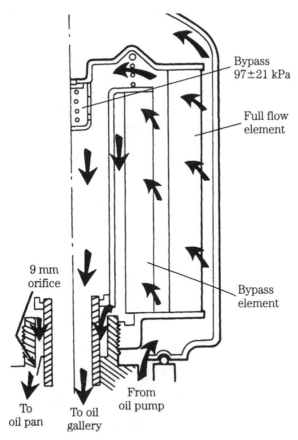

Bypass
97±21 kPa

Full flow
element

Bypass
element

9 mm
orifice

To
oil pan

To oil
gallery

From
oil pump

8-20 Ford series/parallel filter.

bearings—even though preoiled during assembly—suffer for it. The system *must* be pressurized after a new camshaft is installed.

Any repairs involving the turbocharger should be followed by disconnecting the oil return line and cranking until flow is established.

Before start-up after routine oil and filter changes, it is good practice to dry-crank the engine until the panel gauge indicates that oil pressure has been restored.

Lube oil

Oil collects moisture from condensation in the sump and is the repository for the liquid by-products of combustion. These by-products include several acid families that, even in dilute form, attack bearings and friction surfaces. No commercially practical filter can take out these contaminants. In addition oil, in a sense, wears out. The petroleum base does not change, but the additives become exhausted and no longer suppress foam, retard rust, and keep particles suspended. Heavy sludge in the filter is a sure indication that the change interval should be shortened, because the detergents in the oil have been exhausted. In heavy concentrations water emulsifies to produce a white, mayonnaise-like gel that has almost no lubrication qualities.

Oil that has overheated oxidizes and turns black. (This change should not be confused with the normal discoloration of detergent oil.) If fresh oil is added, a reaction might be set up that causes the formation of hard granular particles in the sump, known as "coffee grounds."

Oil change intervals are a matter of specification—usually at every 100 hours—and sooner if the oil shows evidence of deterioration. Most manufacturers are quite specific about the type and brand to be used. Multigrade oils (e.g., 10–30W) are not recommended for some engines, because it is believed they do not offer the protection of single-weight types. Other manufacturers specify SF grades. As a practical matter this specification means that multigrade oils can be used. Brand names are important in diesel service. Compliance with the standards jointly developed by the American Petroleum Institute, the Society of Automotive Engineers, and the American Society for Testing and Materials is voluntary. You have no guarantee that brand X is the equivalent of brand Y, even though the oil might be labeled as meeting the same API-SAE-ASTM standards. General Motors suggests that you discuss your lubrication needs with your supplier and use an oil that has been successful in diesel engines and meets the pertinent military standards. MIL-L-2104B is the standard most often quoted, although some oils have been "doped" with additives to meet the API standards. These additives cause problems with some engines often in the form of deposits around the ring belt. For GM two-cycles zinc should be held in between 0.07% and 0.10% by weight, and sulfated ash to 1.0%. If the lubricant contains only barium detergent-dispersants, the sulfated-ash content can be increased by 0.05%. High-sulfur fuels might call for *low-ash series 3* oils, which do not necessarily meet military low-temperature performance standards. Such oils might not function as well as MIL-L-2104B lubricants in winter operation.

Oil pumps

The Gerotor pump shown in Fig. 8-21 is gear-driven; a more common practice is to drive pumps of this type directly from the crankshaft. Like the Wankel engine, which employs a similar trochoidal geometry, operation is somewhat difficult to visualize on paper and obvious in the hardware. The inner rotor, which has one lobe less than the outer rotor, "walks" as it turns, imparting motion to the outer rotor and simultaneously varying the volume of the working cavity (shaded in the drawing). Volume change translates as a pressure head.

The clearance between the rotors and the end cover is best checked with *Plasti-Gage*. Plasti-Gage consists of rectangular-section plastic wire in various thicknesses. A length of the wire is inserted between the rotors and cover. The working clearance is a function of how much the wire is compressed under assembly torque. The package has a scale printed on it to convert this width to thousandths of an inch. Plasti-Gage is extremely accurate and fast. The only precautions in its use are that the parts must be dry, with no oil or solvent adhering to them, and that the wire must be removed after measurement. Otherwise it might break free and circulate with the oil, where it can lodge in a port. Typical end clearance for a Gerotor pump is on the order of 0.003 in. Of course, all rubbing surfaces should be inspected for deep scratches, and the screen should be soaked in solvent to open the pores.

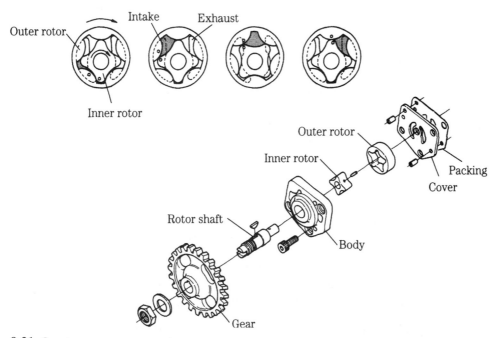

8-21 Gerotor pump. Yanmar Diesel Engine Co. Ltd.

Other critical specifications are the clearance between the outer rotor and the case (Fig. 8-22A) and the approach distance between inner and outer rotors (B).

Camshaft-driven gear-type pumps are the norm (Fig. 8-23). Check the pump for obvious damage—scores, chipped teeth, noisy operation. Then measure the clearance; the closer the better as long as the parts are not in physical contact.

End clearance can be checked by bolting the cover plate up with Plasti-Gage between the cover and gears, or with the aid of a machinist's straightedge. Lay the straightedge on the gear case and measure the clearance between the top of

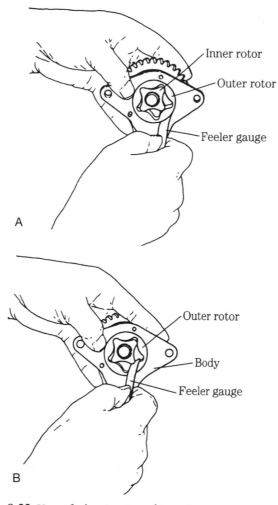

8-22 Use a feeler gauge to determine outer rotor to case clearance (A) and inner to outer rotor clearance. Wear limits for the Yanmar and most other such pumps are 0.15 in. (B).

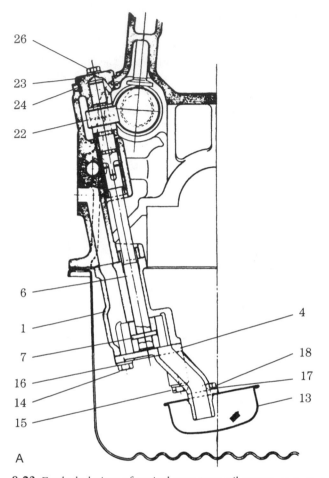

1	Oil pump body
2	Idler shaft
3	Packing (SD22)
4	Oil pump cover
5	Drive gear
6	Drive shaft
7	Pin
8	Driven gear
9	Relief valve
10	Relief valve spring
11	Washer
12	Cotter pin
13	Oil screen
14	Bolt
15	Bolt
16	Washer
17	Lockwasher
18	Bolt
19	Gasket
20	Bolt
21	Lockwasher
22	Driving spindle
23	Driving-spindle support
24	O-ring
25	Bolt
26	Bolt
27	Lockwasher

8-23 Exploded view of typical gear-type oil pump. Marine Engine Div., Chrysler Corp.

the gear and the case with a feeler gage. When in doubt, replace or resurface the cover. Check the diameter of all shafts and replace as needed. The idler gear shaft is typically pressed into the pump body. Use an arbor press to install, and make certain that the shaft is precisely centered in its boss. Allowing the shaft to cant will cause interference and early failure. Total wear between the drive shaft and bushing can be determined with a dial indicator as shown in Fig. 8-24. Move the shaft up and down as you turn it. Check the backlash with a piece of solder or Plasti-Gage between the gears. For most pumps this clearance should not exceed 0.018 in.

In general, it is wise not to tamper with the pressure relief valve, which might be integral with the pump or at some distance from it. If you must open the valve, because of either excessive lube oil pressure or low pressure, observe that the parts are generally under spring tension and can "explode" with considerable force. Check the spring tension and free length against specs, and replace or lap

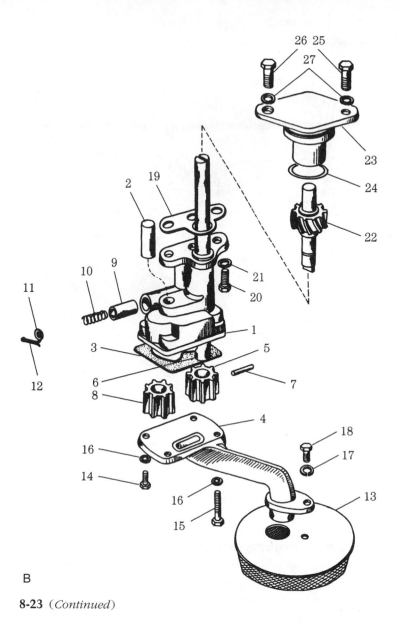

B

8-23 (*Continued*)

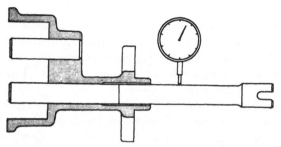

8-24 Determining shaft-bearing wear. Marine Engine Div., Chrysler Corp.

8-25 Oil pressure check. Peugeot

the valve parts as needed. When this assembly has been disturbed it is imperative that the oil pressure be checked, with an accurate instrument such as the gauge shown in Fig. 8-25.

Oil pressure monitoring devices

Because oil pressure is critical, it makes good sense to invest in some sort of alarm to supplement the usual gauge. Gauge failure is rare (erratic operation of mechanical gauges is usually caused by a slug of hardened oil impacted at the gauge fitting), but it does happen. Failure to shut down when pressure is lost will destroy the engine in seconds.

Stewart-Warner monitor Stewart-Warner makes an audiovisual monitor that lights and buzzes when oil pressure and coolant temperatures exceed safe norms. The indicator is a cold-cathode electron tube, which is much more reliable than the usual incandescent bulb. Designed for panel mounting, the 366-T3 is relatively inexpensive insurance.

Onan monitor Onan engines can be supplied with an automatic shutdown feature. The system employs a normally on sensor element and a time delay relay. In conjunction with a 10 W, 1-Ω resistor, this relay allows the engine to be started. When running pressure drops below 13 psi, the sensor opens, causing the fuel rack to pull out.

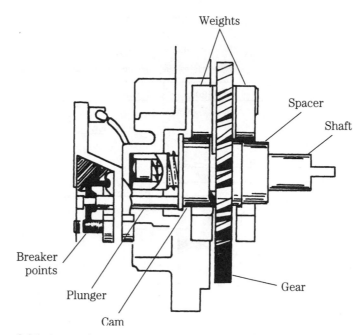

8-26 Onan oil pressure monitor—sensor section.

The sensor (Fig. 8-26) is the most complicated part of the system. It has a set of points that should be inspected periodically, cleaned, and gapped to 0.040 in. When necessary, disassemble the unit to check for wear in the spacer, fiber plunger, and spring-loaded shaft plunger. The spacer must be at least 0.35 in. long. Replace as needed. Check the action of the centrifugal mechanism by moving the weights in their orbits. Binding or other evidence of wear dictates that the weights and cam assembly be replaced. The cam must not be loose on the gear shaft.

Ulanet monitor The George Ulanet Company makes one of the most complete monitoring systems on the market (Fig. 8-27). It incorporates an optional water level sensor, overheat sensor, fuel-pressure sensor, and low- and high-speed oil sensors. For the GM 3-71 and 4-71 series, these sensors are calibrated at 4 psi and 30 psi, respectively.

Crankcase ventilation

The crankcase must he vented to reduce the concentration of acids and water in the oil. A few cases are at atmospheric pressure; others (employing forced-air scavenging) are at slightly higher than atmospheric; and still other systems run the crankcase at a slight vacuum to reduce the possibility of air leaks. The vapors might be vented to the atmosphere or recycled through the intake ports. In any event, the system requires attention to ensure that it operates properly. A clogged mesh element or pipe will cause unhealthy increases in crankcase pressures, forcing oil out

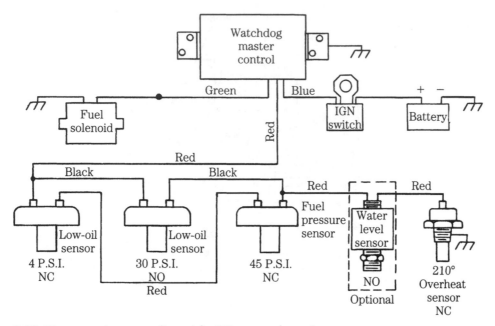

8-27 Ulanet monitor as configured for DD two-cycle engines.

around the gaskets and possibly past the seals, as well as increasing oil consumption. Figure 8-28 shows a breather assembly with check valve to ensure that the case remains at less than atmospheric pressure.

Block casting

Carefully examine the fire deck for pulled head bolt threads, eroded coolant passages, missing or damaged coolant deflectors, and cracks, particularly between adjacent cylinders. In some cases, minor cracks can be ground out and filled. Several dimensional checks also need to be made at this time.

Deck flatness

Figure 8-29 illustrates the basic technique, which involves a series of diagonal and transverse measurements along the length and width of the deck, using a machinist's straightedge and feeler gauge. Block deck surfaces can he milled flat (a process called *decking*), although contemporary design practices give little leeway for corrective machining. As mentioned in Chap. 7, several modern engines cannot be safely decked, and warped blocks must be replaced. However, when it comes to it, most machinists will push these limits and obtain piston-to-cylinder head clearance by shimming the gasket or by selective assembly, fitting the shortest piston and rod sets that can be found. The owner should be aware of these expedients and agree to them.

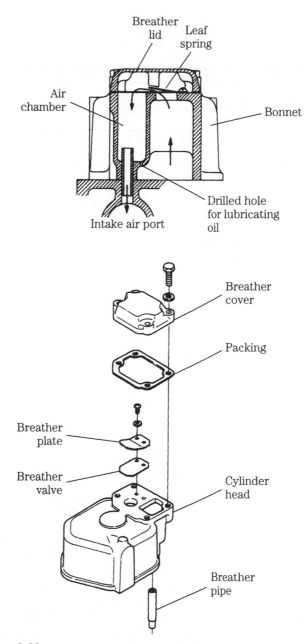

8-28 Yanmar atmospheric crankcase breather.

Piston height

While this is an assembly dimension, an early check is not out of order. Measure piston protrusion or regression with a dial indicator, as described in the previous chapter.

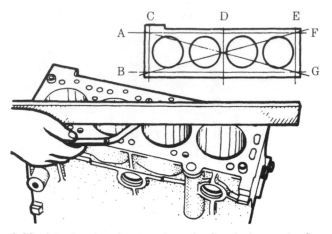

8-29 If the head gasket is to live, the fire deck must be flat and true. Chrysler Corp.

Liner height

A few engines continue to be built with integral cylinder bores, machined directly into the block metal or in the form of a sleeve permanently installed during the casting process. These engines have flat decks, and what will be said here does not apply.

Most diesel engines employ discrete cylinder bore liners, or sleeves, which can be more or less easily replaced. Dry liners press into block counterbores; wet liners insert with a light force fit and come into direct contact with the coolant. Seals confine the coolant to areas adjacent to ring travel (Fig. 8-30). Wet liners simplify foundry work and give better control of water-jacket dimensions. The interface between dry liners and cylinder bores erects a minor, but real, thermal barrier between combustion and coolant. However, the liner is a structural member and eliminates the possibility of coolant leaks into the crankcase or back into the combustion chamber. Liners of either type stand proud of the deck, a practice that gives additional compression to the head gasket in a critical area and, at the same time prestresses the liner (Fig. 8-31). Any liner that does not meet specification must be replaced.

This measurement taken at several points around the circumference of the liner, should be made during initial teardown and after replacement liners are installed. Figure 8-32 illustrates the liner hold-down hardware used in machine shops. Clamps are normally fabricated on site from cold-rolled steel, fitted with hardened washers, and pulled down securely. This technique works for dry liners and wet liners with radial (O-ring type) seals. Liner counterbores—the ledges that establish liner height—can be remachined and shimmed when necessary. The more usual procedure is to shim the liner shoulder, as shown in the illustration.

Some wet liners fit so loosely that seal resiliency can affect liner height when the cylinder head is removed. John Deere and other manufacturers that use this type of liner provide detailed loading instructions. Neither do engines fitted with these liners

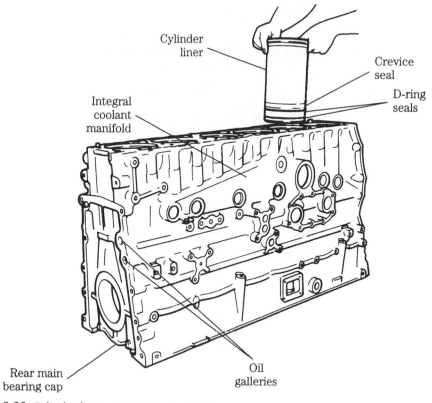

8-30 Cylinder liner—wet type. Detroit Diesel

tolerate moving the crankshaft with the head detached. If this is done, piston-ring friction will raise the liners and possibly damage the O-ring seals.

Boring

Ascertain the amount of bore wear. Most wear occurs near the top of the cylinder at the extreme end of ring travel. This wear is caused by local oil starvation and

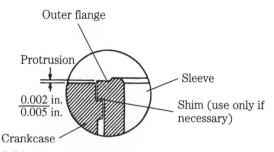

8-31 Liner protrusion measurement is taken at several places around the circumference of the part and averaged. International Harvester

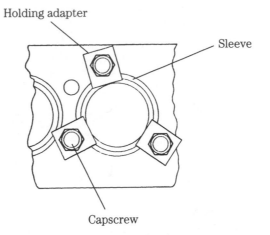

Holding adapter

Sleeve

Capscrew

8-32 Sleeve hold-down hardware. International Harvester

combustion-related acids. Pronounced wear, sometimes taking the form of scuffing, can occur at right angles to the crankshaft centerline on the bore surface that absorbs piston angular thrust. An engine that turns clockwise when viewed from the front will show more wear on the right side of its cylinders than on the left. In addition, axial thrust forces can generate wear in bore areas adjacent to the crankshaft centerline (Fig. 8-33). These forces account for most of the taper and eccentricity exhibited by worn cylinders. Block distortion accounts for the rest.

Some idea of bore condition can be had by inserting a ring into the cylinder with the flat of a piston. The difference in ring end gap between the upper and lower portions of the cylinder, as determined with a feeler gauge, roughly corresponds to cylinder wear. However, such techniques do not substitute for repeated and averaged measurements with a cylinder bore gauge (Fig. 8-34).

Study the surface of the bore under a strong light. Deep vertical scratches usually indicate that the air filter has at one time failed. The causes of more serious

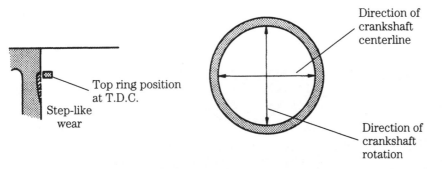

Top ring position
at T.D.C.

Step-like
wear

Direction of
crankshaft
centerline

Direction of
crankshaft
rotation

8-33 Areas of accelerated cylinder wear. Yanmar Diesel Engine Co. Ltd.

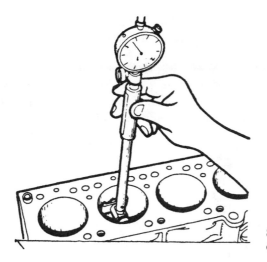

8-34 Using a cylinder gauge.
Chrysler Corp.

damage—erosion from contact with fuel spray, galling from lack of lubrication, rips and tears from ring, ring land, or wrist pin lock failure—will be painfully obvious.

Integral or dry-sleeve bores can be overbored and fitted with correspondingly oversized pistons. When the bore limit is reached, replacement sleeves can be fitted to either type, although the work is considerably easier on an engine that was originally sleeved.

The common practice is to use a boring bar for the initial cuts and finish to size with a hone, preferably an automatic, self-lubricating hone such as the Sunnen CV. The finish will approximate that achieved by the original equipment manufacturer (OEM).

The accuracy of the job can only be as good as the datum—the reference point from which all dimensions, including the bore-to-crankshaft relationship, are taken. Most boring bars index to the deck, on the assumption that the deck is parallel to the main bearing centerline. This is a large assumption. The better tools index to the main bearing saddles. The cylinder with most wear is bored first; how much metal must be removed to clean up this cylinder determines the bore oversize for the engine.

Oversized piston and ring assemblies are normally supplied in increments of 0.010 in. (or 0.25 mm) to the overbore limit. A few manufacturers offer 0.05 in. over pistons for slightly worn cylinders. Occasionally one runs across a 0.015-in. piston and ring set.

The overbore limit varies with sleeve thickness, cylinder spacing (too much overbore compromises the head gasket in the critical "bridge" area between cylinders), and the thickness of the water jacket for engines with integral bores. Jacket thickness depends, in great part, on the quality control exercised by the foundry. Some blocks and whole families of engines have fairly uniform jacket thickness; others are subject to core shift and the unwary machinist can strike water. A final consideration, not of real concern unless the engine is really "hogged out" beyond continence, is obtaining a matching head gasket. The gasket must not be allowed to overhang the bore.

Of course, it is always possible to replace dry liners and to install liners in worn integral bores. The latter operation can be expensive, and most operators would be advised to invest in another block.

Dry sleeves are, by definition, difficult to move; wet sleeves can also stick and some have more propensity for this than others. In short, liner removal and installation tools must be used. Figure 8-35 illustrates a typical combination tool. This or a similar tool must be used for extraction, but a press and a stepped pilot—one diameter matching

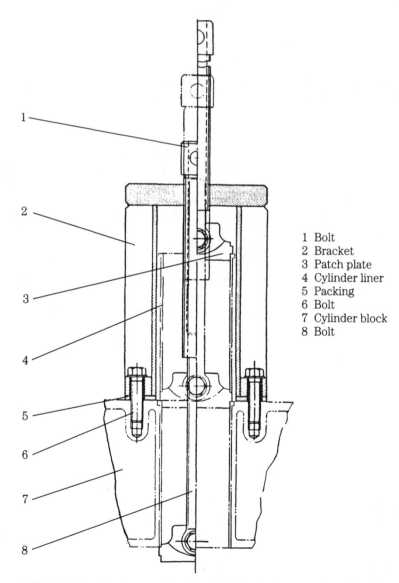

1 Bolt
2 Bracket
3 Patch plate
4 Cylinder liner
5 Packing
6 Bolt
7 Cylinder block
8 Bolt

8-35 Cylinder liner removal and installation tool. Marine Engine Div., Chrysler Corp.

liner inner diameter (ID), the other, liner outer diameter (OD)—is the better choice for installation.

Liner bores and counterbores (the ledge upon which the liner seats) require careful measurement and inspection. Out-of-spec counterbores can sometimes be machined and restored to height with shims, Cumming's fashion.

Wet-liner seals must be lubricated just prior to installation to control swelling. Seal contact areas on the water-jacket ID should be cleaned and inspected under a light. Some machinists oil dry sleeves; others argue against the practice.

Some sleeves, wet or dry, tend to crack at the counterbore area after a few hours of operation. The most common causes have to do with chamfers, either the chamfer on the counterbore or the chamfer on the liner installation pilot.

Detroit Diesel reboring

These two-cycle engines require some special instruction. Many Series 71 engines are cast in aluminum. Early production inserts were a slip fit in the counter-bore; current standards call for the liner to be pressed in. In any event, the block should be heated to between 160° and 180°F in a water tank. Immerse the block for at least 20 minutes.

Counterbore misalignment can affect any engine, although mechanics generally believe that the aluminum block is particularly susceptible to it. You will be able to detect misalignment by the presence of bright areas on the outside circumference of the old liner. The marks will be in pairs—one on the upper half of the liner and the other diagonally across from it, on the lower half. The counterbore should be miked to check for taper and out-of-roundness.

Small imperfections—but not misalignment between the upper and lower deck—can be cleaned up with a hone. Otherwise, the counterbore will have to be machined. Torque the main-bearing caps. Oversize liners are available from the OEM and from outside suppliers such as Sealed Power Corporation.

Liners on early engines projected 0.002–0.006 in. above the block (Fig. 8-36). These low-block engines used a conventional head gasket and shims under the liner

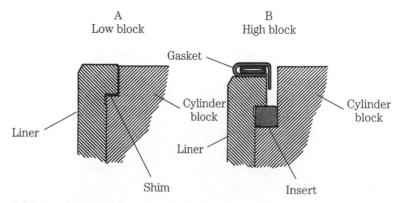

8-36 Detroit Diesel liner arrangements. Sealed Power Corp.

to obtain flange projection. Late-model high-block engines use an insert below the liner. Narrower-than-stock inserts are available to compensate for metal removed from the fire deck, and in various oversized diameters to accommodate larger liners. A 0.002-in. shim is also available for installation under the insert.

Note: Inserts can become damaged in service and can contribute to upper liner breakage.

Series-53 engines employ a wet liner. The upper portion of the liner is surrounded by coolant and sealed with red silicone seals in grooves on the block. Early engines had seals at the top and above the ports. Late-model engines dispense with the lower seal. A second groove is machined at the top of the cylinder to be used in the event of damage to the original.

The seals must be lubricated to allow the liners to pass over them. Do not presoak the seals, because silicone expands when saturated with most lubricants. The swelling tendency is pronounced if petroleum products are used. Lubricate just prior to assembly with silicone spray, animal fat, green soap, or hydraulic brake fluid. Carefully lower the liners into the counterbores, without twisting the seals or displacing them from their grooves.

The eccentricity (out-of-roundness and taper) must be measured before final assembly. On the 110 series you are allowed 0.0015 in. eccentricity. The 53 and 71 engines will tolerate 0.002 in. Eccentricity can often be corrected by removing the liners and rotating them 90° in the counterbores. Do not move the inserts in this operation.

Honing

Honing is used to bring rebored cylinders to size and to remove small imperfections and glaze in used cylinders. Glaze is the hard surface layer of compacted iron crystals formed by the rubbing action of the rings. Most engine manufacturers recommend that the glaze be broken to add in ring seating and to remove the ridge that forms at the upper limit of ring travel. The Perfect Circle people suggest that honing can be skipped if the cylinder is in good shape.

The pattern should be diamond-shaped, as shown in Fig. 8-37, with a 22-32 intersection degree at the horizontal centerline. The cut should be uniform in both directions, without torn or folded metal, leaving a surface free of burnish and imbedded stone particles. These requirements are relatively easy to meet if you have access to an automatic honing machine. However, satisfactory work can be done with a fixed-adjustment hone turned by a drill press or portable drill motor.

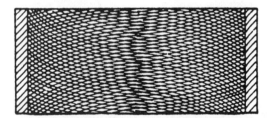

8-37 Preferred crosshatch pattern.

The hone must be parallel to the bore axis. Liners can he held in scrap cylinder blocks or in wood jigs. The spindle speed must be kept low—a requirement that makes it impossible to use a ¼ in. utility drill motor. Suggested speeds are shown below.

Bore diameter (inches)	Spindle (rpm)	Speed strokes (per minute)
2	380	140
3	260	83
4	190	70
5	155	56

Move the hone up and down the bore in smooth oscillations. Do not let the tool pause at the end of the stroke, but reverse it rapidly. Excessive pressure will load the stone with fragments, dulling it and scratching the bore. Flood the stones with an approved lubricant (such as mineral oil), which meets specification 45 SUV at 100°F.

Stone choice is in part determined by the ring material. Most engines respond best to 220–280 grit silicon carbide with code J or K hardness.

Cleaning the bore is a chore that is seldom done correctly. Never use a solvent on a honed bore. The solvent will float the silicon carbide particles into the iron, where they will remain. Instead, use hot water and detergent. Scrub the bore until the suds remain white. Then rinse and wipe dry with paper towels. The bore can be considered "sanitary" when there is no discoloration of the towel. Oil immediately.

Piston rings

Piston rings are primarily seals to prevent compression, combustion, and exhaust gases from entering the crankcase. The principle employed is a kind of mechanical jujitsu—pressure above the ring is conducted behind it to spread the ring open against the cylinder wall. The greater the pressure above the ring, the more tightly the ring wedges against the wall (Fig. 8-38).

The rings also lubricate the cylinder walls. The oil control ring distributes a film of oil over the walls, providing piston and ring lubrication. One or more scraper

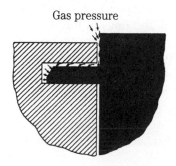

Gas pressure

8-38 A compression ring is a dynamic seal, expanding under the effects of gas pressure.

rings control the thickness of the film, reducing chamber deposits and oil consumption. In addition to sealing and lubrication, the ring belt is the main heat path from the piston to the relatively cool cylinder.

Rings are almost always cast iron, although steel rings are used in some extreme-pressure situations. Cast iron is one of the very few metals that tolerate rubbing contact with the same material. Until a few years ago rings were finished as cast.

Today almost all compression rings are flashed with a light (0.004 in.) coating of chrome. Besides being extremely hard and thus giving good wear resistance, chrome develops a pattern of microscopic cracks in service. These cracks, typically accounting for 2% of the ring's surface, serve as oil reservoirs and help to prevent scuffing. A newer development is to *fill,* or *channel,* the upper compression ring with molybdenum. The outer diameter of the ring is grooved and the moly sprayed on with a hot-plasma or other bonding process. Besides having a very low coefficient of friction and a very high melting temperature, moly gives a piston ring surface that is 15–30% void. It retains more oil than chrome-faced rings and is, at least in theory, more resistant to scuffing.

Rings traditionally have been divided into three types, according to function. Counting from the top of the piston, the first and second rings are *compression* rings, whose task is to control blowby. The middle ring is the *scraper,* which keeps excess oil from the combustion space. The last ring is the *oil* ring, which is serrated to deliver oil to the bore.

This rather neat classification has become increasingly ambiguous with the development of multipurpose ring profiles and the consequent reduction in the number of rings fitted to a piston. Five- and six-ring pistons have given way to three- and four-ring pistons on many of the smaller engines. The function of the middle rings is split between gas sealing and oil control. The lower rings, while primarily operating as cylinder oilers, have some gas-seating responsibilities. Design has become quite subtle, and it is difficult for the uninitiated to distinguish between *compression* and *scraper* rings.

The drawing in Fig. 8-39 illustrates the ring profiles used on the current series of GM Bedford engines. Note the differences in profile among the three. These profiles are typical, but by no means, universal. The Sealed Power Corporation offers several hundred in stock and will produce others on special order.

What this means to the mechanic is that he or she must be very careful when installing rings. Most have a definite *up* and *down,* which might or might not be indicated on the ring. Usually the top side is stamped with some special letter code. Great care must be exercised not to install the rings in the wrong sequence. New rings are packaged in individual containers or in groups that are clearly marked 1 (for first compression), 2, and so forth. Reusable rings should be taken off the piston and placed on a board in the assembly sequence.

Ring wear

The first sign of ring wear is excessive oil consumption, signaled by blue smoke. But before you blame the rings, you should check the bearing clearances at the main

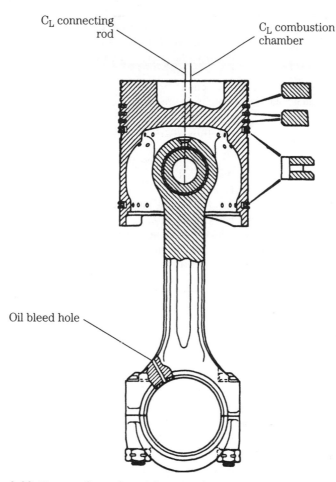

C_L connecting rod

C_L combustion chamber

Oil bleed hole

8-39 Ring configuration and nomenclature. GM Bedford Diesel

and crankpin journals. Bearings worn to twice normal clearance will throw off five times the normal quantity of oil on the cylinder walls. You can make a direct evaluation of oil spill by pressurizing the lubrication system. If appreciable amounts of oil are getting by the rings, the carbon pattern on the piston will be chipped and washed at the edges of the crown.

Check the rings for sticking in their grooves (this can be done on two-cycles from the air box), breaks, and scuffing. The latter is by far the most common malady, and results from tiny fusion welds between the ring material and cylinder walls. Basically it can be traced to lack of lubrication, but the exact cause might require the deductive talents of a Sherlock Holmes. Engineers at Sealed Power suggest these possibilities

Symptom	Possible cause
Overheating	Clogged, restricted, sealed cooling system
	Defective thermostat or shutters
	Loss of coolant
	Detonation
Lubrication failure	Worn main bearings
	Oil pump failure
	Engine lugging under load
	Extensive idle
	Fuel wash on upper-cylinder bores
	Water in oil
	Low oil Level
	Failure pressurize oil system after rebuild
Wrong cylinder finish	Low crosshatch finish
	Failure to hone after reboring
Insufficient clearance	Inadequate bearing clearance at either end of the rod
	Improper ring size
	Cylinder sleeve distortion

Usually inadequate bearing clearance, complicated by a poor fit in the block counterbore, results in overheating. The fundamental cause is often poor torque procedures, or improperly installed sealing rings on wet-sleeved engines. A rolled or twisted sealing ring can distort the sleeve.

Ring breakage is due to abnormal loading or localized stresses. It can be traced to

- Ring sticking—this overstresses the free end of the ring.
- Detonation—this is traceable to the overly liberal use of starting fluid, to dribbling injectors and to out-of-time delivery.
- Overstressing the ring on installation. Usually the ring breaks directly across from the gap.
- Excessively worn grooves, which allow the ring to flex and flutter.
- Ring hitting the ridge at the top of the bore. The mechanic is at fault because this ridge should have been removed.

The last point—involving blame—can be sticky in a shop situation. Mechanics make mistakes the same as everyone else, and the number of mistakes is in part, a function of the complexity of the repair. Few people can overhaul a machine as complicated as a multicylinder engine without making some small error. Assessing blame, if only to correct the situation, is sometimes complicated by having the mechanic who built the engine tear it down. But careful examination of the parts usually points to the fault. For example, rings that have been fitted upside down show reversed wear patterns. Rings that have broken in service are worn on either side of the break, from contact with the cylinder walls. The fracture will be dulled. Ring that have been broken during removal show sharp crystalline breaks without local wear spots.

Pistons

Large diesel engines often employ cast-iron or steel pistons; smaller, high-speed engines generally use aluminum castings. That such pistons survive combustion temperatures that can reach 4500°F and cylinder pressures that, highly supercharged engines, can exceed 2000 psi, is a triumph of engineering over materials. Aluminum has a melting point of 1220°F and rapidly loses strength as this temperature is approached.

Construction

The heavy construction of these pistons—diesel pistons typically weigh half as much again as equivalent SI engine pistons-provides mechanical strength and the heat conductivity necessary to keep the piston crown at about 500°F. The standard practice is to direct a stream of oil to the underside of the crown, usually by means of spray jets. Some designers go a step further, and insulate the crown, which is relatively easy to cool, from the ring belt and skirt. Turn ahead to Fig. 8-48 for an illustration of a heat dam. The cavity above the piston pin slows heat transfer by reducing piston wall thickness. Another approach is to lengthen the thermal path by grooving the area above the ring belt (Fig. 8-40). Yet another approach is to apply a thermal coating to the upper side of the piston crown, thus confining heat to the combustion chamber (Fig. 8-41).

The skirts to aluminum pistons run hotter than cast iron and have a coefficient of expansion that is about twice that of iron. Consequently, light-metal pistons are assembled with fairly generous bore clearances to compensate for thermal expansion, and might be noisy upon starting. Semi-exotic alloys, such as Lo-Ex, or cast-in steel struts, help control expansion and knocking. It is interesting that one of the first experimenters with aluminum pistons, Harvey Marmon, found it necessary to sheathe the piston skirt in an iron "sock."

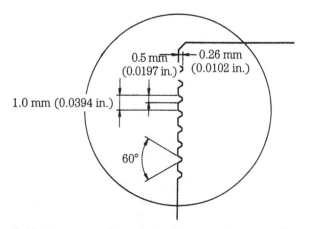

8-40 Fire groves form thermal barriers between the ring band and combustion chamber. Yanmar Diesel Engine Co. Ltd.

Alumite treatment

Front

8-41 Ford 2.3L pistons are insulated with a plasma coating and relieved at areas adjacent to the pins. Note the letter "F" embossed on the relief, which should face the front of the engine when the piston is installed.

Most alloy pistons are cam-ground; when cold the skirts are ovoid, with the long dimension across the thrust faces. As the piston heats and expands, it becomes circular, filling the bore. Other ways of coping with thermal expansion are progressively to reduce piston diameter above the pin and to relieve, or cut back, the skirts in the pin area (also shown in Fig. 8-41). The heavy struts that support the pin bosses transfer heat from the underside of the crown to the skirts.

Critical wearing points include the thrust faces and the sides of the ring grooves. With the exception of certain two-cycle applications, rings are designed to rotate in their grooves and, according to one researcher, reach speeds of about 100 rpm. Rotation is the primary defense against varnish buildup and consequent ring sticking. But it also wears "steps" into the grooves. Piston thrust faces are also subject to rapid wear.

Some manufacturers run the compression ring against a steel insert, cast integrally with the piston. Another approach is to substitute a long-wearing eutectic alloy for the ASE 334 or 335 usually specified. Eutectic alloys consist of clusters of hard silicon crystals distributed throughout an aluminum matrix. As the aluminum wears, silicon— one of the hardest materials known—emerges as the bearing surface. The same mechanism rapidly dulls cutting tools, which is why the cost of eutectic pistons approaches the cost of forgings.

Forged pistons are an aftermarket item, used as a last resort in highly super-charged engines when castings have failed. Forging eliminates voids in the metal and compacts the grain structure at the crown, pin bosses, and ring lands (Fig. 8-42). These pistons have superior hot strength characteristics but require generous running clearances. An engine with forged pistons will be heard from during cold starts.

8-42 Forged pistons gain strength from the uniform grain structure. Sealed Power Corp.

A two-piece piston consists of a piston dome, or ring carrier, element and a skirt element (Fig. 8-43). These parts pivot on the piston pin. Although other manufacturers use this form of the piston, the GM version first appeared on Electromotive railroad engines and, in 1971, replaced conventional trunktype pistons on turbocharged DDA Series 71 engines. It eventually found its way into several other DDA engines, including the Series 60.

Detroit Diesel describes this design as a "crosshead" piston. As the term is usually applied, it refers to a kind of articulated piston used on very large engines. A pivoted extension bar separates the upper piston element and the skirt, which rides against the engine frame. The crosshead isolates the cylinder bores from crankshaft-induced side forces. The DDA design does not relieve the bore of side forces, but it increases the bearing area of the pin and centralizes the load more directly on the conn rod for a reduction in bending forces. The connecting rod, illustrated in the following section, bolts to the underside of the pin, making the upper half of the pin available to support piston-dome thrust.

Failure modes

Piston failure is usually quite obvious. Wear should not be a serious consideration in low-hour engines, because the skirt areas are subject to relatively small forces and have the benefit of surplus lubrication. Excessive wear can be traced to dirty or improperly blended lube oil or inadequate air filtration. Poor cylinder finishing might also contribute to it. Piston collapse or shrinkage is usually due to overheating. If the problem shows itself in one or two cylinders, expect water-jacket stoppage or loose liners.

Combustion roughness or *detonation* damage begins by eroding the crown, usually near the edge. The erosion spreads and grows deeper until the piston "holes." Typically the piston will look as if it were struck by a high-velocity projectile. *Scuffing* and *scoring* (a scuff is a light score) might be confined to the thrust side of the piston. If this is the case, look for the following:

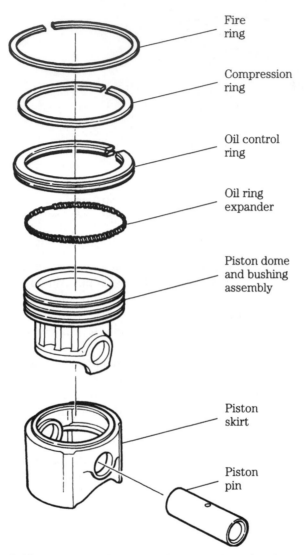

Fire
ring

Compression
ring

Oil control
ring

Oil ring
expander

Piston dome
and bushing
assembly

Piston
skirt

Piston
pin

8-43 Detroit Diesel articulated piston reduces bending
forces on the piston pin.

- Oil pump problems such as clogged screen, excessive internal clearances.
- Insufficient rod bearing clearances, which reduce throw-off, robbing the
  cylinders of oil.
- Lugging.

The probable causes of damage to both sides of the skirt include the ones just
mentioned, plus the following:
- Low or dirty oil.
- Detonation.
- Overheating caused by cooling system failure.

- Coolant leakage into the cylinder.
- Inadequate piston clearance.

Scuffs or scores fanning out 45° on either side of the pinhole mean one of the following conditions:

- Pin fit problems such as too tight in the small end of the rod or in the piston bosses.
- Pinhole damage (see below for installation procedures).

Ring land breakage can be caused by the following:

- Excessive use of starting fluid.
- Detonation.
- Improper ring installation during overhaul.
- Excessive side clearance between the ring and groove.
- Water in cylinder.

Free-floating pins sometimes float right past their lock rings and contact the cylinder walls. Several causes (listed below) have been isolated.

- Improper installation: Some mechanics force the lock rings beyond the elastic limit of the material. In a number of cases, it is possible to install lock rings by finger pressure alone.
- Improper piston alignment: This might be caused by a bent rod or inaccuracies at the crankshaft journal. Throws that are tapered or out of parallel with the main journals will give the piston a rocking motion that can dislodge the lock ring. Pounding becomes more serious if the small-end bushing is tight.
- Excessive crankshaft end play: Fore-and-aft play is transmitted to the lock rings and can pound the grooves open. Again, a too-tight fit at the connecting rod's small end will hasten piston failure.

Servicing

Used pistons can give reliable service in an otherwise rebuilt engine, but only after the most exhaustive scrutiny. Scrape and wire-brush carbon accumulations from the crown, but do not brush the piston flanks. Carbon above the compression ring and on the underside of the crown should be removed chemically. A notch, letter, arrow, or other symbol identifies the forward edge of the piston. These marks are usually stamped on the crown, on the pin boss relief (illustrated in Fig. 8-41), or hidden under the skirt. Make note of the relationship between the leading edge of the piston and the numbered side of the connecting rod.

Lay out the rings in sequence, topsides up, on the bench. Spend some time "reading" the rings—the history of the upper engine is written on them, just as the crank bearings testify to events below. The orientation code—the word *Top* (T) or *Ober.* (O)—will be found on the upper sides of the compression and scraper rings, adjacent to the ring ends.

You might wonder why attention is given to these codes for piston assemblies that, at this stage, are of unknown quality, and for rings that will, in any event, be discarded. The purpose is to become familiar with the concept of orientation as it applies to the particular engine being serviced. New parts might not carry identical codes, but the relationship between coded parts will not change.

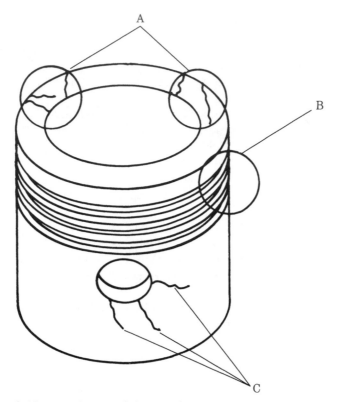

8-44 Typical piston failure modes. ©1990 Deers & Co.

While still attached to the rod and before investing any more time in it, inspect the piston for obvious defects. Reject if the piston is fractured, deeply pitted, scored, or if it exhibits ring land damage (Fig. 8-44). Cracks tend to develop where abrupt changes in cross section act as stress risers. Deep pits usually develop at the edges of the piston; scoring is most likely to develop on the thrust faces. Contact with the cylinder bore occurs at two areas, 90° from the piston-pin centerline. The major thrust face lies in the direction of the crankshaft rotation and is normally the first to score. (Viewed from the front of a clockwise-rotation engine, the major thrust face is on the right.) Machine marks, the light cross-hatching left by the cutting tool, should remain visible over most of the contact area. The next section describes the relationship of wear patterns to crankshaft and rod alignment.

If the piston passes this initial examination, make a micrometric measurement of skirt diameter across the thrust faces. Depending on the supplier, the measurement is made at the lower edge of the skirt, at the pin centerline, or at an arbitrary distance between these points (Fig. 8-45). Piston OD subtracted from cylinder bore ID equals running clearance.

All traces of carbon in the ring grooves and oil spill ports must be removed. Whenever possible, this tedious task should be consigned to the machinist and thought about no more. Figure 8-46 illustrates a factory-supplied plug gauge used to

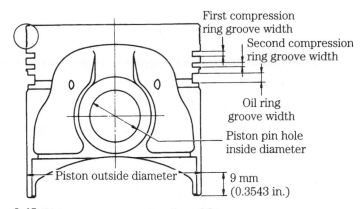

8-45 Piston measurement points. OD measurements may vary with manufacturers. Yanmar Diesel Engine Co. Ltd.

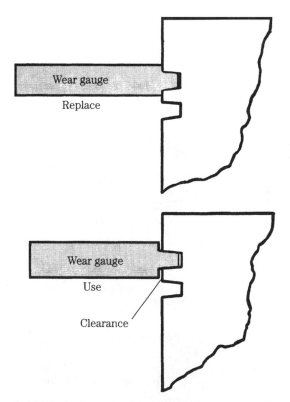

8-46 The best way to determine ring groove wear is with a factory-supplied go-nogo gauge. International Harvester.

measure groove width; the next drawing, Fig. 8-47, shows an alternate method using a *new* ring and a feeler gauge. The latter method is accurate, so long as the gauge bottoms against the back of the groove.

Most engines of this class employ "full-floating" piston pins that, at running temperatures, are free to oscillate on both the rod and piston (Fig. 8-48). Pins secure laterally with one or another variety of snap rings, some of which are flattened on the inboard (or wearing) sides.

Protecting your eyes with safety glasses, disengage and withdraw the snap rings. Although mechanics generally press out (and sometimes hammer out) piston pins, these practices should be discouraged. Instead, take the time to heat the pistons, either with a heat gun or by immersion in warm (160°F) oil. Pins will almost fall out.

While the piston is still warm, check for bore integrity. Insert the pin from each side. If the pin binds at the center, the bore might be tapered; if the bore is misaligned, the pin will click or bind as it enters the far boss (Fig. 8-49).

Other critical areas are illustrated in Fig. 8-45. Measurements are to be made at room temperature. Damage to retainer rings or ring grooves suggests excessive crankshaft end play, connecting-rod misalignment, or crankpin taper. Slide forces generated by a badly worn crankpin or severely bent rod react against the snap rings. Rod or crankpin misalignment might also appear as localized pin and pin-bearing wear, a subject discussed in the following section.

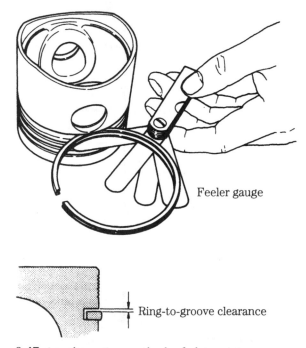

Feeler gauge

Ring-to-groove clearance

8-47 An alternative method of determining groove wear is to measure the width relative to a new ring.
Yanmar Diesel Engine Co. Ltd.

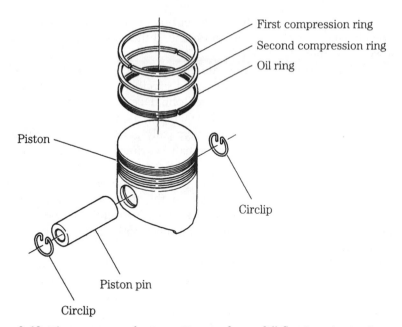

First compression ring

Second compression ring

Oil ring

Piston

Circlip

Circlip

Piston pin

8-48 Like most manufacturers, Yanmar favors full-floating piston pins, pivoting on both the rod and piston, and secured laterally by snap rings ("circlips" in the drawing).

To install, warm the piston, oil the pin and pin bores, and, with the rod in its original orientation, slip the pin home with light thumb or palm pressure. Use *new* snap rings, compressing them no more than necessary. Verify that snap rings seat around their full diameters in the grooves.

A few engines use pressed-in pins, which make a lock on the piston with an interference fit. Support the piston on a padded V-block and press the pin in two

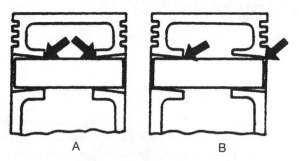

A   B

8-49 This drawing, intended by John Deere to show how piston pin bearing checks (free-floating pins should insert with light thumb pressure; if the pin goes in easily from both sides but binds in the center, the problem might be tapered bores (A); if the pin clicks or binds when contacting the far boss, the piston is warped (B), also illustrated the cavity on the underside of the head that serves as a heat dam.

stages: stop at the point of entry to the lower boss, relieve press force to allow the piston to regain shape, and press the pin home. If the pin is installed in one pressing, the lower boss might be shaved.

What remains is to establish the running clearance. Piston-to-bore clearance is fundamentally a matter of specification, but specifications are never so rigid that they cannot be bent. For example, one John Deere piston/liner combination requires a running clearance of 0.0034–0.0053 in., as measured at the bottom of the piston skirt. A high-volume rebuilder typically goes toward the outside limit, building clearance here and in the crankshaft bearings. A "loose" engine will be more likely to tolerate severe loads as delivered and without the benefit of a break-in period. A custom machinist, who works on one or two engines at a time, often goes in the other direction, aiming at the tightest clearance the factory allows. Besides being aesthetically more satisfying, "tight" engines tend to live long, quiet lives. Of course, such an engine must be carefully run in during the first hours of operation.

In an attempt to stabilize the maintenance process, factory manuals include, wear limits for critical components. The concept of permissible wear is a value judgment, made in an engineering office remote from the world of mechanics, back-ordered parts, and budgetary restraints. Permissible piston-to-liner clearance for the Deere engine is 0.0060 in. Suppose that the running clearance is found to be 0.0055 in.—only 0.0002 in. over the allowable 0.0053 with 0.0030 in. to go before the wear limit is reached. Assuming that the wear is equally distributed over both parts, should the piston or liner or both be replaced? This is only one cylinder of six, none of them worn by identical amounts. Questions like this go beyond mechanics and depend for their answers on the politics of the situation, interpreted in light of the philosophy of maintenance—formal or informal—that characterizes the operation.

Connecting rods

No part is more critical than the connecting rod and none more conservatively designed (Fig. 8-50).

Construction

All diesel engines employ two-piece, H-section rods, heavily faired at the transitions between bearing sections, and forged from medium-carbon steel or from that ubiquitous alloy, SAE 4140. The small end is generally closed and fitted with a replaceable bushing (some automotive OEMs do not catalog replacement bushings, but a component machinist can work around that). The big end carries a two-piece precision insert bearing. A few rods terminate in a slipper, which bolts to the piston pin, as shown in Fig. 8-51. Open construction makes the whole length of the pin available as a bearing surface.

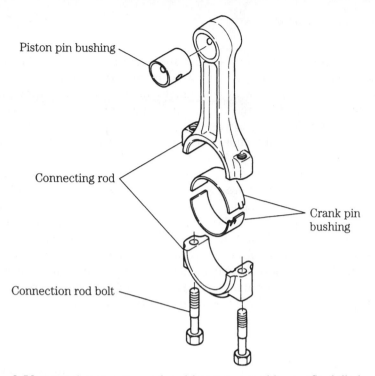

Piston pin bushing

Connecting rod

Crank pin bushing

Connection rod bolt

8-50 Typical connecting rod and bearing assembly. A rifle-drilled oil passage indexes with ports in both the small-end bushing and upper big-end insert. Yanmar Diesel Engine Co. Ltd.

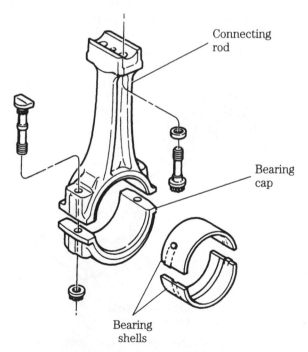

Connecting rod

Bearing cap

Bearing shells

8-51 Detroit Diesel slipper-type connecting rod, used with the articulated piston shown in Fig. 8-44.

Whenever possible, interface between the shank and cap should be perpendicular to the crankshaft centerline. This configuration reduces side forces on the cap. However, the rod must be split at an angle to allow disassembly through the bore when crankpin diameter is large (as in Onan engines) or to prevent contact with other parts (Deere 6076).

Connecting rods have a definite orientation relative to the piston and to the cap. The former is a function of transverse oil ports, drilled in the rod shank; the latter reflects the way connecting rods are manufactured. The relatively small connecting rods that we are dealing with are forged in one piece. Then the cap is separated (sometimes merely by snapping it off), assembled, and honed to size. Reversing the cap or installing a cap from another rod destroys bearing circularity. Typically, a mismatched assembly overheats and seizes within a few crankshaft revolutions. Rod shanks and caps carry match marks, and both parts are identified by cylinder number (Figs. 8-52 and 8-53). Errors can be eliminated by double- and triple-checking match mark alignment and cylinder numbers. Mechanics do well to observe the rule that no more than one rod cap should be disassembled at a time.

Service

The number stamped on the rod shank and cap should correspond to the cylinder number. Sometimes these numbers are scrambled or missing, and the mechanic must supply them. Stamp the correct numbers on the pads provided and, to prevent confusion, deface the originals.

Mike rod journals at several places across the diameter, repeating the measurements on both sides to detect taper (Fig. 8-54). Inertial forces tend to stretch the rod cap and pinch the ends together. When this condition is present, the machinist should mill a few thousandths off the cap interface and, with the cap assembled and

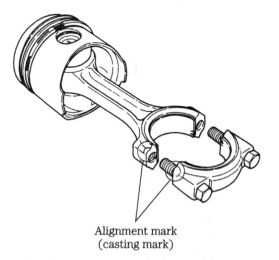

Alignment mark
(casting mark)

8-52 Most rods and caps carry cast match marks as shown on this Yanmar assembly. Cylinder numbers will be stamped on the other side of the rod.

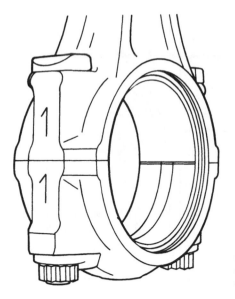

8-53 Rod assembly for No. 1 cylinder on a Detroit Diesel engine.

torqued, machine the journal to size. Typically, this is done with a reamer, although automatic honing machines, such as the Sunnen or the Danish-made AMC, produce a finer, more consistent surface.

The machinist should check the alignment of each connecting rod with a fixture similar to the one shown in Fig. 8-55. Even so, these matters should not be entirely left to the discretion of the machinist. It is always prudent to check the work against the testimony of the engine. Rod and crankpin misalignment produces telltale wear patterns on piston thrust faces, on pin bores, and on the retainers and retainer grooves.

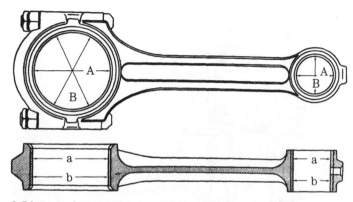

8-54 Journal measurement points. Marine Engine Div., Chrysler Corp.

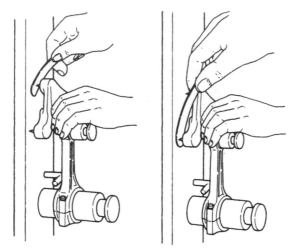

8-55 Rod alignment fixture used to detect bending and torsional misalignment. Yanmar Diesel Engine Co. Ltd.

A bent rod tilts the piston, localizing bearing wear at the points shown in Fig. 8-56, This condition might also be reflected by an hourglass-shaped wear pattern on the thrust faces, as illustrated in Fig. 8-57. A twisted rod imparts a rocking motion to the piston, concentrating wear on the ring band and skirt edges (Fig. 8-58). Crankshaft taper generates thrust, which might scuff the skirt and could drive the pin through one or the other retainer.

Bent or twisted rods generally can be straightened, although some manufacturers warn against the practice.

Most rod bolts can be reused, if visual and magnetic-particle testing fails to reveal any flaws. Some recently developed torque-to-yield bolts can also be reused;

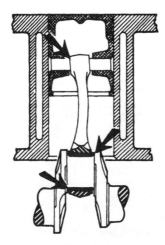

8-56 A bent rod concentrates bearing wear at the points shown. Ford Motor Co.

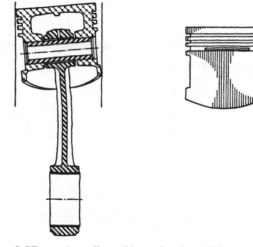

8-57 Another effect of loss of rod parallelism is to tilt the piston and, in extreme cases, produce the wear pattern shown on the left. Sealed Power Corp.

earlier types were sacrificial. Rod-bolt nuts should be replaced, regardless of the fastener type.

Used rods should be Magnafluxed. Interpretation of the crack structure thus revealed requires some judgment. Any rod that has seen service will develop cracks. One must distinguish between inconsequential surface flaws and cracks that can lead to structural failure. The drawing in Fig. 8-59 can serve as a guide.

8-58 A twisted connecting rod causes the piston to rock, a condition that can be signaled by an elliptical wear pattern. Sealed Power Corp.
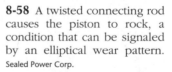

Do not use or attempt to salvage rods with indications over 0.005 in. deep extending over edges of H-section on both sides of flange (shaded areas are most highly stressed)

Start of fatigue crack resulting from overloading (due to hydrostatic lock). Do not attempt to salvage (this type of indication is not visible with bushings in place)

Section A-A

Section B-B

A

A

Example of an indication following longitudinal forging flow lines

Noncritical area

Critical area acceptable limits

B

B

Noncritical area

Indications in noncritical areas are acceptable unless they can be observed as obvious cracks without magnetic inspection

Example of a transverse indication that does not follow longitudinal forging flow lines can be either a forging lap, heat treat crack, or start of a fatigue crack

Longitudinal indications
Following forged flow lines are usually seams and are not considered harmful if less than 1/32 in. deep

Poor practice

Good practice

Grinding notes
Care should be taken in grinding out indications to assure proper blending of ground area into unground surface so as to form a smooth contour

Transverse indications (across flow lines)
Having a maximum length of 1/2 in. which can be removed by grinding no deeper than 1/64 in. are acceptable after their **complete removal.** An exception to this is a rod having an indication which extends over the edge of H-section and is present on both sides of the flange in this case. Maximum allowable depth is 0.005 in. (see section A-A).

8-59 Interpreting Magnaflux indications. Detroit Diesel

In general, longitudinal cracks are not serious enough unless they are 1/32 in. deep, which can be determined by grinding at the center of the crack. Transverse cracks are causes of concern because they can be the first indication of fatigue. If the cracks do not extend over the edges of the H-section, are no more than 1/2 in. long, and less than 1/64 in. deep, they can be ground and feathered. Cracks over the H-section can be removed if 0.005 in. deep or less. Cracks in the small end are cause for rejection.

Small-end bushings are pressed into place with reference to the oil port and are finish-reamed. Loose bushings are a sign that the rod has overheated, and they can cause a major failure by turning and blocking the oil port.

What has been described are standard, industry-wide practices. One can go much further in the quest for a more perfect, less problematic engine. For example, it is good practice to match the weight of piston and rod assemblies to within 10 g or so—even when parts are not to be sent out for balancing. Surplus weight can be ground from the inner edges of the piston skirts. Some mechanics assemble one piston and rod to several crankpins. Differences in deck height between No. 1 and the last cylinder of the bank indicate an alignment problem, either between crankshaft throws or between the crankshaft centerline and deck.

Crankshafts

At this point we have progressed to the heart of the engine, the place where its durability will be finally established.

Construction

Crankshafts for the class of engine under discussion are, for the most part, steel forgings. Materials range from ordinary carbon steels to expensive alloys such as chrome-moly SAE 4140. A number of automotive engines and lightly stressed stationary power plants get by with cast-iron shafts, usually recognized by sharp, well-defined parting lines on the webs and by cored crankpins. Forging blurs the parting lines and mandates solid crankpins, which, however, are drilled for lubrication.

Most crank journals are induction-hardened, a process that leaves a soft, fatigue-resistant core under a hardened "skin." Hardness averages about 55 on the Rockwell scale and extends to a depth of between 0.020–0.060 in. in order to accommodate regrinding. But a cautious machinist will test crankshaft hardness after removing any amount of metal.

A few extreme-duty crankshafts are hardened by a proprietary process known as *Tufftriding*. Wearing quality is comparable to that provided by chrome plating, but unlike chromium, the process does not adversely affect the fatigue life of the shaft.

Service

Remove the crankshaft from the block—a hoist will be required for the heavier shafts—and make this series of preliminary inspections:

- If a main bearing cap or rod has turned blue, discard the shaft, together with the associated cap or rod. The metallurgical changes that have occurred are irreversible. By the same token, be very leery of a crankshaft that is known to have suffered a harmonic balancer failure.
- Check the fit of a new key in the accessory-drive keyway. There should be no perceptible wobble. A competent machinist can rework worn keyways and, if necessary, save the shaft by fabricating an oversized key.

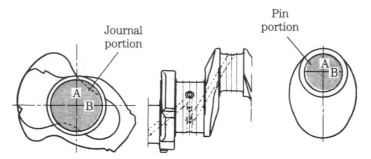

8-60 Crankshaft measurement points. Marine Engine Div., Chrysler Corp.

- Check the timing-gear teeth for wear and chipping. Magnetic-particle testing can be of some value when applied to the hub area, but cannot detect the subsurface cracks that signal incipient gear-tooth failure. Timing gear sprockets and related hardware should be replaced as a routine precaution during major repairs.
- Mike the journals and pins as shown in Fig. 8-60. Compare taper, out-of-roundness, and diameter against factory wear limits.
- If corrective machining does not appear necessary, remove with crocus cloth all light scratches and the superficial ridging left by bearing oil grooves. Tear off a strip of crocus cloth long enough to encircle the journal. Wrap the cloth with a leather thong, crossing the ends, to apply force evenly over the whole diameter of the journal. Work the crocus cloth vigorously, stopping at intervals to check progress. An Armstrong grinder works surprisingly fast.
- Using an EZ-Out or hex wrench, remove the plugs capping the oil passages. Back-drilled sections of these passages serve as chip catchers, and must be thoroughly cleaned. Compressed air, shown in Fig. 8-61, helps but is no substitute for rifle-bore brushes, solvent, and elbow grease. Clean the plugs, seal with Loctite, and assemble.
- Check trueness with one or, preferably, tandem dial indicators while rotating the crank in precision V-blocks (Fig. 8-62).

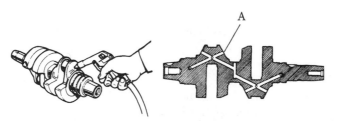

8-61 Blind runs in the crankshaft oiling circuit must be opened, mechanically cleaned, and resealed. Lombardini

Dial gauge

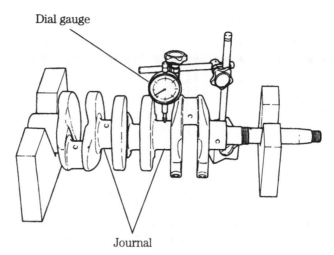

Journal

8-62 Checking crankshaft straightness using the center bearing as the referent. Yanmar Diesel Engine Co. Ltd.

Flaw testing

Flaw testing is generally done with the Magnaflux process, although some shops prefer to use the fluorescent-particle method. Both function on the principle that cracks in the surface of the crankshaft take on magnetic polarity when the crank is put in a magnetic field. Iron particles adhere to the edges of these cracks, making them visible. The fluorescent particle method is particularly sensitive because the metal particles fluoresce and glow under black light.

Most cracks are of little concern because the shaft is loaded only at the points indicated in Fig. 8-63. The strength of the shaft is impaired by crack formations which follow these stresses, as shown in Fig. 8-64. These cracks radiate out at 45° to the crank centerline and will eventually result in a complete break.

Abnormal bending forces are generated by main-bearing bore misalignment, improperly fitted bearings, loose main-bearing caps, unbalanced pulleys, or over-tightened belts. Cracks caused by bending start at the crankpin fillet and progress diagonally across.

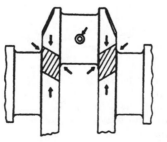

8-63 Most crankshaft loads pass through the webs. Detroit Diesel

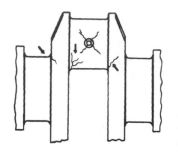

8-64 Typical fatigue crack patterns, all of them Critical. Detroit Diesel.

The distribution of cracks caused by torsional (or twisting) forces is the same as for bending forces. All crankshafts have a natural period of torsional vibration, which is influenced by the length/diameter ratio of the crank, the overlap between crankpins and main journals, and the kind of material used. Engineers are careful to design the crankshaft so that its natural periodicity occurs at a much higher speed than the engine is capable of turning. However, a loose flywheel or vibration damper can cause the crank to wind and unwind like a giant spring. Unusual loads, especially when felt in conjunction with a maladjusted governor, can also cause torsional damage.

Crankshaft grinding

Bearings are available in small oversizes (0.001 and 0.002 in.) to compensate for wear. The first regrind is 0.010 in. Some crankshafts will tolerate as much as 0.040 in., although the heat treatment is endangered at this depth. The crankpin and main-journal fillets deserve special attention. Flat fillets invite trouble, because they act as stress risers. Gently radius the fillets as shown in the left drawing in Fig. 8-65.

All journals and pins should be ground, even if only one has failed. Use plenty of lubricant to reduce the possibility of burning the journal. Radius the oil holes with a stone and check the crankshaft again for flaws with one of the magnetic particle methods.

Surface hardness should be checked after machining and before the crank receives final polishing. Perhaps the quickest and surest way to do this is to use Tarasov etch. Clean the shaft with scouring powder or a good commercial solvent. Wash thoroughly and rinse with alcohol. Apply etching solution No. 1 (a solution of 4 parts nitric acid in 96 parts water). It is important to pour the acid into the water, not vice versa.

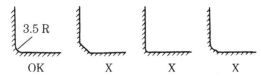

OK   X   X   X

8-65 Fillet profiles. Sharp edges, flat surfaces, and overly wide profiles should be avoided. Rolled, as opposed to ground, fillets extend fatigue life by as much as 60%, but few shops have the necessary equipment. Detroit Diesel

Rinse with clean water and dry. If you use compressed air, see that the system filter traps are clean. Apply solution No. 2, which consists of 2 parts hydrochloric acid in 98 parts acetone. Acetone is highly flammable and has a sharp odor that can produce dizziness or other unpleasant reactions when used in unventilated areas, so allow yourself plenty of breathing room.

The shaft will go through a color change if it has been burned. Areas that have been hardened by excessive heat will appear white; annealed areas turn black or dark gray. Unaffected areas are neutral gray. If any color other than gray is present, the shaft should be scrapped and the machinist should try again, this time with a softer wheel, a slower feed rate, or a higher work spindle speed. Some experimentation might be necessary to find a combination that works.

Cranks that have been Tufftrided must be treated after grinding to restore full hardness. One test is chemical: a 10% solution of copper ammonium chloride and water applied to the crankshaft reacts almost immediately by turning brown if a traditional heat treatment has been applied. There will be no reaction in 10 seconds if the crankshaft is Tufftrided. Another test is mechanical: Tufftriding is applied to the whole crankshaft, not just to the bearing journals. If a file skates ineffectually over the webs without cutting, one can assume that the shaft is Tufftrided.

Any heavy-duty crankshaft, forged or cast, can benefit from Tufftriding, if the appropriate bearing material is specified. This treatment is especially beneficial when the manufacturer has neglected to treat journal fillets. The abrupt change in hardness acts as a stress riser. Treatment should be preceded by heating the crankshaft for several hours at a temperature above 1060°F. It might be necessary to re-straighten the crankshaft.

Camshafts and related parts

Inspect the accessory drive gear train for tooth damage and lash. Crank and camshaft gears are pressed on their shafts and further secured by pins, keys, or bolts.

OHV cam and balance shafts are supported on bushings, which require a special tool, shown in Fig. 8-66, to extract and install. Normally, the machinist services in-block bushings, removing the old bushings and installing replacements that, depending upon engine type, might require finishing. The OEM practice of leaving bushing bores unfinished and reaming installed bushings to size complicates the rebuilding process. It is also possible for bushings to spin in their bosses, a condition that can be corrected by sleeving.

Camshaft lobes are the most heavily loaded parts of the engine, with unit pressures in excess of 50,000 psi at idle. Valve lifters, whether hydraulic or mechanical, are normally offset relatively to the lobes, so that contact occurs over about half of the cam surface. This offset, together with a barely visible convexity ground on the lifter foot tends to turn the lifter as it reciprocates. As the parts wear, the contact area increases and eventually spreads over the whole width of the lobe. Thus, if you find a camshaft with one or more "widetrack" lobes, scrap the cam, together with the complete lifter set. Although the condition is rare in diesel engines, overspeeding and consequent valve float might batter the flat toes of the lobes into points. Figure 8-67,

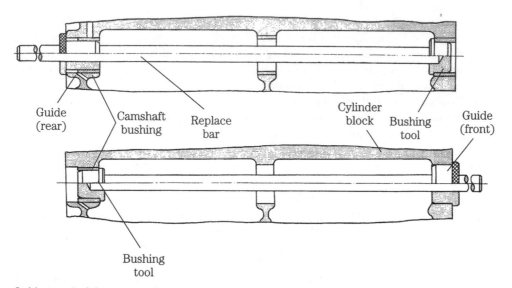

Guide (rear)   Camshaft bushing   Replace bar   Cylinder block   Bushing tool   Guide (front)

Bushing tool

8-66 Camshaft bearing tool. Marine Engine Div., Chrysler Corp.

which illustrates lobe wear measurements, inadvertently illustrates this condition. The same two-point measurements should be made on the journals.

Normally, hydraulic valve lifters are replaced during a rebuild; attempting to clean used lifters is a monumental waste of time. This means that the camshaft must also be replaced, because new lifters will not live with a used camshaft (and vice versa). Assemble with the special cam lubricant provided, smearing the grease over each lobe and lifter foot. Oil can be used on the journals.

Replacement hydraulic lifters normally are charged with oil as received, and ample time must be allowed for bleed when screwing down the rocker arms on OHV engines. Some valves will be open; if the rockers are tightened too quickly, pushrods will bend. As mentioned earlier, the low bleed-down rates of Oldsmobile 350 lifters make rocker assembly an exercise in patience; most other rockers can be assembled in a few minutes of careful wrenching.

Break in the camshaft exactly per maker's instructions. Otherwise, it will score.

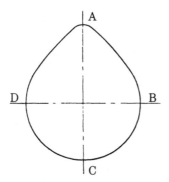

8-67 Cam lobe measurement points. Navista.

Harmonic balancers

The harmonic balancer, or vibration dampener, mounts on the front of the crankshaft where it muffles torsional vibration. Power comes to the crankshaft as a series of impulses that cause the shaft to twist, first in the direction of rotation and then against normal rotation. The shaft winds and unwinds from a node point near the flywheel. This movement can, unless dampened, quickly break the shaft.

Most harmonic balancers consist of an outer, or driven, ring bonded by means of rubber pads to the hub, which keys to the crankshaft. The rubber medium dampens crankshaft accelerations and deceleration, transferring motion to the outer ring at average crankshaft velocity. Some balancers drive through silicon-based fluid that exerts the same braking effect.

It is difficult to test a harmonic balancer in a meaningful way. Rubberized balancers can be stressed in a press and the condition of the rubber observed. Cracks or separation of the bonded joint means that the unit should be replaced. Fluid-filled balancers are checked for external damage and fluid leaks. Some machinists equate noise when the balancer is rolled on edge with failure.

Detroit Diesel's advice is best: replace the balancer whenever the engine is rebuilt.

Crankshaft bearings

All modern engines are fitted with two-piece precision insert bearings at the crankpin, as illustrated in Figs. 8-50 and 8-51. Most employ similar two-piece inserts, or shells, at the main journals, although full-circle bearings are appropriate for barrel-type crankcases. No bearing can be allowed to spin in its carrier.[1] Full-circle bearings pressed into their carriers or locked by pins or cap screws. Two-piece shells secure with a tab and gain additional resistance to spinning from residual tension. Bearing shells are slightly oversized, so that dimension A in Fig. 8-68 is greater than dimension B. The difference is known as the *crush height*. Torquing the cap equalizes the diameters, forcing the inserts hard against their bosses.

Thrust bearings, usually in the form of flanges of main inserts, limit crankshaft fore-and-aft movement (Fig. 8-69). So long as the crankpin is not tapered, connecting-rod thrust loads are insignificant and transfer through rubbing contact between the rod big end and the crankpin flanges.

Insert-type bearings are made up in layer-cake fashion of a steel backing and as many as five tiers of lining material. Although their numbers are dwindling, perhaps half of the high-speed engines sold in this country continue to use copper-lead bearings, the best known of which is the Clevite (now Michigan)-77. The facing surface of this bearing consists of 75% copper, 24% lead, and a 1% tin overplate. Daimler-Benz, Perkins, and several other manufacturers have followed the example of Caterpillar and specify aluminum-based bearings, which tolerate acidic oils better

---

[1]Henry Ford made an exception to the "no-spin" rule; his early V-8s used floating crankpin bearings, babbitted on both sides. This technique cut bearing speed in half, but introduced a complication in the form of big-end connecting rod wear.

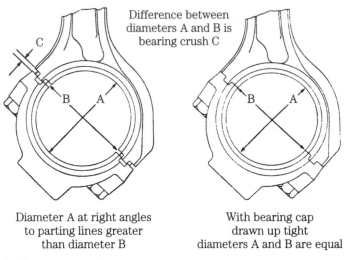

Difference between
diameters A and B is
bearing crush C

Diameter A at right angles
to parting lines greater
than diameter B

With bearing cap
drawn up tight
diameters A and B are equal

8-68 Bearing crush. Navistar

than copper-lead types. One material commonly used is Alcoa 750, an alloy of almost 90%-pure aluminum, 6.5% tin, alloyed with zinc and trace amounts of silicon, nickel, and copper.

Normal bearing wear should not exceed 0.0005 in. per 1000 hours of continuous operation. The operative word is "continuous"—frequent startups, cold loads, and chronic overloads can subtract hundreds of hours from the expected life. In practice, main bearings outlast crankpin bearings by about three to one, although No. 1 main, which receives side loads from accessories, can fail early.

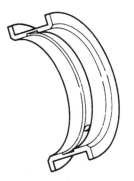

8-69 Thrust bearing configuration for two Yanmar engines. Other makes use two-piece thrust washers.

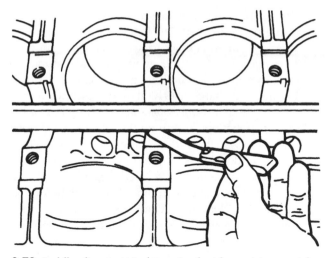

8-70 Saddle alignment is determined with precision straight-edge and feeler gauge. Ford Motor co.

According to Michigan Bearing, 43% of failings can be attributed to dirt, 15% to oil starvation, 13% to assembly error, and 10% to misalignment. Overloading, corrosion, and miscellaneous causes account for the remainder. In other words, the mechanic contributes to most failures, by either assembling dirt into the engine, reversing nonsymmetrical bearing shells or bearing caps, or failing to provide adequate lubrication during initial start-up. These matters are discussed in the next section.

Bearing alignment problems come about from improperly seated rod caps or from warped main-bearing saddles. The condition can sometimes be spotted during disassembly as accelerated wear on outboard or center mainbearing shells. The mechanic should verify alignment by spanning the saddles with a precision straight-edge (Fig. 8-70). Loss of contact translates as block warp.

Bearing alignment can be restored, but at the cost of raising the crankshaft centerline a few thousandths of an inch. The effect of chain-driven camshafts is to retard valve timing by an almost imperceptible amount; but gear trains are not so forgiving and the machinist might be forced to resort to some fairly exotic (and expensive) techniques to maintain proper tooth contact. The "cleanest" solution is to resize the bearing bosses by a combination of metalizing and honing.

Assembly—major components

Building up the major components—crankshafts, pistons, and rods—is a special kind of activity, that can be neither forced nor hurried.

Although remote from our subject, a certain insight into it can be gained from the experience of U.S. Naval gunners almost a century ago. As the century closed, five warships fired at a hulk from a mile distance for 25 minutes and registered two hits. This was an average performance, achieved by taking approximate aim and waiting for the roll of the ship to bring the sights into alignment with the target. Meanwhile an Englishman by the name of Timms had developed another system of aiming, which involved constant correction, in an attempt to keep the sights on the target however the ship moved. His explanations that the gunner would naturally adjust to ship's motion went unheeded by Navy brass until he got the ear of Teddy Roosevelt, who put him in charge of gunnery training. In 1910, the test described above was repeated, this time using the new method and only one ship. The number of hits doubled.

Timms scored by concentrating directly on the problem and not on some ritualized technique that was supposed to yield an automatic solution. There is a great deal of room in the mechanic's trade for this kind of pragmatic thinking. For example, the whole business of measurement could be reformed, beginning with the establishment of a datum line (such as the crankshaft centerline) from which deck flatness, cylinder bore centerlines, and bearing clearances would be generated. And much thought could be given to substituting a direct measurement of clamping force for the indirect system currently used. When a capscrew is tightened, about 90% of the torque load appears as friction between the underside of the fastener head and the work piece. A 5% miscalculation in friction values represents a 50% clamping force error.

Glaze breaking

Cast-iron bores develop a hard glaze in service, which must be "broken" or roughened, with a hone to facilitate ring sealing. Consequently, worn cylinder bores must be honed whenever piston rings are replaced. Remachined cylinders must also be honed, but this is usually the province of the machine shop.

Most mechanics use a brush hone, sized to cylinder diameter, such as the 120-grit BRM Flex-Hone pictured in Fig. 8-71. Obtaining the desired Crosshatch pattern

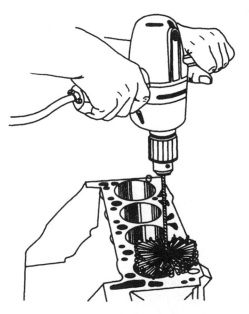

8-71 Most mechanics break glaze with a brush, or ball, hone.

(shown earlier in Fig. 8-37) depends on a four-sided relationship among the ring material, spindle speed, bore diameter, and stroke frequency. For what it is worth, chrome rings running in a 4-in. cylinder need the finish produced by a 280-grit stone, rotated 190 rpm and reciprocated 70 times a minute. Such precision is unobtainable, but a little practice will produce an acceptable surface.

It is important to use plenty of lubricant, such as mineral oil or PE-12, available in aerosol cans from specialty tool houses. Keep the hone moving—it cuts as it rotates—and do not pause at the ends of the strokes. Hone for 3 or 4 seconds and stop to inspect the bore. Crosshatched grooves should intersect at angles of 30–45° relative to the bore centerline. If the pattern appears flat, reduce the spindle speed or increase the stroke rate. The job should require no more than 15 or 20 seconds to complete. Withdraw the hone while it is still turning.

Few mechanics take the time to clean a bore properly. Solvents merely float abrasive hone fragments deeper into the metal. Detergent, hot water, and a scrub brush are the prescription. Scrub until the suds are white, and wipe dry with paper shop towels. Continue to scrub and wipe until the towels no longer discolor. According to TRW, half of early ring failures can be traced to abrasive particles left in the bore.

Laying the crank

This discussion applies to engines with two-piece main-bearing shells; full-circle bearings require a slightly different technique, as outlined in the appropriate shop manual.

1. Although it is late in the day, make one final inspection of the block. Critical areas include saddle alignment for two-piece GM block castings, cracks that are likely to be associated with saddle webbing and liner flanges, and main bearing cap-to-saddle fits. Most caps are buttressed laterally by interference fits with their saddles. Fretting on contact surfaces can, if not corrected, result in crankshaft failure. Some mechanics make it a practice to chase main-bearing bolt threads; if you opt to do this, use the best, most precise tap that money will buy.
2. Verify that main-bearing saddles are clean and dry.
3. Unwrap one set of main-bearing inserts. Bearing size will be stamped on the back of each insert. All machinists can tell stories about mislabeled bearings, but certainly it would give one pause if the crankshaft were turned 0.010 in. and the inserts were marked "Std."
4. Identify the upper insert, which always is ported for crankshaft lubrication. A few lower shells carry a superfluous port; most are blanked off and grooved.
5. Roll the upper insert into place on the web, leading with the locating tab, as shown in Fig. 8-72. Make absolutely certain that the tab indexes with the web recess. When installed correctly, the ends of the insert stand equally proud above the parting face.
6. Verify that the oil passage drilled in the web indexes with the oil port in the insert. A glance will suffice, but some shops go a step further and run a drill bit through the insert and into the web.

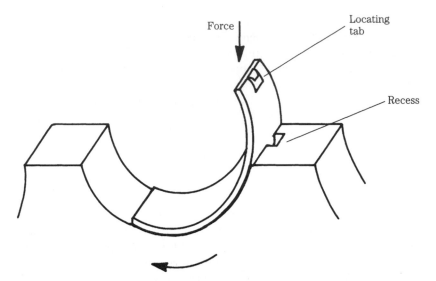

Force

Locating tab

Recess

8-72 Main bearing inserts roll into place. Slip the bearing into the cap or saddle, leading with the smooth edge. Then, with thumbs providing the force, push the insert home, seating the locating tab in its recess.

7. Install the remaining upper shells, one of which might be flanged for thrust. Check each insert for indicated size and, once they are installed, verify that oiling passages are open.
8. Where appropriate, install the upper half of the rear crankshaft seal, per manufacturer's instructions. Rope seals must be rolled into place, using a bar sized to match crankshaft journal diameter.
9. Saturate the bearings and seal with clean lube oil. (Some mechanics prefer to use a grease formulated for bearings assembly, such as Lubriplate No. 105. Grease is very appropriate when the engine will not be returned to immediate service.)
10. Wipe down the crankshaft with paper shop towels. Coat journals and crankpins with a generous amount of lubricant.
11. Lower the shaft gently and squarely into position on the webs. Exercise care not to damage thrust flanges.
12. Install lower bearing shells into the caps, lubricating as before.
13. Mount the caps in the correct orientation and sequence. Lightly oil cap bolt threads.
14. Torque the caps, working from the center cap outward. Conventional bolts make up in three steps—1/3, 2/3, and 3/3 torque; torque-to-yield bolts are run down to a prescribed torque limit and rotated past that limit by a set amount. After each cap is pulled down, turn the crankshaft to detect possible binds.
    - *Radius ride,* a condition recognized by bright edges on the bearing shells and caused by excessively large crankshaft fillets left after regrinding. Return the crankshaft to the machinist.

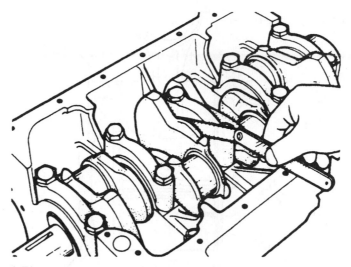

8-73 Lever the crankshaft through its full range of axial movement before measuring thrust bearing clearance. The same measurement can be made with dial indicator, registering off of the crankshaft nose. Chrysler Corp.

- Dirt on the bearing face or OD.
- Insufficient running clearance, a condition that can be determined with Plasti-Gage (see below).
- Bent crankshaft or misaligned bearing saddles.

15. Using a pry bar, lever the crankshaft toward the front of the engine, then pull back (Fig. 8-73). Measure crankshaft float with a feeler gauge between the thrust bearing face and crankshaft web.

Piston and rod installation

It is assumed that piston and rod assemblies have already been made up. If the machinist has not already done so, install the rings on the piston. Rings are packaged with detailed instructions, which supersede those in the factory manual. Here, it is enough to remind you that:

- The time required to check ring gap is well spent, because rings, like bearing inserts, are sometimes mislabeled. Using the flat of the piston as a pilot, insert each compression ring into the cylinder and compare ring gap with the specification (Fig. 8-74).
- Ring orientation is important; according to one manufacturer, a reversed scraper ring increases oil consumption by 500%.
- Using the proper tool, expand the ring just enough to slip over the piston (Fig. 8-75).
- Stagger ring gaps, as detailed by the manufacturer.

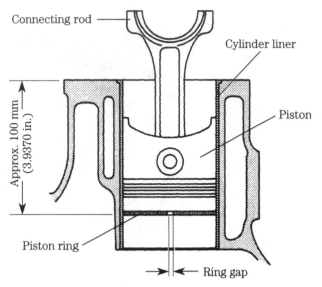

8-74 Measure ring gap relative to the lower, and least worn part of the cylinder. Yanmar Diesel Engine Co. Ltd.

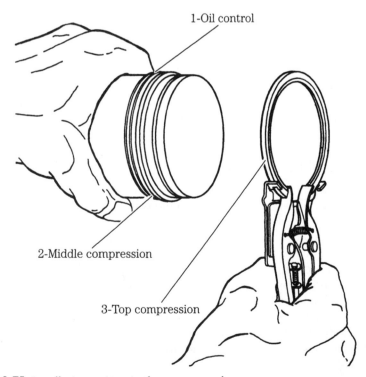

8-75 Install piston rings in the sequence shown. Kohler of Kohler

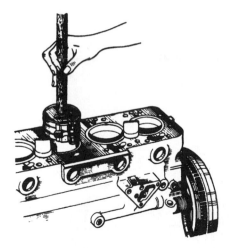

8-76 Piston installation should be a gentle process, involving no more than the force exerted by one hand. Peugeot

At this point, you are ready to install the piston assemblies. Follow this procedure:

1. Remove the bearing cap from No. 1 rod and piston assembly
2. Mount an upper bearing shell on the rod, indexing oil ports.
3. Coat the entire bearing surface with fresh lube oil. Repeat the process for the cap.
4. Slip lengths of fuel line over the rod bolt ends to protect the crankshaft journals during installation.
5. Turn the crankshaft to bottom dead center on No. 1 crankpin. Saturate the crankpin with oil.
6. Install a ring compressor over the piston. The bottom edge of the compressor should be a ½ in. or so below the oil ring. Tighten the compressor bands just enough to overcome ring residual tension.
7. With the block upright and the leading side of the piston toward the front of the engine, place the compressor and captive piston over No. 1 cylinder bore. A helper should be stationed below to guide the rod end over the crankpin. While holding the compressor firmly against the fire deck, press the piston out of the tool and into the bore. As shown in Fig. 8-76, thumb pressure should be sufficient. Stop if the piston binds and reposition the piston in the compressor.
8. Install the lower bearing shell in the cap; verify that the locating tab indexes with its groove.
9. Coat the bearing contact surface with lube oil.
10. Working from below, carefully remove the hoses from the bolt ends and pull the rod down over the crankpin. Make up the rod cap, making sure match-marks align.
11. Torque the rod bolts to spec (Fig. 8-77).
12. Using a hammer handle or brass knocker, gently tap the sides of the big end journal. The rod should move along the length of the crankpin.

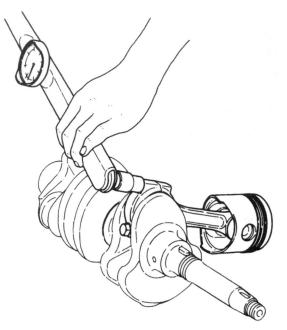

8-77 Using an accurate torque wrench, pull the rod bolts down evenly to spec. Yanmar Diesel Engine Co. Ltd.

13. Rotate the crankshaft a few revolutions to detect possible binds and the bore scratches that mean a broken ring. Resistance to turning will not be uniform: piston speed increases at mid-stroke, with a corresponding increase in crankshaft drag.
14. Repeat this operation for remaining pistons.
15. Press or bolt on piston cooling jets, aiming them as the factory manual indicates.

Plastigage

Old-time mechanics checked timing gears by rolling a piece of solder between the gear teeth. The width of the solder represented gear lash. (This technique has generally been replaced by direct measurement with a dial indicator.) Plastic gauge wire represents an application of the same principle to journal bearings. Perfect Circle and several other manufacturers supply color-coded gauge wire together with the necessary scales that translate wire width into bearing clearance. Green wire responds to the normal clearance range of 0.0001–0.0003 in.; red goes somewhat higher—0.0002–0.0006; blue extends to 0.0009 in. Accuracy compares to that obtained with a micrometer.

1. Remove a bearing cap and wipe the lubricant off both the cap and the exposed portion of the journal.
2. Lay a strip of gauge wire longitudinally on the bearing, about ¼ in. off-center, as shown in Fig. 8-78.

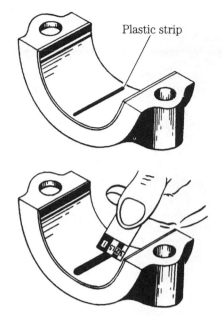

Plastic strip

8-78 Plastic gauge wire is an inexpensive and quite accurate method of determining bearing clearance. Ditroit Diesel

Note: Invert the engine or raise the crankshaft with a jack under an adjacent counterweight before gauging main-bearing clearances. Otherwise, the weight of the crankshaft and flywheel will compress the wire and give false readings.

3. Assemble the cap, torquing the hold-down bolts to specifications.
4. Without turning the crankshaft, remove the cap. Using the scale printed on the gauge-wire envelope, read bearing clearance as a function of wire width. Clearance should fall within factory assembly specifications. If it does not, remove the crankshaft and return it to the machinist. Scrape or wipe off the gauge wire with a rag soaked in lacquer thinner. Reoil the bearing and journal before final assembly.

Balancing

Precision balancing is an optional service, difficult to justify in economic terms, but worth having done if the goal is to build the best possible engine. Balanced engines run smoother, should last longer, and sometimes exhibit a small gain in fuel economy.

But do not expect wonders; balancing is a palliative, arrived at by compromise, and effective over a fairly narrow rpm band. No amount of tweaking can take all of the shake, rattle, and roll out of recip engines.

There are three steps in the balancing process. First the technician equalizes piston and pin weights to within $1\frac{1}{2}$ g (Fig. 8-79). If the technician works out of a large inventory, this can sometimes be achieved by selective assembly. Otherwise, machine work is in order; heavy pistons are lightened by turning the inner edge of

8-79 Balancing begins with match-weighing piston assemblies.

the skirts, and when absolutely necessary, aluminum slugs can be pressed into the hollow piston pins to add weight to light assemblies.

The next step is to match-weight the rods using a precision scale and a special rod adapter, such as shown in Fig. 8-80. The Stewart-Warner adapter makes it possible to distinguish between rotating (or lower-end) and reciprocating (upper-end) masses; when such equipment is not available, the technicians assume that the lower half of the rod rotates and that the upper end describes a purely reciprocating path. However arrived at, rod reciprocating and rotating masses are equalized to a 1½ g tolerance by removing metal from the balance pads (shown as the extensions on the ends of the DDA rod in the photograph and more clearly back in Fig. 8-59, where they are labeled "noncritical" areas).

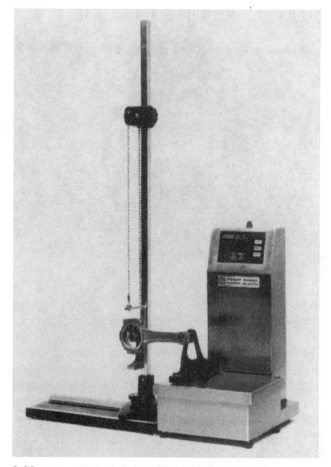

8-80 Stewart-Warner fixture splits rod weight into the rotating mass of piston and rod assemblies.

The technician computes the total rotating mass and bolts an equivalent mass to the crankshaft (Fig. 8-81). The shaft is then mounted on the balancer and rotated at the desired speed, which should correspond to actual operating conditions. Balance, or matching counterweight mass to rotating mass, can normally be achieved by drilling the counterweights. Most shops work to a 0.5 oz.-in. tolerance; the equipment will support 0.2 oz.-in. accuracy. Once this is done, the process is repeated for the harmonic balancer and clutch assembly.

Oil seals

Oil seals can be a headache, especially if one leaks just after an overhaul. Premature failure is almost always the mechanic's fault and can be traced to improper installation.

8-81 The crankshaft is dynamically balanced relative to the rotating mass of piston and rod assemblies.

Rope or strip seals are still used at the aft end of the crankshaft on some engines. These seals depend on the resilience of the material for wiping action and so must be installed with the proper amount of compression. Remove the old seal from the grooves and install a new one with thumb pressure. (You might want to coat the groove—but not the seal face—with stickum.) The seal should not be twisted or locally bound in the groove.

Now comes the critical part. Obtain a mandrel of bearing boss diameter (journal plus twice the thickness of the inserts) and, with a soft-faced mallet, drive the seal home. Using this tool as a ram, you can cut the ends of the seal flush with the bearing parting line. Without one of these tools you will have to leave some of the seal protruding above the parting line in the elusive hope that the seal will compact as the caps are tightened. Undoubtedly some compaction does take place, but the seal ends turn down and become trapped between the bearing and cap, increasing the running clearance. Lubricate the seal with a grease containing molybdenum disulfide to assist in break-in.

Synthetic seals, usually made of Neoprene, are used on all full-circle applications and on the power takeoff end of many crankshafts as well. These seals work in a manner analogous to piston rings. They are preloaded to bear against the shaft and designed so that oil pressure on the wet side increases the force of contact.

These seals must be installed with the proper tools. If the seal must pass over a keyway, obtain a seal protector (a thin tube that slides over the shaft) or at least cover the keyway with masking tape. Drive the seal into place with a bar of the correct diameter. The numbered side is the driven side in most applications. The steep side of the lip profile is the wet side. It is good practice to use a nonhardening sealant on the back of metallic seals. Plastic coated seal cams are intended to conform to irregularities in the bore and do not need sealant.

More elaborate seals require special one-of-a-kind factory tools. The better engines often incorporate wear sleeves over the shafts, either as original equipment

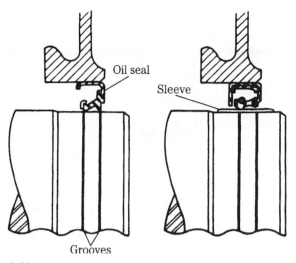

8-82 Wear sleeves extend crankshaft life. When retro-fitted, an oversized seal must be used to compensate for the increase in journal diameter. Detroit Diesel

or as a field option. Figure 8-82 shows the use of a wear sleeve on a Detroit Diesel crankshaft. The sleeve makes an interference fit over the shaft and is further secured with a coating of shellac. Worn sleeves are cut off with a chisel or peened to stretch the metal. The latter method is preferred because there is less chance of damaging the shaft.

9
CHAPTER

Air systems

Most diesels have open manifolds without the restriction imposed by a throttle plate. As a result, these engines ingest about 2.5 ft³ of air per horsepower. Much of the air merely cycles through the engine without taking part in combustion. But atmospheric air, even at its most pristine, contains mineral dust and other solids that, unless removed, result in rapid upper cylinder wear. Research has shown that the most destructive particulates in the air-inlet system are between 10 and 20 μm in diameter. A micron is one millionth of a meter or 4/100,000 in.

Air cleaners

The ideal air cleaner would trap all solids, regardless of size. Unfortunately, such a device does not exist, at least in practical form. Some particulates get through, which is why engines tend to wear rapidly in dusty environments, no matter how carefully maintained.

Construction

Older engines are often fitted with oil-bath filters that combine oil-wetted filtration with inertial separation (Fig. 9-1). Air enters at the top of the unit through the precleaner, or cyclone. Internal vanes cause the air stream to rotate, which tends to separate out the larger and heavier particulates. The air then passes through the central tube to the bottom of the canister, where it reverses direction. Some fraction of the remaining particulates fail to make the U-turn and end in the oil reservoir. Most of those that remain are trapped in the oiled mesh.

These devices depend, in great part, upon the velocity of the incoming air to centrifuge out heavier particles. At low engine speeds, the particles remain in the air stream, and filtration depends solely upon the oil-wetted mesh. Under ideal circumstances filtering efficiency is no more than 95% and, in practice, can be much less.

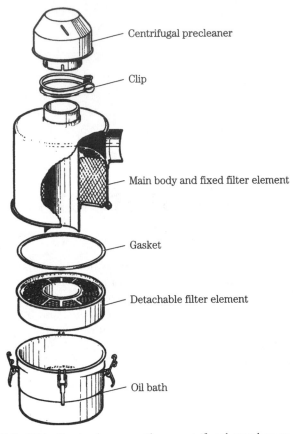

Centrifugal precleaner

Clip

Main body and fixed filter element

Gasket

Detachable filter element

Oil bath

9-1 Oil-bath air cleaner with a centrifugal precleaner. These units have a peak efficiency of about 95%, which falls off rapidly at low engine speeds.

If overfilled, subjected to high flow-rates, or tilted much off the horizontal, oil-bath cleaners bleed oil into the intake manifold. The oil can plug the aftercooler and raise exhaust temperatures enough to cause early turbocharger failure.

Figure 9-2 shows a replaceable paper-element filter of the type used on automobiles and light trucks. When used for industrial engines, the paper element is often combined with a prefilter and a centrifugal precleaner. At their most sophisticated, paper-element filters can have an efficiency of 99.99%. Efficiency improves when the filter is lightly impacted with dust. Engines operated in extremely dusty environments also benefit from an exhaust-powered ejector mounted upstream of the filter.

The filter should be mounted horizontally on the manifold to reduce the possibility of dirt entry when the element is changed. Filters that mount vertically, such as the one shown in Fig. 9-2, should include a semi-permanent *safety* element below the main filter. This element functions to trap dust when the filter is removed for servicing.

All engines should have some sort of filter monitoring device that, if not supplied by the manufacturer, can be fabricated by the simple expedient of plumbing a

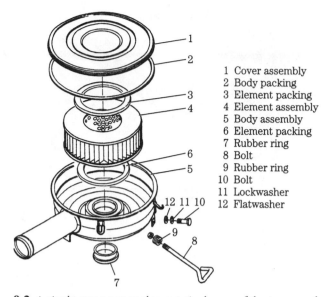

1 Cover assembly
2 Body packing
3 Element packing
4 Element assembly
5 Body assembly
6 Element packing
7 Rubber ring
8 Bolt
9 Rubber ring
10 Bolt
11 Lockwasher
12 Flatwasher

9-2 A single-stage paper-element air cleaner of the type used for automotive and marine applications. Two-stage paper filters achieve efficiencies of as high as 99.99%, which sounds impressive, but air-borne abrasives continue to be the primary cause of engine wear. Marine Engine Div., Chrysler Corp.

vacuum gauge immediately downstream of the filter. When new, paper-element filters impose a pressure drop of about 6.0 in./H_2O and the inlet ducting usually adds about 3.0 in./H_2O. Engine performance falls off noticeably when the total system pressure drop exceeds 15 in./H_2O.

Service the filter only when the air-restriction indicator trips, since each time the element is removed some dust enters the system. Before removing the element wipe off all dust from the inside of the filter housing. Do not blow out paper elements with compressed air: a tear in the element, so small that it may not be visible to the eye, exposes the engine to massive amounts of dust intrusion.

And before we leave the subject, it should be remarked that "high-performance" filters can provide marginal power increases by reducing pumping losses. But the reduction in pressure drop often comes at the cost of reduced filtration efficiency. Any air filter worthy of its name should have its efficiency certified by a third party under the protocols of the SAE air cleaner test code J726.

Rebuilt engines almost always have shorter lives than new engines. For years it was believed that factory inspection and assembly procedures gave new engines the edge. But evidence is accumulating that abrasives entering from leaks in the flexible tubing couplings downstream of the filter are the culprit (Fig. 9-3a). These couplings, which see boost temperatures of 300°–400°F (149°–204°C), harden with age and are rarely replaced. Nor is it possible to detect leaks by external inspection of the tubing and hose couplings. However, the presence of leaks will be revealed by streaks in the dust film that collects on inlet-tubing inner diameters (IDs). The origin of the streak marks the point of dirt entry.

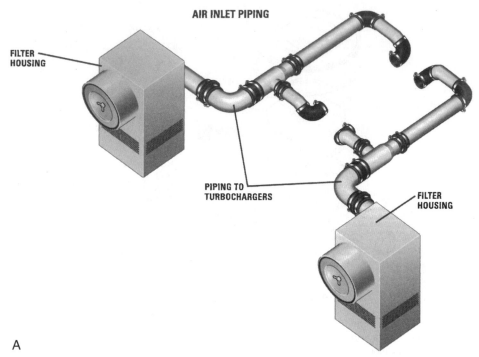

AIR INLET PIPING

FILTER HOUSING

PIPING TO TURBOCHARGERS

FILTER HOUSING

A

9-3A. Elastomer tubing couplings downstream of the filter are a major and nearly always overlooked source of dust entry. Caterpillar Inc.

Flexible couplings should be replaced periodically and as inexpensive insurance for rebuilt engines. Secure the coupling with SAE type F clamps—not worm-gear plumber's clamps—that provide a 360° seal.

Turbochargers

A turbocharger is an exhaust-powered supercharger, that unlike conventional super-chargers, has no mechanical connection to the engine (Figs. 9-3b and 9-4). The exhaust stream, impinging against the turbine (or "hot") wheel, provides the energy to turn the compressor wheel. For reasons that have to do with the strength of materials, turbo boost is usually limited to 10 or 12 psi. This is enough to increase engine output by 30—40%.

Turbocharging represents the easiest, least expensive way to enhance performance. It is also something of a "green" technology, because the energy for compression would otherwise be wasted as exhaust heat and noise (Fig. 9-5). On the other hand, the interface between sophisticated turbo machinery, turning at speeds as great as 140,000 rpm and at temperatures in excess of 1000°F, and the internal combustion engine is not seamless.

Unless steps are taken to counteract the tendency, turbochargers develop maximum boost at high engine speeds and loads. The turbine wheel draws energy from exhaust gas velocity and heat, qualities that increase with piston speed and load. The

B

9-3B. John Deere 6068T turbocharged engine. One of the appeals of turbocharging is its apparent simplicity. In this instance, a 25-lb turbocharger boosted output to 175 hp for a gain of 45 hp over the naturally aspirated version of the same engine. Torque went from 335 lb-ft to 473 lb-ft.

compressor section behaves like other centrifugal pumps, in that pumping efficiency is a function of impeller speed. At low speeds, the clearance between the rim of the impeller and the housing shunts a large fraction of the output. At very high rotational speeds, air takes on the characteristics of a viscous liquid and pumping efficiency approaches 100%. In its primitive form, a turbocharger acts like the apprentice helper, who loafs most of the day and, when things get busy, becomes too enthusiastic.

Another innate, but not necessarily uncorrectable, characteristic of turbocharged engines is the lag, or flat spot, felt during snap acceleration. Perceptible time is required to overcome the inertia of the rotating mass. By the same token, the wheels continue to coast for a few seconds after the engine stops.

Background

The exhaust-driven supercharger was first demonstrated in 1915, but remained impractical until the late 1930s, when the U.S. Army, working closely with General Electric, developed a series of liquid-cooled and turbocharged aircraft engines. The expertise in metallurgy and high temperature bearings gained in this project made GE a leader in turbocharging and contributed to its success with jet engines.

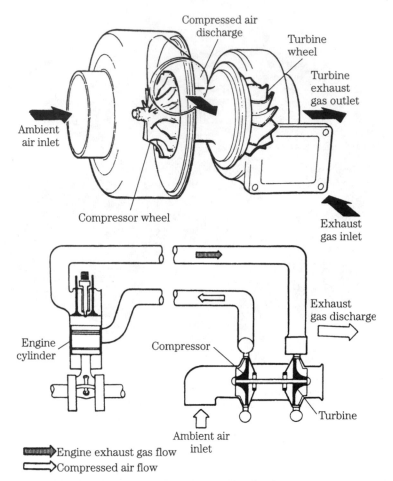

Compressed air
discharge

Turbine
wheel

Turbine
exhaust
gas outlet

Ambient
air inlet

Compressor wheel

Exhaust
gas inlet

Exhaust
gas discharge

Engine
cylinder

Compressor

Ambient air
inlet

Turbine

Engine exhaust gas flow
Compressed air flow

9-4 A turbocharger is an exhaust-driven centrifugal pump, operating at six-figure speeds and typically generating 10–12 psi of supercharge.

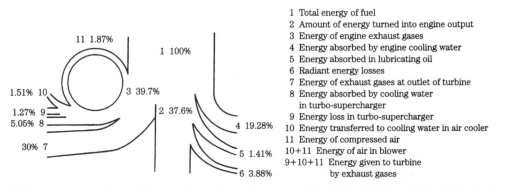

11 1.87%   1 100%

1.51% 10
1.27% 9
5.05% 8   2 37.6%   3 39.7%

30% 7   4 19.28%

5 1.41%

6 3.88%

1 Total energy of fuel
2 Amount of energy turned into engine output
3 Energy of engine exhaust gases
4 Energy absorbed by engine cooling water
5 Energy absorbed in lubricating oil
6 Radiant energy losses
7 Energy of exhaust gases at outlet of turbine
8 Energy absorbed by cooling water
    in turbo-supercharger
9 Energy loss in turbo-supercharger
10 Energy transferred to cooling water in air cooler
11 Energy of compressed air
10+11 Energy of air in blower
9+10+11 Energy given to turbine
           by exhaust gases

9-5 Turbocharger heat balance chart. Note that even with turbocharging 30% of the fuel energy goes out the exhaust.

As applied to aircraft, turbocharging was a kind of artificial lung that normalized manifold pressure at high altitudes. A wastegate deflected exhaust gases away from the turbine at low altitudes, where the dense air would send manifold pressures and engine power outputs to dangerously high levels. In emergencies, P-51 pilots were instructed to push their throttles full forward, an action that broke a restraining wire, closed the wastegate, richened the mixture, and initiated alcohol injection for a few seconds of turboboost. The broken wire was a telltale sign, signaling the ground crew to rebuild the engine.

Turbochargers continue to be used on reciprocating aircraft engines and enjoy some currency in high performance automobiles, thanks to electronic controls that minimize detonation. But diesel engines are the real success story.

Turbocharging addresses the fundamental shortcoming of compression ignition, which is the delay between the onset of injection and ignition. That delay, terminated by the sudden explosion of puddled fuel, results in rapid cylinder pressure rise, rough running and incomplete combustion with attendant emissions. Turbocharging (and supercharging generally) increase the density of the air charge. Ignition temperature develops early during the compression stroke and coincides more closely with the onset of injection. Fuel burns almost as quickly as delivered for a smoother running, cleaner, and more economical engine.

Applications

Depending on how it is accomplished, turbocharging can have three quite distinct effects on performance. If no or little additional fuel is supplied, power output remains static, but emissions go down. Surplus air lowers combustion temperatures and provides internal cooling. Such engines should be more durable than their naturally aspirated equivalents.

The high-boost/low-fuel approach to turbocharging is limited to large stationary and marine plants. Makers of small, high-speed engines are more concerned with maximum power or mid-range torque.

Supplying additional fuel in proportion to boost yields power, which can translate as fuel savings for engines that run under constant load. Some gains in fuel efficiency (calculated on a hp/hour basis) accrue from turbocharging, but the real advantage comes about when high supercharge pressures allow for lower piston speeds. Large marine engines, developing a 1000 hp and more per cylinder, use this approach to achieve thermal efficiencies of 50%. On a more familiar scale, the naturally aspirated International DT-414 truck engine produced 157 hp at 3000 rpm. The addition of a large turbocharger boosted output to 220 hp, for a gain of 40%. Once in the fairly narrow power band, the trucker could save fuel by selecting a higher gear.

The third approach is more characteristic of automotive and light truck engines, which operate under varying loads and rarely, if ever, develop full rated power. What is wanted is torque.

The Navistar 7.3L, developed for Ford pickups and Econoline vans, is perhaps the best demonstration of the way a turbocharger can be throttled for torque production. In its naturally aspirated form, the engine develops 185 hp at 3000 rpm and

360 ft./lb. of torque at 1400 rpm. A Garrett TC43 turbocharger boosts power output marginally to 190 hp; but torque goes up 17.8% to 385 ft/lb. In normal operation the wastegate remains closed, directing exhaust gases to the turbine, until 1400 rpm. At that point, which corresponds to the torque peak, the hydraulically actuated wastegate begins to open, shunting exhaust away from the turbine.

Serious applications of turbocharging, whether for peak power or mid-range torque, impose severe mechanical and thermal loads, which should be anticipated in the design stage. Insofar as they work as advertised, add-on turbocharger kits—which rarely involve more than a rearrangement of the plumbing—are a buyer-beware proposition. The 7.9L is considered a very rugged engine in its naturally aspirated form. But turbocharging called for hundreds of engineering changes, including shot-peened connecting rods, oversized piston pins, Inconel exhaust valves, and special Zollner pistons, with anodized crowns.

Construction

Figure 9-6 illustrates an air-cooled Ishikawajima-Harima turbocharger of the type found on engines in the 100–150-hp range. The inset shows the water-cooled version of the same turbocharger, plumbed into the engine cooling system. All modern small-engine turbochargers follow these general patterns.

Floating bushings, located in the central bearing chamber, support the shaft. These bushings, like the connecting rod bearings specified for the original Ford V-8, float on the ID and OD. Thus, bushing speed is half that of shaft speed, which is to say that the bushings can reach speeds of 60,000 or 70,000 rpm. A floating thrust bushing contains axial motion.

Also note the way the impeller wheels cantilever from the bushings, so that the masses of the rotating assembly are concentrated near the ends of the shaft. This "dumb-bell" configuration requires precise wheel balance and extremely accurate shaft alignment.

The bearing section is lubricated and cooled by engine oil, generally routed through external pipes or hoses. Shaft seals keep oil from entering the turbine and compressor sections.

The turbocharger is usually the last component to receive oil pressure and might continue to rotate after the engine stops. To ensure an adequate oil supply to the bearings, operators should idle the engine for at least 30 seconds upon starting and for the same period before shutdown.

Wastegate

All turbocharger installations incorporate some form of boost limitation; otherwise, boost would rise with load until the engine destroyed itself.

Turbocharger geometry, sometimes abetted by inlet restrictions and designed-in exhaust backpressure, limit boost on constant-speed engines. Automotive and light truck engines have surplus turbocharging capability for boost at part throttle. These applications employ a wastegate—a kind of flap valve—that automatically opens at a preset level of boost to shunt exhaust gas around the turbine.

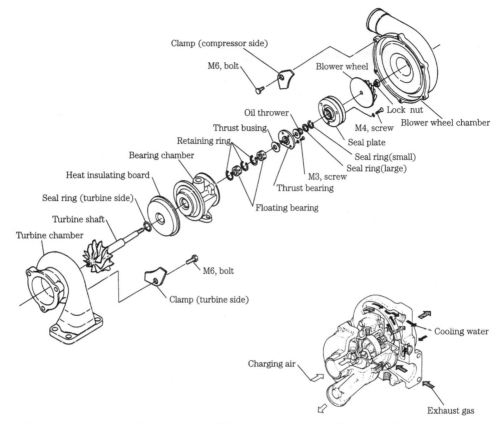

Clamp (compressor side)

M6, bolt

Blower wheel

Oil thrower

Thrust busing

Retaining ring

Bearing chamber

Heat insulating board

Seal ring (turbine side)

Turbine shaft

Turbine chamber

Floating bearing

Thrust bearing

M3, screw

Seal ring(large)

Seal ring(small)

Seal plate

M4, screw

Lock nut

Blower wheel chamber

M6, bolt

Clamp (turbine side)

Cooling water

Charging air

Exhaust gas

9-6 Air- or water-cooled turbocharger of the type used on small Japanese diesels. Water cooling may prolong bearing life and should reduce the coking experienced during hot shutdowns, when the turbocharger continues to rotate after the engine stops.

Most wastegates are controlled by a diaphragm, open to the atmosphere on one side and to manifold pressure on the other (Fig. 9-7). The Ford unit shown also incorporates a relief valve. Normally the wastegate opens at 10.7 psi; should it fail to do so, the relief valve opens at 14 psi and, because it is quite noisy, alerts the driver to the overboost condition.

Other wastegates are spring-loaded and usually include an adjustment, appropriately known as the "horsepower screw." As usually configured, tightening the screw increases available boost. Therefore, exercise restraint.

Test wastegate operation by loading the engine while monitoring rpm and manifold pressure. If the installation does not include a boost gauge, connect a 0–20 psi pressure gauge at any point downstream of the compressor. The diaphragm-sensing line (on units so-equipped) serves as a convenient gauge point. Note, however, that the gauge must be connected with a tee fitting to keep the wastegate functional. High gear acceleration from 2500 rpm or so should generate sufficient load to open the gate.

The control philosophy discussed in the preceding paragraphs implies that the turbocharger is used for power enhancement. When mid-range torque is the object,

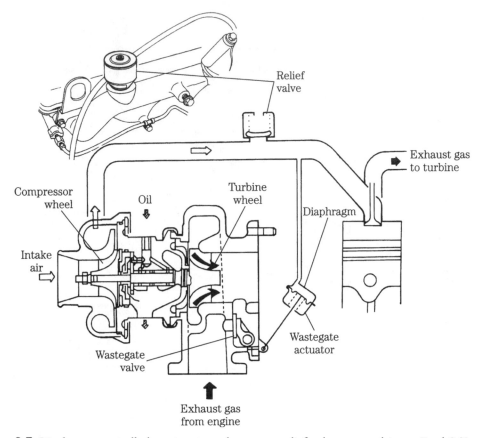

Relief valve

Exhaust gas to turbine

Compressor wheel

Oil

Turbine wheel

Diaphragm

Intake air

Wastegate actuator

Wastegate valve

Exhaust gas from engine

9-7 Diaphragm-controlled wastegate and pressure-relief valve, as used in on Ford 2.3L diesel engines.

the wastegate opens early, at the engine speed corresponding to peak torque output. Most of these applications employ computer-controlled wastegates, whose response is conditioned by manifold pressure, engine rpm, coolant temperature, and other variables. No generalized test procedure has been developed for these devices.

Aftercoolers

Compressing air raises its temperature, which reduces change density and tends to defeat the purpose of supercharging. Sophisticated turbocharger installations include a heat exchanger, or aftercooler, between the compressor outlet and intake manifold. The cooling medium can be air, engine coolant, or—for marine applications—water. The plate and tube seawater-fed aftercooler used with the Yanmar 4LH-HTE boosts output to 135 hp, or 30 hp more than an identical engine without aftercooling.

Air-cooled heat exchangers work most efficiently when mounted in front of the radiator. Engine coolant should be taken off at the pump discharge and returned to some point low on the water jacket.

Air-cooled units require little attention, other than an occasional dust off. Liquid-cooled heat exchangers should be cleaned as needed to remove scale and fouling, and periodically tested by blowing low-pressure (25-psi maximum) air through the tubes. Water intrusion into the intake tract can be expensive.

Routine maintenance

The first priority is to obtain actual performance data, particularly with reference to turbocharger behavior. The full story would require a dynamometer to extract, but one can gain useful insight by observing the changes in manifold pressure under working loads that, at some point in the test, should be great enough to cause the wastegate to open. The rise in oil temperature and variations in crankcase pressure supply additional parts of the picture.

Do not operate a turbocharged engine unless the air cleaner (or spark arrestor) is in place and the intake-side ducting secure. The compressor acts as a vacuum cleaner, drawing in foreign, objects which will severely damage the unit and might cause it to explode. The troubleshooting chart (Table 9-1) makes reference to coast-down speed. If you feel it is necessary to observe compressor rotation, cover the turbocharger inlet with a screen to at least keep fingers and other large objects out of the mechanism.

When dismantling a turbocharger and related hardware, make a careful tally of all fasteners, lockwashers, and small parts removed. Be absolutely certain that all are accounted for before starting the engine. Immediately shut down the engine if the turbocharger makes unusual noise or vibrates.

Turbocharging (and supercharging generally) put severe stress on lubrication, air inlet, crankcase ventilation, and exhaust systems.

Lubrication system

Elevated combustion pressure contaminates the oil with blowby gases and promotes oxidation by raising crankcase oil temperature. That fraction of the oil diverted to the turbocharger can undergo 80°F temperature rise in its passage over the bushings. Change lube oil and filter(s) frequently.

The cost and disposal problems associated with filters make reusable filters attractive for fleet operators. Racor, a division of Parker Hannifin, manufactures a series of liquid filters with washable, stainless-steel elements and a TattleTale light that alerts the operator when the filter needs to be cleaned.

Frequently inspect the turbocharger and its oiling circuitry for evidence of leaks that, if neglected, can draw down the crankcase.

Air inlet system

Dust particles, entering through a poorly maintained filter or through leaks in the ducting rapidly erode the compressor wheel. Leaks downstream of the compressor cost engine power and waste fuel.

Crankcase ventilation system

This system removes combustion residues and, in the process, subjects the crankcase to a slight vacuum. In normal operation, fresh air enters through the breather filter and crankcase vapors discharge to the atmosphere (prepollution engines) or to the

Table 9-1. Turbocharger fault diagnosis

Engine lacks power, black smoke in exhaust

Symptom	Probable causes	Corrective actions
1. Insufficient boost pressure; compressor wheel coasts to a smooth stop when engine is shut down; turns easily by hand.	1. Clogged air filter element.	Clean or replace element.
	2. Restriction in air intake.	Remove restriction.
	3. Air leak downstream of compressor.	Repair leak.
	4. Insufficient exhaust gas energy	
	• Exhaust leaks upstream of turbocharger.	Repair leak.
	• Exhaust restriction downstream of turbocharger.	Remove restriction.
2. Insufficient boost pressure; compressor wheel does not turn or judders during coastdown; drags or binds when turned by hand.	1. Carbon accumulations on turbine-shaft oil seals.	Disassemble, clean turbocharger and change oil and filter.
	2. Bearing failure, traceable to	
	• Normal wear.	Rebuild turbo.
	• Insufficient oil.	Rebuild turbo. Inspect oil supply/return circuit and repair as necessary to restore full flow. Change engine oil and filter.
	• Excessive oil temp.	Rebuild turbo. Clean oil cooler (if fitted); change engine oil and filter. Monitor oil temp.
	• Turbo rotating assy out of balance.	Rebuild and balance turbo.
	• Bad operating practices— application of full throttle upon startup and/or hot shutdown.	Rebuild turbo. Train operators. A prelube system may extend turbo life in harsh operating environments.
	3. Rotating assembly makes rubbing contact with case. Leading causes are	
	• Bearing failure (see above).	Rebuild turbo. Determine cause of failure and correct.
	• Entry of foreign matter into compressor.	Rebuild turbo. Inspect air cleaner and repair as necessary.
	• Excessive turbo speed.	Rebuild turbo. Turbo rpm is a function of exhaust gas temperature (engine load). Check air filter for restrictions, bar over engine to verify that crankshaft rotates freely. If necessary, adjust governor to reduce engine power output.
	• Improper assembly.	Repair turbo, review-service procedures.

(Continued)

Table 9-1. Turbocharger fault diagnosis (*Continued*)

Excessive oil consumption, blue or white exhaust smoke

Symptom	Probable cause	Corrective actions
1. Condition may be accompanied by oil stains in the inlet and/or outlet ducting, or, in extreme eases, by oil drips from the turbocharger housing.	1. High restriction in the air inlet. 2. Worn turbocharger seals, possibly associated with bearing failure and/or wheel imbalance.	Clean or repalce filter element. Repair turbo.

Abnormal turbocharger lag

Symptom	Probable cause	Corrective actions
1. Engine power output exhibits a pronounced "flat spot" during cceleration. Fuel system appears to function normally.	1. Carbon buildup on compressor wheel and housing.	Disassemble and clean turbo.

Unusual noise or vibration

Symptom	Probable cause	Corrective actions
1. "Chuffing" noise, often most pronounced during acceleration.	1. Surging caused by restriction at compressor discharge nozzle.	Disassemble and clean turbo.
2. Knocking or squeal.	1. Rotating elements in contact with housing because of bearing failure or impact damage.	Replace or rebuild turbo.
3. Speed-sensitive vibration.	1. Loose turbocharger mounts 2. Rotating elements in contact with housing. 3. Severe turbine shaft imbalance.	Tighten Replace or rebuild turbo. Rebuild and balance.

intake side of the turbo compressor. The latter arrangement, known as positive crankcase ventilation (PCV) is the norm.

Under severe load, blowby gases accumulate faster than they can be vented and escape through the breather filter. These flow reversals, which occur more frequently in turbocharged engines, tend to clog the filter. Restrictions at the filter allow corrosive gases to linger in the crankcase. A partially functional filter can also pressurize the crankcase under the severe blowby conditions that accompany heavy loads. Oil seals and gaskets leak as a consequence.

Under light loads, reduced flow through the breather depressurizes the crankcase. Low crankcase pressures encourage oil to migrate into the turbocharger and collect in the aftercooler on engines with positive crankcase ventilation. Because oil is a fairly good thermal insulator, the efficiency of the aftercooler suffers. Tests of a Detroit Diesel engine, conducted by Diesel Research, Inc., established that oil migration resulting from a 40% efficient breather cost 6% of engine output. The engine in question had 4000 hours on the clock.

Exhaust system

Restrictions downstream of the turbine—crimped pipes, abrupt changes in direction, clogged mufflers—reduce turbo efficiency and, in extreme cases, represent enough load to induce boost. However, exhaust leaks between the engine and turbocharger are more typical. The pipe rusts out (especially if wrapped in insulation), fatigues, or cracks under the 1000°F-plus temperature. If OEM parts do not give satisfactory service, you might try calling a Flexonics applications engineer (312-837-1811). The company makes a line of exhaust tubing, fitted with metal bellows to absorb thermal expansion.

Turbocharger inspection

The seven-step inspection procedure outlined here was adapted, with modifications, from material supplied by John Deere.

1. *Turbo housing* Before disconnecting the oil lines, examine external surfaces of the housing for oil leaks, which would almost certainly mean turbo seal failure.
2. *Compressor housing inlet and wheel* Inspect the compressor wheel for erosion and impact damage. Erosion comes about because of dust intrusion; impact damage is prima facie evidence of negligence. Carefully examine the housing ID and compressor blade tips for evidence of rubbing contact, which means bearing failure.
3. *Compressor housing outlet* Check the compressor outlet for dirt, oil, and carbon accumulations. Dirt points to a filtration failure; oil suggests seal failure, although other possibilities exist, such as clogged turbo-oil return line or crankcase breather. Carbon on the compressor wheel might suggest some sort of combustion abnormality, but the phenomenon is also seen on healthy engines. I can only speculate about the cause.
4. *Turbine housing inlet* Inspect the inlet ports for oil, heavy carbon deposits, and erosion. Any of these symptoms suggest an engine malfunction.
5. *Turbine housing outlet and wheel* Examine the blades for impact damage. Look for evidence of rubbing contact between the turbine wheel and the housing, which would indicate bearing failure.
6. *Oil return port* The shaft is visible on most turbochargers from the oil return port. Excessive bluing or coking suggests lubrication failure, quite possibly caused by hot shutdowns.
7. *Bearing play measurements* Experienced mechanics determine bearing condition by feel, but use of a dial indicator gives more reliable results. Note that measurement of radial, or side-to-side, bearing clearance involves moving the shaft from one travel extreme to another, 180° away (Fig. 9-8A). Hold the shaft level during this operation, because a rocking motion would muddy the results. Axial motion is measured as travel between shaft thrust faces (Fig. 9-8B). In very general terms, subject to correction by factory data for the unit in question, we would be comfortable with 0.002 in. radial and 0.003 in. axial play. Note that Schwitzer and small foreign types tend to be set up tighter.

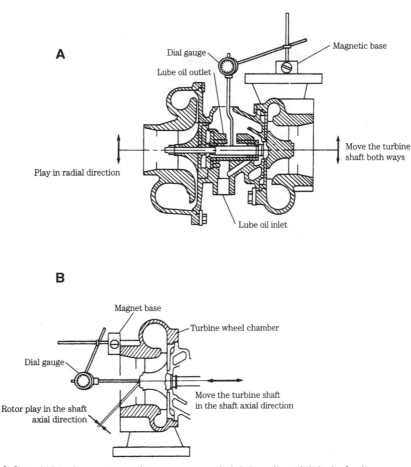

9-8 A dial indicator is used to measure radial (A) and axial (B) shaft play, specified as total indicator movement. Take off the radial measurement at a point near the center of the shaft, with shaft held level throughout its range of travel.
Yanmar Diesel Engine Co., Ltd.

Overhaul

Only the most general instructions can be provided here, because construction details, wear limits, and torque specifications vary between make and model. It should also be remarked that American mechanics do not, as a rule, attempt turbocharger repairs. The defective unit is simply exchanged for another one.

However, there are no mysteries or secret rites associated with turbocharger work. Armed with the necessary documentation and the one indispensable special tool—a turbine-shaft holding fixture—any mechanic can replace the bearings and seals, which is what the usual field overhaul amounts to.

The holding fixture secures the integral turbine wheel and shaft during removal and installation of the compressor nut. Figure 9-9 illustrates plans for one such fixture, used on several Detroit Diesel applications. Dimensions vary, of course, with turbocharger make and model.

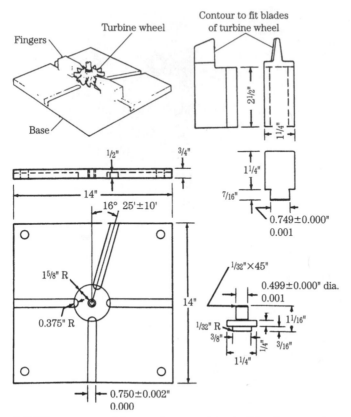

9-9 Whenever possible, purchase a turbine-holding fixture rather than attempting to fabricate one. Dimensions are critical. Detroit Diesel

Factory-supplied documentation should alert you to any peculiarities of the instrument, such as a left-handed compressor-nut thread or the need to heat the compressor wheel prior for removal. Special precautions include the following:

- Do not use a wire wheel or any other sort of metallic tool on turbo wheels or shaft. Remove carbon deposits with a soft plastic scraper and one of the various solvents sold for this purpose. Light scratches left by wheel contact on the turbo housings can be polished out with an abrasive.
- Do not expose the turbine shaft to bending forces of any magnitude. Pull the shaft straight out of its bearings. Use a double u-joint between the socket and the wrench when removing and torquing the compressor-wheel nut.
- Do send out the rotating assembly for shaft alignment and balancing (Fig. 9-10).
- Do exercise extreme cleanliness during all phases of the operation.
- Do prelubricate the bearings with motor oil and prefill the bearing housing prior to starting the engine.
- Do make certain that all fasteners and small tools are accounted for and not lurking within the turbocharger or its plumbing.

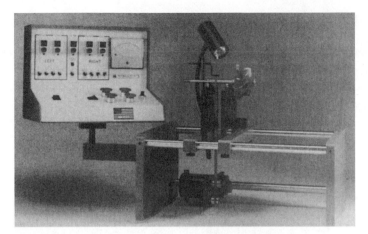

9-10 Heins Balancing Systems makes specialized tools for turbocharger rebuilders, including this precision balancer.

Variable geometry turbine

Variable geometry turbine (VGT) turbochargers include a throttling mechanism on the exhaust inlet to generate boost at engine speeds just above idle. With boost comes torque, which results in low-rpm flexibility unmatched since the days of steam. But torque is a side-effect—VGT is an emissions-control expedient, intended to reduce NOx levels in the exhaust.

Until 2004, when EPA Tier 2 regulations went into effect, EGR was required only at high speeds, when exhaust-gas pressure exceeded manifold pressure. All that was needed was a connection between the exhaust pipe and the intake manifold, and exhaust flow would take care of itself. Under the present rules, EGR must be available across the rpm band.

A conventional turbocharger, sized for maximum power output, cannot provide the low-speed boost necessary for exhaust gases to overcome manifold pressure. BMW uses two conventional turbochargers—a small unit that develops boost just off idle and a larger turbocharger for maximum power production. Other manufacturers get around the manifold-pressure problem by throttling the intake with a butterfly valve. At low speeds, the valve closes, creating a vacuum behind it. But this expedient costs power.

The VGR generates boost at all speeds. Figure 9-11 illustrates the construction of adjustable-vane type used on motor vehicles. Pivoted vanes (3), controlled by movement of the adjuster ring (4), surround the turbine housing. At low speeds, the vanes swing closed like a Venetian blind. This flow obstruction increases the velocity of the exhaust gases striking the turbine wheel. The vanes also direct the gas stream toward the outer edge of the wheel for improved leverage. Exhaust-gas volume and energy increase with engine speed. At high speeds, the control vanes swing open to take advantage of the energy now available. In their full-open position, control vanes also function as a boost limiter, to prevent turbocharger overspeeding.

Variable geometry turbine systems are closely monitored for boost, exhaust backpressure, and compressor inlet and outlet temperature, so that failure generates one

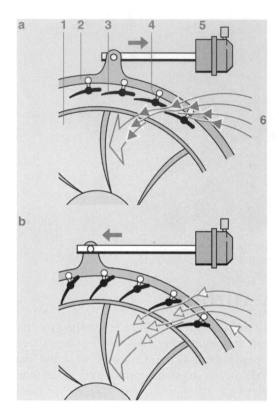

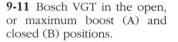

9-11 Bosch VGT in the open, or maximum boost (A) and closed (B) positions.

or more trouble codes. European automobiles control vane pitch and turbo boost with a stepper motor. U.S. heavy truck VGTs are controlled by lube-oil or air pressure acting on a spring-loaded diaphragm. A rod transfers diaphragm movement to the turbo adjustment ring. Replacement actuators include instructions for adjustment, an operation that requires a scanner capable of cycling the unit and a precisely regulated source of pressure. Fig. 9-12 shows a Detroit Diesel air-operated VGT actuator.

9-12 Detroit Diesel air-operated VGT actuator.

10
CHAPTER

Electrical fundamentals

Most small diesel engines are fitted with an electric starter, battery, and generator. The circuit might include glow plugs for cold starting and electrically operated instruments such as pyrometers and flow meters. The diesel technician should have knowledge of electricity.

This knowledge cannot be gleaned from the hardware just looking at an alternator will not tell you much about its workings. The only way to become even remotely competent in electrical work is to have some knowledge of basic theory. This chapter is a brief, almost entirely nonmathematical, discussion of the theory.

Electrons

Atoms are the building blocks of all matter. These atoms are widely distributed; if we enlarged the scale to make atoms the size of pinheads, there would be approximately one atom per cubic yard of nothingness. But as tiny and as few as they are, atoms (or molecules, which are atoms in combination) are responsible for the characteristics of matter. The density of a substance, its chemical stability, thermal and electrical conductivity, color, hardness, and all its other characteristics are fixed by the atomic structure.

The atom is composed of numerous subatomic particles. Using high-energy disintegration techniques, scientists are discovering new particles almost on an annual basis. Some are reverse images of the others; some exist for only a few millionths of a second. But, we are only interested in the relatively gross particles whose behavior has been reasonably well understood for generations.

In broad terms the atom consists of a nucleus and one or more electrons in orbit around it. The nucleus has at least one positively charged *proton* and might have one or more electrically neutral *neutrons*. These particles make up most of the atom's mass. The orbiting *electrons* have a negative charge. All electrons, as far as we know, are identical. All have the same electrical potency. Their orbits are balanced by centripetal force and the pull of the positive charge of the nucleus (Fig. 10-1).

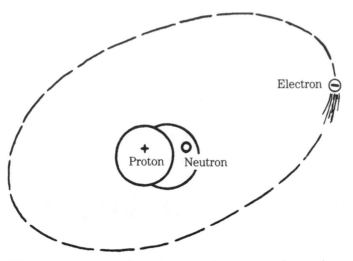

10-1 Representation of an electron orbiting around a nucleus consisting of a proton and a neutron.

A fundamental electrical law states unlike charges attract, like charges repel. Thus two electrons, both carrying negative charges, repel each other. Attraction between opposites and the repulsion of likes are the forces that drive electrons through circuits.

Circuits

The term *circuit* comes from a Latin root meaning *circle.* It is descriptive because the path of electrons through a conductor is always in the form of a closed loop, bringing them back where they started.

A battery or generator has two *terminals,* or posts. One terminal is negative. It has a surplus of electrons available. The other terminal has a relative scarcity of electrons and is positive.

A *circuit* is a pathway between negative and positive terminals. The relative ease with which the electrons move along the circuit is determined by several factors. One of the most Important is the nature of the conductor. Some materials are better conductors than others. Silver is at the top of the list, followed by copper, aluminum, iron, and lead. These and other conductors stand in contrast to that class of materials known as insulators. Most gases, including air, wood, rubber, and other organic materials, most plastics, mica, and slate are insulators of varying effectiveness.

Another class of materials is known as *semiconductors.* They share the characteristics-of both conductors and insulators and under the right conditions can act as either. Silicon and germanium semiconductors are used to convert alternating current (AC) to direct current (DC) at the alternator and to limit current and voltage outputs. These devices are discussed in more detail later in this chapter.

The ability of materials to pass electrons depends on their atomic structures. Electrons flood into the circuit from the negative pole. They encounter other electrons already in orbit. The newcomers displace the orbiting electrons (like repels like) toward the positive terminal and, in turn, are captured by the positive charge on the nucleus (unlike attracts). This game of musical chairs continues until the circuit is broken or until the voltages on the terminals equalize. Gold, copper, and other conductors give up their captive electrons with a minimum of fuss. Insulators hold their electrons tightly in orbit and resist the flow of current.

You should not have the impression that electrons flow through various substances in a go/no-go manner. All known materials have some reluctance to give up their orbiting electrons. This reluctance is lessened by extreme cold, but never reaches zero. And all insulators 'leak' to some degree. A few vagrant electrons will pass through the heaviest, most inert insulation. In most cases the leakage is too small to be significant. But it exists and can be accelerated by moisture saturation and by chemical changes in the insulator.

Circuit characteristics

Circuits, from the simplest to the most convoluted, share these characteristics:
- Electrons move from the negative to the positive terminal. Actually the direction of electron movement makes little difference. In fact, for many years it was thought that electrons moved in the opposite direction—from positive to negative.
- The circuit must be complete and unbroken for electrons to flow. An incomplete circuit is described as *open*. The circuit might be opened deliberately by the action of a switch or a fuse, or it might open of its own accord as in the case of a loose connection or a broken wire. Because circuits must have some rationale besides the movement of electrons from one pole to the other, they have *loads*, which convert electron movement into heat, light, magnetic flux, or some other useful quantity. These working elements of the circuit are also known as *sinks*. They absorb, or sink, energy and are distinguished from the sources (battery and the generator). A circuit in which the load is bypassed is described as *shorted*. Electrons take the easiest, least resistive path to the positive terminal.
- Circuits can feed single or multiple loads. The arrangement of the loads determines the circuit type.

Series and parallel circuits

A series circuit has its loads connected one after the other like beads on a string (Fig. 10-2A). There is only one path for the electrons between the negative and positive terminals. In a string of Christmas tree lights, each lamp is rated at 10V; 11 in. series required 110V. If any lamp failed, the string went dark.

Pure series circuits are rare today. (They can still be found as filament circuits in some AC/DC radios and in high-voltage applications such as airport runway lights.) But switches, rheostats, relays, fuses, and other circuit controls are necessarily in series with the loads they control.

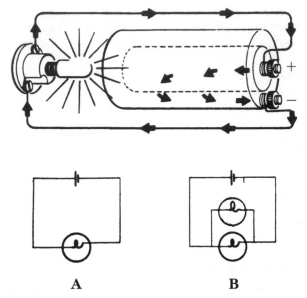

A   **B**

10-2 Basic series circuit (top) and its schematic (A). In
B, a second lamp has been added with the first.

Parallel circuits have the loads arranged like the rungs of a ladder, to provide
multiple paths for current (Fig. 10-2). When loads are connected in parallel, we say
they are *shunt.* Circuits associated with diesel engines usually consist of parallel
loads and series control elements. The major advantage of the parallel arrangement
from a serviceman's point of view is that a single load can open without affecting
the other loads. A second advantage is that the voltage remains constant throughout
the network.

Single- and two-wire circuits

However the circuit is arranged—series, parallel, or in some combination of
both—it must form a complete path between the positive and negative terminals of
the source. This requirement does not mean the conductor must be composed
entirely of electrical wire. The engine block, transmission, and mounting frame are
not the best conductors from the point of view of their atomic structures. But because
of their vast cross-sectional area, these components have almost zero resistance.

In the single-wire system the battery is *grounded*, or as the British say, *earthed*
(Fig. 10-3). These terms seem to have originated from the power station practice of
using the earth as a return. With the exception of some Lucas and CAV systems, most
modem designs have the negative post grounded. The 'hot' cable connects the positive post to the individual loads, which are grounded. Electrons flow from the negative terminal, through the loads, and back to the battery via the wiring.

The single-wire approach combines the virtues of simplicity and economy. But
it has drawbacks. Perhaps the most consequential is the tendency for the connections to develop high resistances. One would think that a heavy strap bolted to the

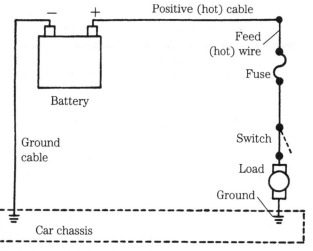

10-3 The principle of grounding.

block would offer little resistance. Unfortunately this is not the case. A thin film of oil, rust, or a loose connection is enough to block the flow of current.

Corrosion problems are made more serious by weather exposure and *electrolysis*. Electrolysis is the same phenomenon as occurs during electroplating. When current passes through two dissimilar metals (e.g., copper and cast iron) in the presence of damp air, one of the metals tends to disintegrate. In the process it undergoes chemical changes that makes it a very poor conductor.

Another disadvantage of the single-wire system, and the reason it is not used indiscriminately on aircraft and ocean-going vessels, is the danger of short circuits. Contact between an uninsulated hot wire and the ground will shunt the loads down circuit.

The *two-wire system* (one wire to the load, and a second wire from it to the positive terminal) is preferred for critical loads and can be used in conjunction with a grounded system. Electrolytic action is almost nil, and shorts occur only in the unlikely event that two bare wires touch.

Electrical measurements

Voltage is a measure of electrical pressure. In many respects, it is analogous to hydraulic pressure. The unit of voltage is the volt, abbreviated V. Thus we speak of a 12-V or 24-V system. Prefixes, keyed to the decimal system, expand the terms so we do not need to contend with a series of zeros. The prefix K stands for *kilo*, or 1000. A 50-kV powerline delivers 50,000V. At the other end of the scale, *milli* means $\frac{1}{1000}$, and one millivolt (1 mV) is a thousandth of a volt.

The *ampere*, shortened sometimes to *amp* and abbreviated *A* is a measure of the quantity of electrons flowing past a given point in the circuit per second. One ampere represents the flow of 6.25×10^{18} electrons per second. Amperage is also referred to as *current intensity* or *quantity*. From the point of view of the loads on the circuit, the amperage is the *draw*. A free-running starter motor might draw 100A, and three times as much under cranking loads.

Resistance is measured in units named after G. S. Ohm. Ohms are expressed by the last letter of the Greek alphabet, omega (Ω). Thus we might speak of a 200Ω resistance.

The resistance of a circuit determines the amount of current that flows for a given applied voltage. The resistance depends on the atomic structure of the conductor—how tightly the electrons are held captive in their orbits—and on certain physical characteristics. The broader the cross-sectional area of the conductor, the lesser opposition to the current. And the longer the path formed by the circuit between the poles of the voltage source, the more is the resistance. Think of these two dimensions in terms of ordinary plumbing. The *resistance* to the flow of a liquid in a pipe is inversely related to its diameter (decreases as diameter increases) and directly related to its length. Resistance in the pipe produces heat, exactly as does resistance in an electrical conductor.

Resistance is generally thought of as the electrical equivalent of friction—a kind of excise tax that we must pay to have electron movement. There are, however, positive uses of resistance. Resistive elements can be deliberately introduced in the circuit to reduce current in order to protect delicate components. The heating effect of resistance is used in soldering guns and irons and in the glow plugs employed as starting aids in diesel engines.

Ohm's law

The fundamental law of simple circuits was expressed by the Frenchman G. S. Ohm in the early 1800s in a paper on the effects of heat on resistance. The law takes three algebraic forms, each based on the following relationship: a potential of 1V drives 1A through a resistance of 1Ω. In the equations the symbol E stands for *electromotive force*, or, as we now say, volts. An I represents *intensity*, or current, and R is resistance in ohms.

The basic relationship is expressed as:

$$I = \frac{E}{R}$$

This form of the equation states that current in amperes equals voltage in volts divided by resistance in ohms. If a circuit with a resistance of 6Ω, is connected across a 12V source, the current is 2A (12/6 = 2). Double the voltage (or halve the resistance), and the current doubles. Figure 10-4 shows this linear (straight-line graph) relationship.

Another form of Ohm's law is

$$R = \frac{E}{I}$$

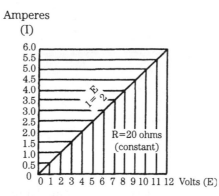

10-4 Relation between amperage and voltage, resistance constant.

or resistance equals voltage divided by current. The 12V potential of our hypothetical circuit delivers 2A, which means the resistance is 6Ω. The relationship between amperage and resistance with a constant voltage is illustrated in Fig. 10-5.

Another way of expressing Ohm's law is

$$E = IR$$

or voltage equals current multiplied by resistance. Two amperes through 6Ω requires a potential of 12V. Double the resistance, and twice the voltage is required to deliver the same amount of current (Fig. 10-6).

Various memory aids have been devised to help students remember Ohm's law. One involves an Indian, an eagle, and a rabbit. The Indian (I) sees the eagle (E) flying over the rabbit (R); this gives the relationship $I = E/R$. The eagle sees both the Indian and the rabbit in the same level, or $E = IR$. And the rabbit sees the eagle over the Indian, or $R = E/I$. A visual aid is shown in Fig. 10-7. The circle is divided into three segments. To determine which form of the equation to use, put your finger on the quantity you want to solve for.

All of this might seem academic, and in truth, few mechanics perform calculations with Ohm's law. It is usually easier to measure all values directly with a meter. Still, it is very important to have an understanding of Ohm's law. It is the best description of simple circuits we have.

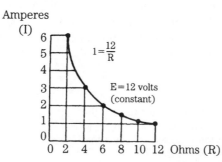

10-5 Relation between amperage and resistance, voltage constant.

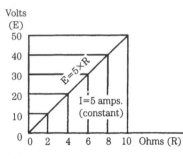

10-6 Relation between voltage and resistance, amperage constant.

For example, suppose the wiring is *shorted:* electrons have found a more direct (shorter) path to the positive terminal of the battery. Ohm's law tells us certain facts about the nature of shorts, which are useful in troubleshooting. First, a short increases the current in the affected circuit, because current values respond inversely to resistance. This increase will generate heat in the conductor and might even carbonize the insulation. Voltage readings will be low because the short has almost zero resistance. Now suppose we have a partially open circuit caused by a corroded terminal. The current through the terminal will be reduced, which means that the lights or whatever other load is on the circuit will operate at less-than-peak output. The terminal will be warm to the touch, because current is transformed to heat by the presence of resistance. Voltage readings from the terminal to ground will be high on the source side of the resistance and lower than normal past the bottleneck.

Direct and alternating current

Diesel engines might employ direct or alternating currents. The action of *direct,* or *unidirectional,* current can be visualized with the aid of the top drawing in Fig. 10-8. *Alternating current* is expressed by the opposed arrows in the lower drawing. Flow is from the negative to the positive poles of the voltage source, but the poles exchange identities, causing the current reversal. The positive becomes negative, and the negative becomes positive.

The graph in Fig. 10-9 represents the rise, fall, and reversal of alternating current. Because the amplitude changes over time, alternating voltage and current values require some qualification. The next drawing illustrates the three values most often used.

Peak-to-peak values refer to the maximum amplitude of the voltage and amperage outputs in both directions. In Fig. 10-10 the peak-to-peak value is 200 V_{p-p} (or 200 A_{p-p}). The half-cycle (alternation) on top of the zero reference line is considered to be positive; below the line alternation is negative.

10-7 Ohm's law: Cover the unknown with your finger to determine which of the three equations to use.

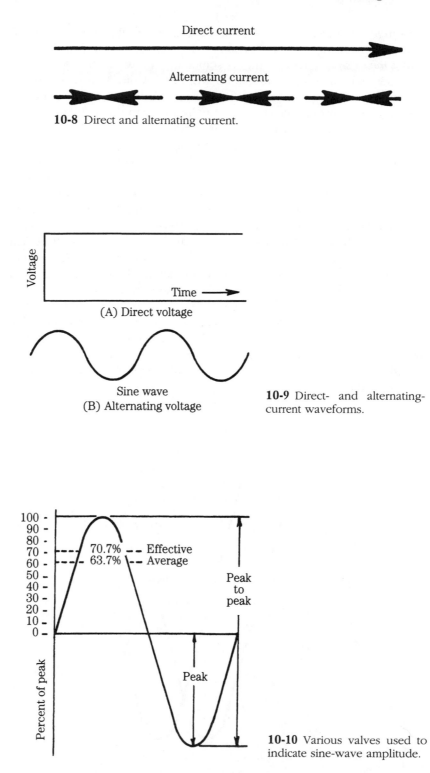

Direct current

Alternating current

10-8 Direct and alternating current.

(A) Direct voltage

Sine wave
(B) Alternating voltage

10-9 Direct- and alternating-current waveforms.

70.7% -- Effective
63.7% --- Average

Peak
to
peak

Peak

Percent of peak

10-10 Various valves used to indicate sine-wave amplitude.

The *average,* or *mean,* value represents an average of all readings. In Fig. 10-10 it is 63.7 units.

The *root mean square* (rms) value is sometimes known as the *effective* value. One rms ampere has the same potential for work as 1-A direct current. Unless otherwise specified, alternating current values are *rms* (effective) values and are directly comparable to direct current in terms of the work they can do. Standard meters are equipped with appropriate scales to give rms readings.

The illustration depicts one complete cycle of alternating current. The number of cycles completed per second is the frequency of the current. One cycle per second is the same as one *hertz* (1 Hz). Domestic household current is generated at 60 Hz. The alternating current generators used with diesel engines are variable-frequency devices because they are driven by the engine crankshaft. At high speed a typical diesel alternating current generator will produce 500–600 Hz.

In addition to these special characteristics, alternating current outputs can be superimposed upon each other (see Fig. 10-11). The horizontal axis (X-axis) represents zero output—it is the crossover point where alternations reverse direction. In this particular alternator 360° of rotation of the armature represents a complete output cycle from maximum positive to maximum negative and back to maximum positive. The X-axis can be calibrated in degrees or units of time (assuming that the alternator operates at a fixed rpm). Note that there are two sine waves shown. These waves are out of phase; E_1 leads E_2 by 90° of alternator rotation. Alternators used on diesel installations usually generate three output waves 120° apart. Multiphase outputs are smoother than single-phase outputs and give engineers the opportunity

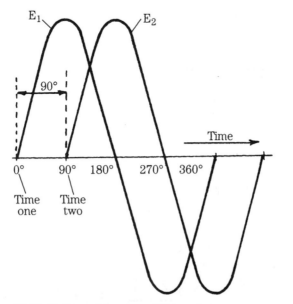

10-11 Voltage sine waves 90° apart.

to build multiple charging circuits into the alternator. Circuit redundancy gives some assurance against total failure; should one charging loop fail, the two others continue to function.

Magnetism

Electricity and magnetism are distinct phenomena, but related in the sense that one can be converted into the other. Magnetism can be employed to generate electricity, and electricity can be used to produce motion through magnetic attraction and repulsion.

The earth is a giant magnet with magnetic poles located near the geographic poles (Fig. 10-12). Magnetic *fields* extend between the poles over the surface of the earth and through its core. The field consists of *lines of force*, or *flux*. These lines of force have certain characteristics, which, although they do not fully explain the phenomenon, at least allow us to predict its behavior. The lines are said to move from the north to the south magnetic pole just as electrons move from a negative to a positive electrical pole. Magnetic lines of force make closed loops, circling around and through the magnet. The poles are the interface between the internal and external paths of the lines of force. When encountering a foreign body, the lines of force tend to stretch and snap back upon themselves like rubber bands. This characteristic is important in the operation of generators.

Lines of force penetrate every known substance, as well as the emptiness of outer space. However, they can be deflected by soft iron. This material attracts and focuses flux in a manner analogous to the action of a lens on light. Lines of force 'prefer' to travel through iron, and they digress to take advantage of the *permeability* (magnetic conductivity) of iron (Fig. 10-13).

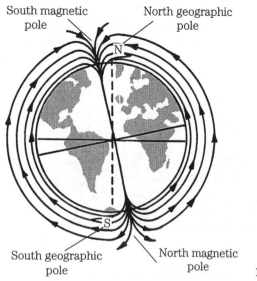

South magnetic pole North geographic pole

South geographic pole North magnetic pole

10-12 Earth's magnetic field.

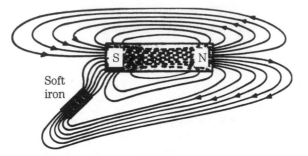

10-13 Effects of soft iron on a magnetic field.

The permeability of iron is exploited in almost all magnetic machines. Generators, motors, coils, and electromagnets all have iron cores to direct the lines of force most efficiently.

When free to pivot, a magnet aligns itself with a magnetic field, as shown in Fig. 10-14. *Unlike magnetic poles attract, like magnetic poles repel.* The south magnetic pole of the small magnet (compass needle) points to the north magnetic pole of the large bar magnet.

Electromagnets

Electromagnets use electricity to produce a magnetic field. Electromagnets are used in starter solenoids, relays, and voltage-and current-limiting devices. The principle upon which these magnets operate was first enunciated by the Danish researcher Hans Christian Oersted. In 1820, he reported a rather puzzling phenomenon: A compass needle when placed near a conductor, deflected as the circuit was made and broken. Subsequently it was shown that the needle reacted because of the presence of a magnetic field at right angles to the conductor. The field exists as long as current flows, and its intensity is directly proportional to the amperage.

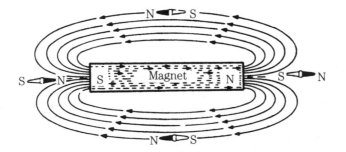

Bar magnet

10-14 Compass deflections in a magnetic field.

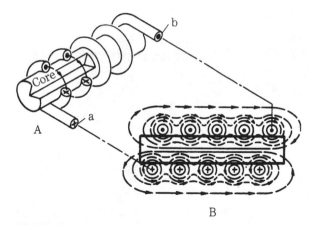

10-15 Electromagnet construction showing how magnetic fields mutually reinforce one another.

The field around a single strand of conductor is too weak to have practical application. But if the conductor is wound into a coil, the weak fields of the turns reinforce each other (Fig- 10-15). Further reinforcement can be obtained by inserting an iron bar inside the coil to give focus to the lines of force. The strength of the electromagnet depends on the number of turns of the conductor, the length/width ratio of the coil, the current strength, and the permeability of the core material.

Voltage sources

Starting and charging systems employ two voltage sources. The generator is electromagnetic in nature and is the primary source. The battery, which is electrochemical in nature, is carried primarily for starting. Also, in case of overloads, the battery can send energy into the circuit.

Generator principles

In 1831 Michael Faraday reported that he had induced electricity in a conductor by means of magnetic action. His apparatus consisted of two coils, wound over each other but not electrically connected. When he made and broke the circuit to one coil, momentary bursts of current were induced in the second. Then Faraday wound the coils over a metal bar and observed a large jump in induced voltage. Finally he reproduced the experiment with a permanent magnet. Moving a conductor across a magnetic field produced voltage. The intensity of the voltage, depended on the strength of the field, the rapidity of movement, and the angle of movement. His generator was most efficient when the conductor cut the lines of force at right angles.

The rubber band effect mentioned earlier helps one to visualize what happens when current is induced in the conductor. A secondary magnetic field is set up,

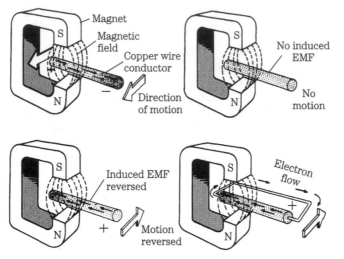

10-16 Voltage produced by magnetism.

which resists the movement of the conductor through the primary field. The greater the amount of induced current, the greater the resistance, and the more power required to overcome it.

Alternator

Figure 10-16 illustrates Faraday's apparatus. Note that there must be relative movement between the magnetic field and conductor. Either can be fixed as long as one can move. Note also that the direction of current is determined by the direction of movement. Of course, this shuttle generator is hardly efficient; the magnet or conductor must be accelerated, stopped, and reversed during each cycle.

The next step was to convert Faraday's laboratory model to rotary motion (Fig. 10-17). The output alternates with the position of the armature relative to the fixed magnets. In Fig. 10-17A the windings are parallel to the lines of force and the output is zero. Ninety degrees later the output reaches its highest value because the armature windings are at right angles to the field. At 180° of rotation (C) the output is again zero. This position coincides with a polarity shift. For the remainder of the cycle, the movement of the windings relative to the field is reversed.

To determine the direction of current flow from a generator armature, position the left hand so that the thumb points in the direction of current movement and the forefinger points in the direction of magnetic flux (from the north to the south pole). Extend the middle finger 90° from the forefinger; it will point to the direction of current. As convoluted as the lefthand rule for generators sounds, it testifies to the fact that polarity shifts every 180° of armature rotation. This is true of all dynamos, whether the machine is DC generator or AC generator (alternator).

The frequency in hertz depends on the rotational speed of the armature and on the number of magnetic poles. Figure 10-18 illustrates a four-pole (two-magnet) alternator, which for a given rpm, has twice the frequency of the single-pole machine in

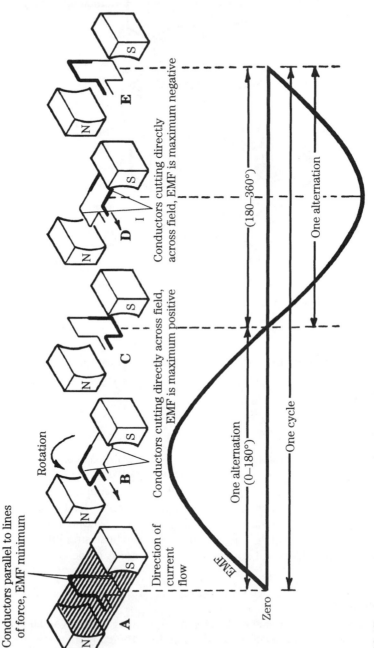

Conductors parallel to lines of force, EMF minimum

Rotation

Direction of current flow

A

B

Conductors cutting directly across field, EMF is maximum positive

C

D

Conductors cutting directly across field, EMF is maximum negative

E

One alternation (0–180°)

(180–360°)

One cycle

One alternation

EMF

Zero

10-17 Alternating current generator (or alternator).

287

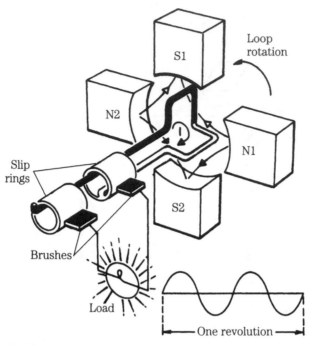

10-18 Four-pole alternator.

the previous illustration. Normally, alternators supplied for diesel engines have four magnetic poles and are designed to peak at 6000 rpm. Their frequency is given by:

$$f = \frac{P \times \text{rmp}}{120}$$

where *f* is frequency in hertz and P is the number of poles. Thus, at 6000 rpm, a typical alternator should deliver current at a frequency of 2000 Hz.

The alternators depicted in Figs. 10-17 and 10-18 are simplifications. Actual alternators differ from these drawings in two important ways. First, the fields are not fixed, but rotate as part of an assembly called the *rotor*. Secondly, the fields are not permanent magnets; instead they are electromagnets whose strength is controlled by the voltage regulator. Electrical connection to the rotor is made by a pair of brushes and slip rings.

DC generator

Direct-current (DC) generators develop an alternating current (AC) that is mechanically rectified by the commutator—brush assembly. The commutator is a split copper ring with the segments insulated from each other. The brushes are carbon bars that are spring-loaded to bear against the commutator. The output is pulsating direct current, as shown in Fig. 10-19. The pulse frequency depends on the number of armature loops,

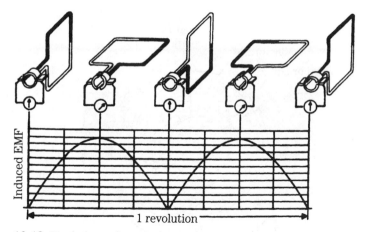

10-19 Single-loop direct current generator.

field poles, and rpm. A typical generator has 24–28 armature loops and 4 poles. The fields are electromagnets and are energized by 8–12% of the generator output current. The exact amount of current detailed for this *excitation* depends on the regulator, which in turn, responds to the load and state of charge of the generator. A few surviving generators achieve output regulation by means of a third brush.

Third-brush generator

The third brush is connected to the field coils as shown in Fig. 10-20. At low rotational speeds the magnetic lines of force bisect the armature in a uniform manner. But as speed and output increase, the field distorts. The armature generates its own field, because a magnetic field is created at right angles to a conductor when current flows through it. The resulting field distortion, sometimes called *magnetic whirl*, places the loops feeding the third brush in an area of relative magnetic weakness. Consequently, less current is generated in these loops, and the output to the fields is lessened. The fields become correspondingly weak, and the total generator output as defined by the negative and positive brushes remains stable or declines.

The output depends on the position of the third brush, which can be moved in or out of the distorted field to adjust output for anticipated loads. Moving the brush in the direction of armature rotation increases the output; moving it against the direction of rotation reduces the output. When operated independently from the external circuit, the brushes must be grounded to protect the windings. Many of these generators have a fuse in series with the fields.

dc motors

One of the peculiarities of a direct current generator is that it will 'motor' if the brushes are connected to a voltage source. In like manner a direct current motor will "gen" if the armature is rotated by some mechanical means. Some manufacturers of

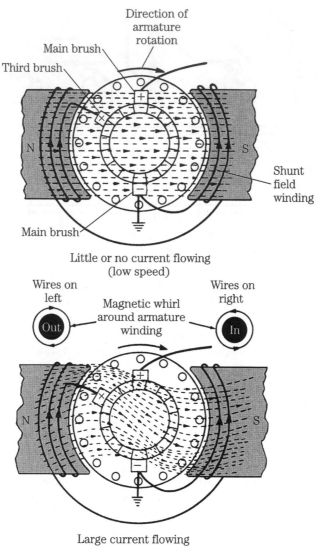

10-20 Third-brush generator. A speed and output increase, the field's excitation current drops because of magnetic whirl.

small-engine accessories have taken advantage of this phenomenon to combine both functions in a single housing. One such system is the *Dynastart* system employed on many single-cylinder Hatz engines (shown schematically in Fig. 10-21).

Figure 10-22 illustrates the motor effect. In the top sketch the conductor is assumed to carry an electric current toward you. The magnetic flux around it travels in a clockwise direction. Magnetic lines of force above the conductor are distorted and stretched. Because the lines of force have a strong elastic tendency to shorten, they push against the conductor. Placing a loop of wire in the field (bottom sketch)

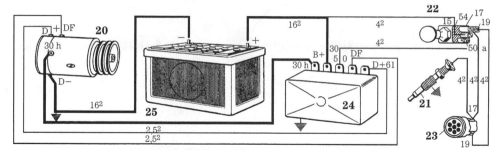

10-21 Combination generator and starter motor. Teledyne Wisconsin Motor

instead of a single conductor doubles the motor effect. Current goes in the right half of the loop and leaves at the left. The interaction of the fields causes the loop (or, collectively, the armature) to turn counterclockwise. Reversing the direction of the current would cause torque to be developed in a clockwise direction.

Starter motors are normally wired with the field coils in series with the armature (Fig. 10-23). Any additional load added to a series motor will cause more current in the armature and correspondingly more torque. Because this increased current must pass through the series field, there will be a greater flux. Speed changes rapidly with load. When the rotational speed is low, the motor produces its maximum torque. A starter might draw 300A during cranking, more than twice that figure at stall.

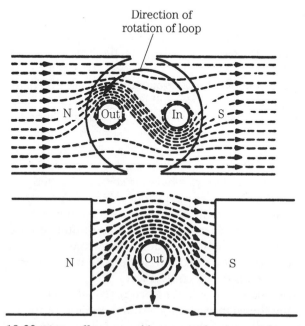

10-22 Motor effect created by current bearing conductor in a magnetic field.

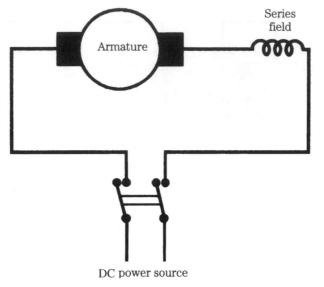

10-23 Series field and armature connections typical of starter motors.

Storage batteries

Storage batteries accumulate electrical energy and release it on demand. The familiar lead-acid battery was invented more than 100 years ago by Gaston Plante. It suffers from poor energy density (watt-hours per pound) and poor power density (watts per pound). The average life is said to be in the neighborhood of 360 complete charge-discharge cycles. During charging, the lead-acid battery shows an efficiency of about 75%; that is, only three-quarters of the input can be retrieved.

Yet it remains the only practical alternative for automotive, marine, and most stationary engine applications. Sodium-sulfur, zinc-air, lithium-halide, and lithium-chlorine batteries all have superior performance, but are impractical by reason of cost and, in some cases, the need for complex support systems.

The lead-acid battery consists of a number of cells (hence the name *battery*) connected in series. Each fully charged cell is capable of producing 2.2V. The number of cells fixes the output; 12V batteries have six cells; 24V batteries have 12. The cells are enclosed in individual compartments in a rubberoid or (currently) high-impact plastic case. The compartments are sealed from each other and, with the exception of Delco and other "zero maintenance" types, open to the atmosphere. The lower walls of the individual compartments extend below the plates to form a sediment trap. Filler plugs are located on the cover and can be combined with wells or other visual indicators to monitor the electrolyte level.

The cells consist of a series of lead plates (Fig. 10-24) connected by internal straps. Until recently the straps were routed over the top of the case, making convenient test points for the technician. Unfortunately, these straps leaked current and were responsible for the high self-discharge rates of these batteries.

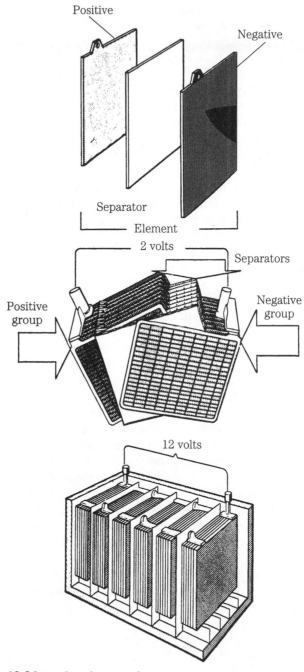

10-24 Lead-acid storage battery construction.

The plates are divided into positive and negative groups and separated by means of plastic or fiberglass sheeting. Some very large batteries, which are built almost entirely by hand, continue to use fir or Port Orford cedar separators. A few batteries intended for vehicular service feature a loosely woven fiberglass padding between the separators and positive plates. The padding gives support to the lead filling and reduces damage caused by vibration and shock.

Both sets of plates are made of lead. The positive plates consist of a lead grid-work that has been filled with lead oxide paste. The grid is stiffened with a trace of antimony. Negative plates are cast in sponge lead. The plates and separators are immersed in a solution of sulfuric acid and distilled water. The standard proportion is 32% acid by weight.

The level of the electrolyte drops in use because of evaporation and hydrogen loss. (Sealed batteries have vapor condensation traps molded into the roof of the cells.) It must be periodically replenished with distilled water. Nearly all storage batteries are shelved dry, and filled upon sale. Once the plates are wetted, electrical energy is stored in the form of chemical bonds.

When a cell is fully charged the negative plates consist of pure sponge lead (Pb in chemical notation), and the positive plates are lead dioxide (PbO_2, sometimes called lead peroxide). The electrolyte consists of water (H_2O) and sulfuric acid (H_2SO_4). The fully charged condition corresponds to drawing A in Fig. 10-25. During discharge both sponge lead and lead dioxide become lead sulfate ($PbSO_4$). The percentage of water in the electrolyte increases because the SO_4 radical splits off from the sulfuric acid to combine the plates. If it were possible to discharge a lead-acid battery completely, the electrolyte would be safe to drink. In practice, batteries cannot be completely discharged in the field. Even those that have sat in junkyards for years still have some charge.

During the charge cycle the reaction reverses. Lead sulfate is transformed back into lead and acid. However, some small quantity of lead sulfate remains in its crystalline form and resists breakdown. After many charge-discharge cycles, the residual sulfate permanently reduces the battery's output capability. The battery then is said to be *sulfated*.

The rate of self-discharge is variable and depends on ambient air temperature the cleanliness of the outside surfaces of the case, and humidity. In general it averages 1% a day. Batteries that have discharged below 70% of full charge might sulfate. Sulfation becomes a certainty below the 70% mark. In addition to the possibility of damage to the plates, a low charge brings the freezing point of the electrolyte to near 32°F.

Temperature changes have a dramatic effect on the power density. Batteries are warm-blooded creatures and perform best at room temperatures. At 10°F the battery has only half of its rated power.

Switches

Switches neither add energy to the circuit nor take it out. Their function is to give flexibility by making or breaking circuits or by providing alternate current paths. They are classified by the number of *poles* (movable contacts) and *throws* (closed positions). Figure 10-26 illustrates four standard switch types in schematic form.

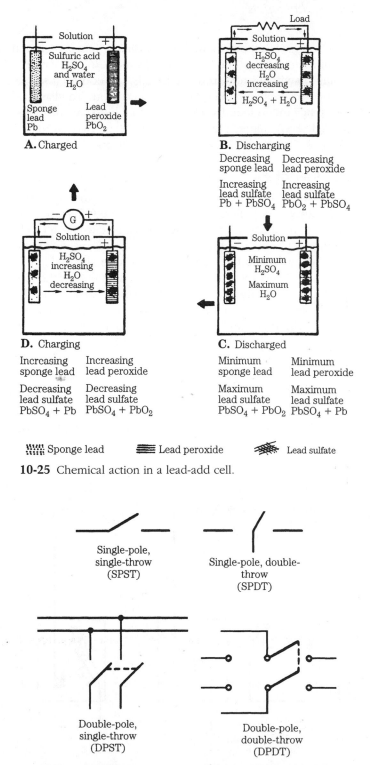

A. Charged

B. Discharging

Decreasing sponge lead   Decreasing lead peroxide

Increasing lead sulfate $Pb + PbSO_4$ Increasing lead sulfate $PbO_2 + PbSO_4$

D. Charging

Increasing sponge lead   Increasing lead peroxide

Decreasing lead sulfate $PbSO_4 + Pb$ Decreasing lead sulfate $PbSO_4 + PbO_2$

C. Discharged

Minimum sponge lead   Minimum lead peroxide

Maximum lead sulfate $PbSO_4 + PbO_2$ Maximum lead sulfate $PbSO_4 + Pb$

Sponge lead   Lead peroxide   Lead sulfate

10-25 Chemical action in a lead-add cell.

Single-pole, single-throw (SPST)

Single-pole, double-throw (SPDT)

Double-pole, single-throw (DPST)

Double-pole, double-throw (DPDT)

10-26 Schematic symbols of commonly used switches.

A single-pole, single-throw switch is like an ordinary light switch in that it makes and breaks a single circuit with one movement from the rest position. Single-pole, double-throw (SPDT) switches have three terminals, but control a single circuit with each throw. As one is completed, the other is opened. You can visualize an SPDT as a pair of SPST switches in tandem. Double-pole, double-throw switches control two circuits with a single movement. Think of the DPDT as two DPST switches combined, but working so that when one closes the other opens.

Self-actuating switches

Not all switching functions are done by hand. Some are too critical to trust to the operator's alertness. In general two activating methods are used. The first depends on the effect of heat on a bimetallic strip or disc to open or close contacts. Heat can be generated through electrical resistance or, as is more common, from, the engine coolant. The bimetallic element consists of two dissimilar metals bonded back-to-back. Because the coefficient of expansion is different for the metals, the strip or disc will deform inward, in the direction of the least expansive metal. Figure 10-27 illustrates this in a flasher unit. Other switches operate by the pressure of a fluid acting against a diaphragm. The fluid can be air, lube oil, or brake fluid. The most common use of such switches is as oil pressure sensors (Fig. 10-28).

Relays

Relays are electrically operated switches. They consist of an electromagnet, a movable armature, a return spring, and one or more contact sets. Small currents excite the coil, and the resulting magnetic field draws the armature against the coil.

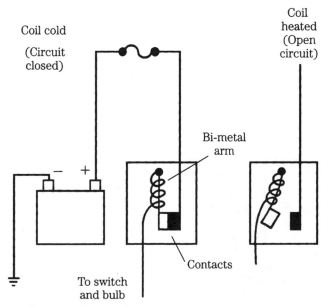

10-27 Bimetallic switch operation.

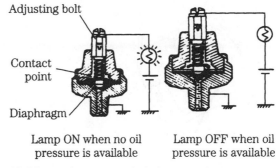

Lamp ON when no oil pressure is available

Lamp OFF when oil pressure is available

10-28 Diaphragm switch (oil pressure).

This linear movement closes or opens the contacts. The attractiveness of relays is that small currents—just enough to excite the electromagnet—can be used to switch large currents. You will find relays used to activate charging indicator lamps (Fig. 10-29), some starter motors, horns, remotely controlled air trips, and the like.

Solenoids are functionally similar to relays. The distinction is that the movable armature has some nonelectrical chore, although it might operate switch contacts in the bargain. Many starter motors employ a solenoid to lever the pinion gear into engagement with the flywheel. The next chapter discusses several of these starters.

Circuit protective devices

Fuses and fusible links are designed to vaporize and open the circuit during overloads. Some circuits are routed through central fuse panels. Others can be protected by individual fuses spliced into the circuit at strategic points.

Fusible links are appearing with more frequency as a protection for the battery-regulator circuit (Fig. 10-30). Burnout is signaled by swollen or discolored insulation over the link. New links can be purchased in bulk for soldered joints, or in precut lengths for use with quick-disconnects. Before replacing the link, locate the cause of failure, which will be a massive short in the protected circuit.

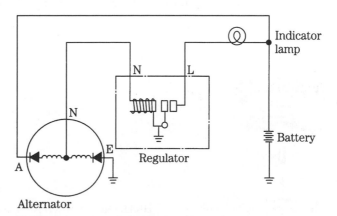

10-29 Relay-tripped charging indicator lamp.

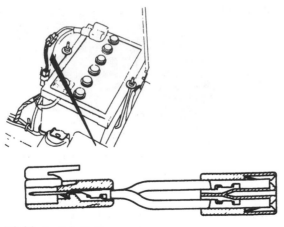

10-30 Fusible link.

On the other hand, a blown fuse is no cause for alarm. Fuses fatigue and become resistive with age. And even the best-regulated charting circuit is subject to voltage and current spikes, which can take out a fuse but otherwise are harmless. Chronic failure means a short or a faulty regulator.

Capacitors (condensers)

Two centuries ago it was thought that Leyden jars for storing charges actually condensed electrical "fluid." These glass jars, wrapped with foil on both the inner and outer surfaces, became known as *condensers*. The term lingers in the vocabulary of automotive and marine technicians, although *capacitor* is more descriptive of the devices' capacity for storing electricity.

A capacitor consists of two plates separated by an insulator or *dielectric* (Fig. 10-31). In most applications the plates and dielectric are wound on each other to save space.

A capacitor is a kind of storage tank for electrons. In function it is similar to an accumulator in hydraulic circuits. Electrons are attracted to the negative plate by the close proximity of the positive plate. Storing the electrons requires energy, which is

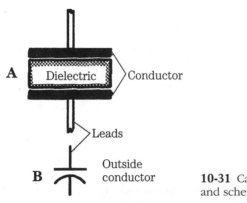

10-31 Capacitor construction and schematic symbol.

released when the capacitor is allowed to discharge. Shorting the plates together sends free electrons out of the negative plate and to the positive side. There is a to-and-fro shuttle of electrons between the plates until the number of electrons on the plates is equalized. The number of oscillations depends on circuit resistance and *reactance* (or the reluctance of the capacitor to charge because of the mutual repulsion of electrons on the negative plate).

Capacitors block direct current, but pass alternating current. The electrons do not physically penetrate the dielectric. The distortions in the orbits of the electrons in the dielectric displace or attract electrons at the plates. The effect on alternating current in a circuit is often as though the dielectric did not exist at least at the higher frequencies.

Capacitors are found at the generator or alternator, where they shunt high-frequency alternating current to ground to muffle radio interference.

Solid-state components

Although the transistor was invented in the Bell Laboratories in 1947, it and its progeny remain a mystery to most mechanics. These components are used in all alternators to rectify alternating current to direct current and can be found in current regulators produced by CAV, Lucas, Bosch, and Delco. In the future they will have a much wider use and possibly will substitute for relays.

Solid-state components are made of materials called semiconductors, which fall midway in the resistance spectrum between conductors and insulators. In general these components remain nonconductive until subjected to a sufficient voltage.

Silicon and germanium are the most-used materials for semiconductors. Semiconductor crystals for electronics are artificial crystals, grown under laboratory conditions. In the pure state these crystals are electrically just mediocre conductors. But if minute quantities of impurities are added in the growth state, the crystals acquire special electrical characteristics that make them useful for electronic devices. The process of adding impurities is known as *doping*.

When silicon or germanium, is doped with arsenic or phosphorus, the physical structure remains crystalline. But for each atom of impurity that combines with the intrinsic material, one free electron becomes available. This material has electrons to carry current and is known as *n*-type semiconductor material.

On the other hand the growing crystals might be doped with indium, aluminum, boron, or certain other materials and result in a shortage of electrons in the crystal structure. "Missing" electrons can be thought of as electrical *holes* in the structure of the crystal. These holes accept electrons that blunder by, but the related atom quickly releases the newcomer. When a voltage is applied to the semiconductor, the holes appear to drift through the structure as they are alternately filled and emptied. Crystals with this characteristic are known as *p*-type materials, because the holes are positive charge carriers.

Diode operation

When *p*-type material is mated with *n*-type material, we have a *diode*. This device has the unique ability to pass current in one direction and block it in the other. When the diode is not under electrical stress, *p*-type holes in the p-material and electrons in the *n*-material complement each other at the interface of the two materials. This interface, or junction, is electrically neutral, and presents a barrier to the charge carriers. When we apply negative voltage to the *n*-type material, electrons are forced out of it. At the same time, holes in the *p*-type material migrate toward the direction of current flow. This convergence of charge carriers results in a steady current in the circuit connected to the diode. Applying a negative voltage to the *n*-type terminal so the diode will conduct is called *forward biasing*.

If we reverse the polarity—that is, apply positive voltage to the *n*-type material—electrons are attracted out of it. At the same time, holes move away from the junction on the *p* side. Because charge carriers are thus made to avoid the junction, no current will pass. The diode will behave as an insulator until the reverse-bias voltage reaches a high enough level to destroy the crystal structure.

Diodes are used to convert alternating current from the generator to pulsating direct current. An alternator typically has three pairs of diodes mounted in heat sinks (heat absorbers).

Diodes do have some resistance to forward current and so must be protected from heat by means of shields and sinks. In a typical alternator the three pairs of diodes that convert alternating current to pulsating direct current are pressed into the aluminum frame. Heat generated in operation passes out to the whole generator.

Solid-state characteristics

As useful as solid-state components are, they nevertheless are subject to certain limitations. The failure mode is absolute. Either the device works or it doesn't. And in the event of failure, no amount of circuit juggling or tinkering will restore operation. The component must be replaced. Failure might be spontaneous—the result of manufacturing error compounded by the harsh environment under the hood—or it might be the result of faulty service procedures. Spontaneous failure usually occurs during the warranty period. Service-caused failures tend to increase with time and mileage as the opportunities for human error increase.

These factors are lethal to diodes and transistors

High inverse voltage resulting from wrong polarity Jumper cables connected backward or a battery installed wrong will scramble the crystalline structure of these components.

Vibration and mechanical shock Diodes must be installed in their heat sinks with the proper tools. A 2 × 4 block is not adequate.

Short circuits Solid-state devices produce heat in normal operation. A transistor, for example, causes a 0.3–0.7V drop across the collector-emitter terminals. This drop, multiplied by the current, is the wattage consumed. Excessive current will overheat the device.

Soldering The standard practice is for alternator diodes to be soldered to the stator leads. Too much heat at the connection can destroy the diodes.

11
CHAPTER

Starting and generating systems

The diesel engines we are concerned with are almost invariably fitted with electric starter motors. A number of engines, used to power construction machinery and other vehicles that are expected to stand idle in the weather, employ a gasoline engine rather than an electric motor as a starter. A motor demands a battery of generous capacity and a generator to match.

Starting a cold engine can be somewhat frustrating, particularly if the engine is small. The surface/volume ratio of the combustion space increases disproportionately as engine capacity is reduced. The heat generated by compression tends to dissipate through the cylinder and head metal. In addition, cold clearances might be such that much of the compressed air escapes past the piston rings. Other difficulties include the effect of cold on lube and fuel oil viscosity. The spray pattern coarsens, and the drag of heavy oil between the moving parts increases.

Starting has more or less distinct phases. Initial or breakaway torque requirements are high because the rotating parts have settled to the bottom of their journals and are only marginally lubricated. The next phase occurs during the first few revolutions of the crankshaft. Depending on ambient temperature, piston clearances, lube oil stability, and the like, the first few revolutions of the crankshaft are free of heavy compressive loads. But cold oil is being pumped to the journals, which collects and wedges between the bearings and the shafts. As the shafts continue to rotate, the oil is heated by friction and thins, progressively reducing drag. At the same time, cranking speed increases and compressive loads become significant. The engine accelerates to *firing speed*. The duration from breakaway to firing speed depends on the capacity of the starter and battery, the mechanical condition of the engine, lube oil viscosity, ambient air temperature, the inertia of the flywheel, and the number of cylinders. A single-cylinder engine is at a disadvantage because it cannot benefit from the expansion of other cylinders. Torque demands are characterized by sharp peaks.

Starting aids

It is customary to include a cold-starting position at the rack. This position provides extra fuel to the nozzles and makes combustion correspondingly more likely.

Lube oil and water immersion heaters are available that can be mounted permanently on the engine. Lube oil heaters are preferred and can be purchased from most engine builders. Good results can be had by heating the oil from an external heater mounted below the sump. Use an approved type to minimize the fire hazard. Alternatively, one can drain the oil upon shutdown and heat it before starting. The same can be done with the coolant, although temperatures in both cases should be kept well below the boiling temperature of water to prevent distortion and possible thermal cracking.

If extensive cold weather operation is intended or if the engine will be stopped and started frequently, it is wise to add one or more additional batteries wired in parallel. Negative-to-negative and positive-to-positive connections do not alter the output voltage, but add the individual battery capacities.

Once chilled beyond the cloud point, diesel fuel enters the gelling stage. Flow through the system is restricted, filter efficiency suffers, and starting becomes problematic. Racor is probably the best known manufacturer of fuel heaters, which are available in a variety of styles. Several combine electric resistance elements with a filter, to heat the fuel at the point of maximum restriction. Another type incorporates a resistance wire in a flexible fuel line.

Makers of indirect injection engines generally fit glow plugs as a starting aid (Fig. 11-1). These engines would be extremely difficult to start without some method of heating the air in the prechamber. A low-resistance filament (0.25–1.5 Ω, cold) draws heavy current to generate 1500°F at the plug tip. Early types used exposed filaments, which sometimes broke off and became trapped between the piston and chamber roof with catastrophic effects on the piston and (when made of aluminum) the head. Later variants contain the filament inside of a ceramic cover, which eliminates the problem. However, ceramic glow plugs are quite vulnerable to damage when removed from the engine and must be handled with extreme care.

In all cases, glow plugs are wired in parallel and controlled by a large power relay. Test filament continuity with an ohmmeter.

Primitive glow-plug systems are energized by a switch, sometimes associated with a timer, and nearly always in conjunction with a telltale light. The more sophisticated systems used in contemporary automobiles automatically initiate glow-plug operation during cranking and, once the engine starts, gradually phase out power.

Heating body   Terminal

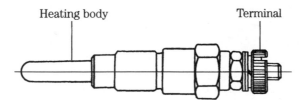

11-1 Sheathed-type glow plug. Marine Engine Div., Chrysler Corp.

Two types of circuits are encountered, both built around a solid-state module with an internal clock. The pulsed system opens the glow-plug power circuit for progressively longer intervals as the engine heats and the timer counts down. In the Ford/Navistar version of this circuit, glow-plug resistance varies with tip temperature, so that the plugs themselves function as heat sensors. Note that these low-resistance devices self-destruct within seconds of exposure to steady-state battery voltage. Pulsed glow plugs can be tested with a low-voltage ohmmeter and plug operation can be observed by connecting a test lamp between the power lead and the glow-plug terminal. Normally, if the circuit pulses, it can be considered okay; when in doubt, consult factory literature for the particular engine model.

Most manufacturers take a less ambitious approach, and limit glow-plug voltage by switching a resistor into the feed circuit. During cold starts, a relay closes to direct full battery voltage to the glow plugs; as the engine heats (a condition usually sensed at the cylinder-head water jacket), the first relay opens and a second relay closes to switch in a large power resistor. Power is switched off when the module times out.

Starting fluid can be used in the absence of intake air heaters. In the old days a mechanic poured a spoonful of ether on a burlap rag and placed it over the air intake. This method is not the safest nor the most consistent; too little fluid will not start the engine, and too much can cause severe detonation or an intake header explosion. Aerosol cans are available for injection directly into the air intake. Use as directed in a well-ventilated place.

More sophisticated methods include pumps and metering valves in conjunction with pressurized containers of starting fluid. Figure 11-2 illustrates a typical metering valve. The valve is tripped only once during each starting attempt, to forestall explosion. Caterpillar engines are sometimes fitted with a one-shot starting device consisting of a holder and needle. A capsule of fluid is inserted in the device and the needle pierces it, releasing the fluid.

The starter motor should not be operated for more than a few seconds at a time. Manufacturers have different recommendations on the duration of cranking, but none suggests that the starter button be depressed for more than 30 seconds. Allow a minute or more between bouts for cooling and battery recovery.

Wiring

Figure 11-3 illustrates a typical charging/starting system/in quasi-realistic style. The next drawing (Fig. 11-4) is a true schematic of the same system, encoded in a way that conveys the maximum amount of information per square inch.

Neither of these drawings is to scale and the routing of wires has been simplified. The technician needs to know where wires terminate, not what particular routes they take to get there. When routing does become a factor, as in the case of electronic engine control circuitry, the manufacturer should provide the necessary drawings.

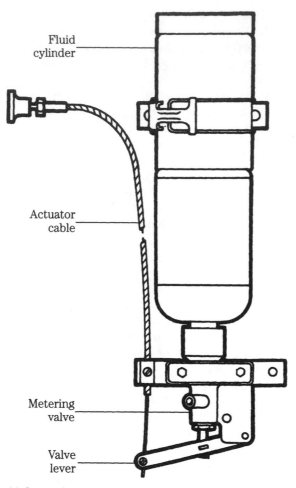

11-2 Quick-Start Unit. Detroit Diesel

Color coding

As is customary, the schematic in Fig. 11-4 indicates the color of the insulation and the size of the various conductors. Color coding has not been completely standardized, but most manufacturers agree that black should represent ground. This does not mean that colors change in a purely arbitrary fashion. In the example schematic, red denotes the positive side of the battery (the 'hot' wire); white/blue is associated with the temperature switch; orange with the tachometer, and so on.

Two-color wires consist of a base, or primary, color and a tracer. The base color is always first in the nomenclature. Thus, white/blue means a wire with a blue stripe.

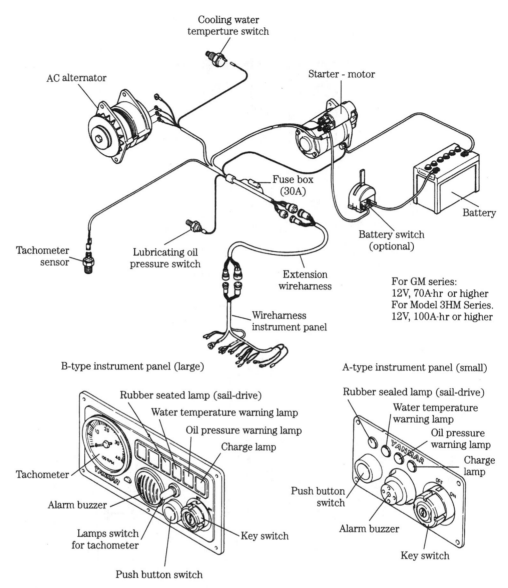

11-3 Wiring layout for a Yanmar Marine application.

Wiring repairs

The material that follows applies to simple DC starting, charging, and instrumentation circuits. Cutting or splicing wiring harnesses used with electronic engine management systems can do strange things to the computer.

The first consideration when selecting replacement wire is its current-carrying capacity, or gauge. In the context of diesel engines, two more or less interchangeable

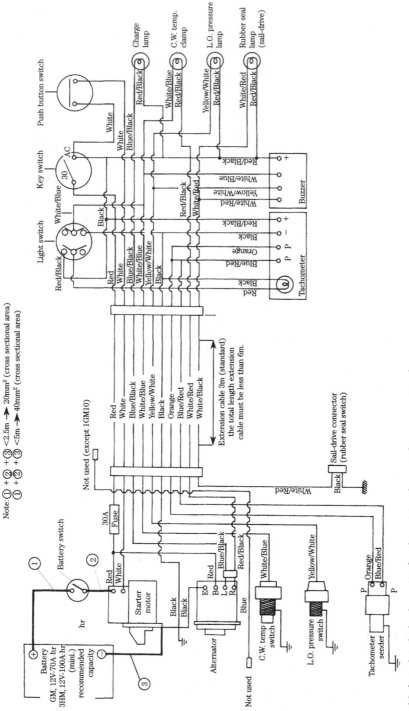

11-4 A detailed schematic for the wiring layout shown previously.

Table 11-1. SAE/metric wire sizes

| SAE wire size | Sectional area | | Outer dia. | | Conductor resistance (Ω/km) | Allowable continuous load current (A) |
	Circular mils	mm²	in.	mm		
20	1,094	0.5629	0.040	1.0	32.5	7
18	1,568	0.8846	0.050	1.2	20.5	9
16	2,340	1.287	0.060	1.5	14.1	12
14	3,777	2.091	0.075	1.9	8.67	16
12	5,947	3.297	0.090	2.4	5.50	22
10	9,443	5.228	0.115	3.0	3.47	29
8	15,105	7.952	0.160	3.7	2.28	39
6	24,353	13.36	0.210	4.8	1.36	88
4	38,430	20.61	0.275	6.0	0.871	115
2	63,119	35.19	0.335	8.0	0.510	160
1	80,010	42.73	0.375	8.6	0.420	180
0	100,965	54.29	0.420	9.8	0.331	210
2/0	126,882	63.84	0.475	10.4	0.281	230
3/0	163,170	84.96	0.535	12.0	0.211	280
4/0	207,740	109.1	0.595	13.6	0.135	340

Source: Marine Engine Div., Chrysler Corp.

standards apply: the Society of Automotive Engineers (SAE) and Japanese Industrial Standard (JIS).

For most wire, the smaller the SAE gauge number, the greater the cross-sectional area of the conductor (See Table 11.1). Thus, #10 wire will carry more current than #12. The schema reverses when we get into heavy cable: 4/0 is half again as large as 2.0 and has a correspondingly greater current capacity. Note that one must take these current values on faith.

The JIS standard eliminates much confusion by designating wire by type of construction and cross-sectional area. Thus, JIS AV5 translates as automotive-type (stranded) wire with a nominal conductor area of 5 mm². The Japanese derive current-carrying capacity from conductor temperature, which cannot exceed 60°C (140°F). Ambient temperature and wire type affect the rating, as shown back in Table 4-2.

Vinyl-insulated, stranded copper wire is standard for engine applications. Teflon insulation tolerates higher temperatures than vinyl and has better abrasion resistance. But Teflon costs more and releases toxic gases when burned. In no case should you use Teflon-insulated wire in closed spaces.

All connections should be made with terminal lugs. Solder-type lugs (Fig. 11-5A) can provide mechanically strong, low-resistance joints and are infinitely preferable to the crimp-on terminals shown in Fig. 11-5B. Insulate with shrink tubing. When shrink tubing is impractical (as when insulating a Y-joint), use a good grade of vinyl electrician's tape. The 3-M brand costs three times more than the imported variety and is worth every penny.

Table 11-2. **Allowable amperage and voltage drop (JIS AV wire).**

Ambient temp.	30°C		40°C		50°C	
Allowable amperage/ Voltage drop	A	mv/m	A	mv/m	A	mv/m
Nominal section (mm²)						
0.5	11	414	9	338	6	226
0.85	15	356	12	284	8	190
1.25	19	310	15	245	10	163
2	25	251	20	201	14	140
3	34	216	28	178	19	121
5	46	185	37	149	26	104
8	60	158	49	129	34	90
15	82	129	66	104	46	72
20	109	110	86	87	61	62
30	152	90	124	73	87	51
40	170	83	139	68	98	48
50	195	75	159	61	113	43
60	215	70	175	57	124	40
85	254	62	207	51	146	36
100	294	56	240	46	170	33

Figure 11-6 illustrates how stranded wire is butt spliced. Cut back the insulation ¾ in. or so, and splay the strands apart. Push the wires together, so that the strands interleave, and twist. Apply a small amount of solder to the top of the joint, heating from below.

Soldering

Most mechanics believe they know how to solder, but few have received any training in the art. Applied correctly, solder makes a molecular bond with the base metal. Scrape conductors bright to remove all traces of oxidation.

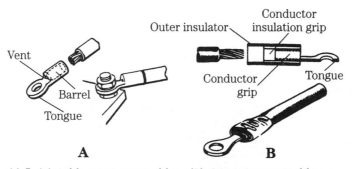

11-5 (A) Solder-type terminal lug. (B) Crimp-on terminal lug.

11-6 Splice with stranded conductor (A) strands splayed, (B) interleaved, (C) twisted.

Use a good grade of low-temperature, rosin-core solder, such as Kessler "blue," which consists of 60% tin and 40% lead. A 250-W gun should be adequate for all but the heaviest wiring.

The tip of a nonplated soldering iron should be dressed with a file down to virgin copper and tinned, or coated with molten solder. Silver solder is preferable for tinning because it melts at a higher temperature than lead-based solder and so protects the tip from corrosion. Retighten the tip periodically.

The following rules were developed from experience and from a series of experiments conducted by the military:

- Use a minimum amount of solder.
- Wrapping terminal lugs and splicing ends with multiple turns of wire does not add to the mechanical strength of the joint and increases the heat requirement. Wrap only to hold the joint while soldering.
- Heat the connection, not the solder. When the parts to be joined are hot enough, solder will flow into the joint.
- Do not move the parts until the solder has hardened. Movement while the solder is still plastic will produce a highly resistive "cold" joint.
- Use only enough heat to melt the solder. Excessive heat can damage nearby components and can crystallize the solder.
- Allow the joint to air cool. Dousing a joint with water to cool it weakens the bond.

Starter circuits

Before assuming that the motor is at fault, check the battery and cables. The temperature-corrected hydrometer reading should be at least 1.240, and no cell should vary from the average of the others by more than 0.05 point. See that the battery terminal connections are tight and free of corrosion.

Excessive or chronic starter failure might point to a problem that is outside the starter itself. It could be caused by an engine that is out of tune and that consequently requires long cranking intervals.

Starter circuit tests

There are several methods that you can use to check the starting-circuit resistance. One method is to open all the connections, scrape bright, and retighten. Another method requires a low-reading ohmmeter of the type sold by Sun Electric and other suppliers for the automotive trades. But most mechanics prefer to test by voltage drop.

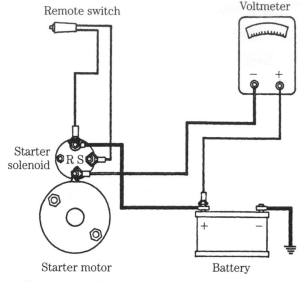

11-7 Hot side voltage drop test.

Connect a voltmeter as shown in Fig. 11-7. The meter shunts the positive, or hot battery post and the starter motor. With the meter set on a scale above battery voltage, crank. Full battery voltage means an open in the circuit.

If the starter functions at all, the reading will be only a fraction of this. Expand the scale accordingly. A perfect circuit will give a zero voltage drop because all current goes to the battery. In practice some small reading will be obtained. The exact figure depends on the current draw of the starter and varies between engine and starter motor types. As a general rule, subject to modification by experience, a 0.5V drop is normal. Much more than this means: (1) resistance in the cable, (2) resistance in the connections (you can localize this by repeating the test at each connection point), or (3) resistance in the solenoid.

Figure 11-8 shows the connections for the ground-side check. A poor ground, and consequent high voltage on the meter, can occur at the terminals, the cable, or between the starter motor and engine block. If the latter is the case, remove the motor and clean any grease or paint from the mounting flange.

Starter motors

Starter motors are series-wound; i.e., they are wound so that current enters the field coils and goes to the armature through the insulated brushes. Because a series-wound motor is characterized by high no-load rpm, some manufacturers employ limiting coils in shunt with the fields. The effect is to govern the free-running rpm and prolong starter life should the starter be energized without engaging the flywheel.

The exploded view in Fig. 11-9 illustrates the major components of a typical starter motor. The frame (No. 1) has several functions. It locates the armature and fields, absorbs torque reaction, and forms part of the magnetic circuit.

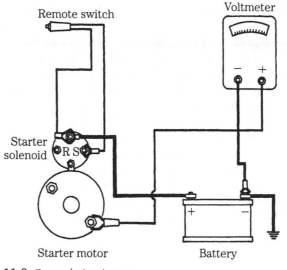

11-8 Ground circuit test.

The field coils (No. 2) are mounted on the pole shoes (No. 3) and generate a magnetic field, which reacts with the field generated in the armature to produce torque. The pole pieces are secured to the frame by screws.

The armature (No. 4) consists of a steel form and a series of windings, which terminate at the commutator bars. The shaft is integral with it and splined to accept the starter clutch.

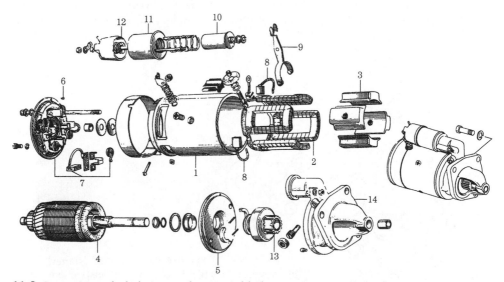

11-9 Starter in exploded view and as assembled. Lehman Manufacturing Co., Inc.

The end plates (5 and 6) locate the armature by means of bronze bushings. The commutator end plate doubles as a mounting fixture for the brushes, while the power takeoff side segregates the starter motor from the clutch.

The insulated (hot) brushes (No. 7) provide a current path from the field coils through the commutator and armature windings to the grounded brushes (No. 8).

Engagement of this particular starter is done by means of a yoke (No. 9), which is pivoted by the solenoid plunger (No. 10) in response to current flowing through the solenoid windings (No. 11). Movement of the plunger also trips a relay (No. 12) and energizes the motor. The pinion gear (No. 13) meshes with the ring gear on the rim of the flywheel. The pinion gear is integral with an overrunning clutch.

The starter drive housing supports the power takeoff end of the shaft and provides an accurately machined surface for mounting the starter motor to the ending block or bellhousing.

Brushes

Before any serious work can be done, the starter must be removed from the engine, degreased, and placed on a clean bench. Disconnect one or both cables at the battery to prevent sparking; disconnect the cable to the solenoid and the other leads that might be present (noting their position for assembly later); and remove the starter from the flywheel housing. Starters are mounted with a pair of cap screws or studs.

Remove the brush cover, observing the position of the screw or snap, because wrong assembly can short the main cable or solenoid wire (Fig. 11-10). Hitachi starters do not have an inspection band as such. The end plate must be removed for access to the brushes and commutator.

Brushes are sacrificial items and should be replaced when worn to half their original length. The rate of wear should be calculated so that the wear limit will not be reached between inspection periods. Clean the brush holders and commutator with a preparation intended for use on electrical machinery. If old brushes are used,

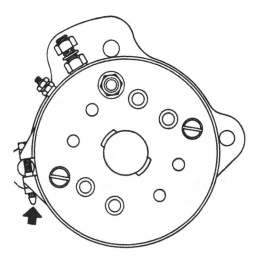

11-10 CAV CA-45C starter; band location in inspection.
GM Bedford Diesel

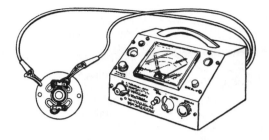

11-11 Testing brush holder insulation. Tecumseh Products Co.

lightly file the flanks at the contact points with the holders to help prevent sticking. New brushes are contoured to match the commutator, but should be fitted by hand. Wrap a length of sandpaper around the commutator—do not use emery cloth—and turn in the normal direction of rotation. Remove the paper and blow out the dust.

Try to move the holders by hand. Most are riveted to the end plate and can become loose, upsetting the brush-commutator relationship. With an ohmmeter, test the insulation on the hot-side brush holders (Fig. 11-11). There should be no continuity between the insulated brush holders and the end plate. Brush spring tension is an important and often overlooked factor in starter performance. To measure it, you will need an accurate gauge such as one supplied by Sun Electric. Specifications vary between makes and models, but the spring tension measured at the free (brush) end of the spring should be at least $1\frac{1}{2}$ lb. Some specifications call for 4 lb.

The commutator bars should be examined for arcing, scores, and obvious eccentricity. Some discoloration is normal. If more serious faults are not apparent, buff the bars with a strip of 000 sandpaper.

Armature

Further disassembly requires that the armature shaft be withdrawn from the clutch mechanism. Some starters employ a snap ring at the power takeoff end of the shaft to define the outer limit of pinion movement. Others use a stop nut (Fig. 11-12, No. 49). The majority of armatures can be withdrawn with the pinion and overrunning clutch in place. The disengagement point is at the yoke (Fig. 11-9, No. 9) and sleeve on the clutch body. Remove the screws holding the solenoid housing to the frame and withdraw the yoke pivot pin. The pin might have a threaded fastener with an eccentric journal, as in the case of Ford designs. The eccentric allows for wear compensation. Or it might be a simple cylinder, secured by a flanged head on one side and a cotter pin or snap ring on the other. When the pin is removed there will be enough slack in the mechanism to disengage the yoke from the clutch sleeve, and the armature can be withdrawn. Observe the position of shims—usually located between the commutator and end plate—and, on CAV starters, the spring-loaded ball. This mechanism is shown in Fig. 11-12 as 18 and 34. It allows a degree of end float so that the armature can recoil if the pinion and flywheel ring gear do not mesh on initial contact.

The armature should be placed in a jig and checked for trueness because a bent armature will cause erratic operation and might, in the course of long use, destroy the flywheel ring gear. Figure 11-13 illustrates an armature chucked between lathe

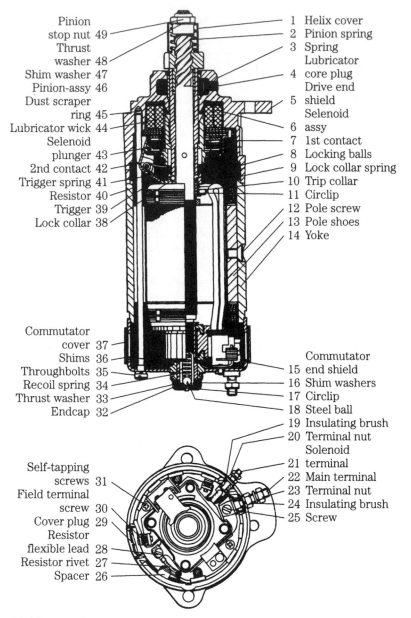

Pinion
stop nut 49
Thrust
washer 48
Shim washer 47
Pinion-assy 46
Dust scraper
ring 45
Lubricator wick 44
Selenoid
plunger 43
2nd contact 42
Trigger spring 41
Resistor 40
Trigger 39
Lock collar 38

1 Helix cover
2 Pinion spring
3 Spring
  Lubricator
4 core plug
  Drive end
5 shield
  Selenoid
6 assy
7 1st contact
8 Locking balls
9 Lock collar spring
10 Trip collar
11 Circlip
12 Pole screw
13 Pole shoes
14 Yoke

Commutator
cover 37
Shims 36
Throughbolts 35
Recoil spring 34
Thrust washer 33
Endcap 32

Commutator
15 end shield
16 Shim washers
17 Circlip
18 Steel ball
19 Insulating brush
20 Terminal nut
   Solenoid
21 terminal
22 Main terminal
23 Terminal nut
24 Insulating brush
25 Screw

Self-tapping
screws 31
Field terminal
screw 30
Cover plug 29
Resistor
flexible lead 28
Resistor rivet 27
Spacer 26

11-12 Typical CAV starter. GM Bedford Diesel.

centers. The allowable deflection at the center bearing is 0.002 in., or 0.004 in. on the gauge. With the proper fixtures and skill with an arbor press, an armature shaft can be straightened, although it will not be as strong as it was originally and will be prone to bend again. The wiser course is to purchase a new armature.

Make the same check on the commutator. Allowable out-of-roundness is 0.012–0.016 in., or less, depending on the rate of wear and the intervals between inspection periods. Commutators that have lost their trueness or have become pitted should be

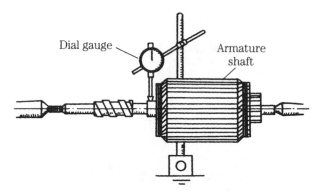

11-13 Checking armature shaft defection. Marine Engine Div, Chrysler Corp.

turned on a lathe. The cutting tool must be racked more for copper than for steel. Do not allow the copper to smear into the slots between the segments. Chamfer the end of the commutator slightly. Small imperfections can be removed by chucking the commutator in a drill press and turning against a single-cut file.

It is necessary that the insulation (called, somewhat anachronistically, *mica*) be buried below the segment edges; otherwise, the brushes will come into contact with the insulation as the copper segments wear. Undercutting should be limited to 0.015 in. or so. Tools are available for this purpose, but an acceptable job can be done with a hacksaw blade (flattened to fit the groove) and a triangular file for the final bevel cut (Fig. 11-14).

Inspect the armature for evidence of overheating. Extended cranking periods, dragging bearings, chronically low battery charge, or an under-capacity starter will cause the solder to melt at the commutator-armature connections. Solder will be splattered over the inside of the frame. Repairs might be possible because these connections are accessible. Continued overheating will cause the insulation to flake and powder. Discoloration is normal, but the insulation should remain resilient.

Armature insulation separates the nonferrous parts (notably the commutator segments) from the ferrous parts (the laminated iron segments extending to the outer diameter of the armature and shaft). Make three tests with the aid of a 120V continuity lamp or a *megger* (meter for very high resistances). An ordinary ohmmeter is useless to measure the high resistances involved. In no case should the lamp light up or resistance be less than 1 MΩ.

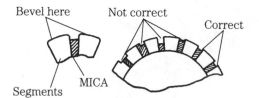

11-14 Undercutting mica. Marine Engine Div., Chrysler Corp.

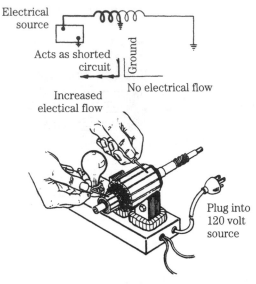

11-15 Continuity check between commutator bars and armature segments.

WARNING: Exercise extreme care when using 120V test equipment.
- Test between adjacent commutator segments.
- Test between individual commutator segments and the armature form (Fig. 11-15).
- Test between armature or commutator segments and the shaft.

It is possible for the armature windings to short, thus robbing starter torque. Place the armature on a growler and rotate it slowly while holding a hacksaw blade over it as shown in Fig. 11-16. The blade will be strongly attracted to the armature segments because of the magnetic field introduced in the windings by the growler.

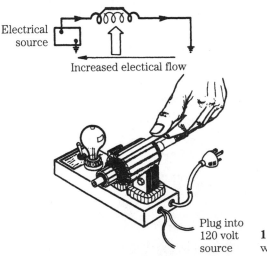

11-16 Checking for shorts with a growler.

But if the blade does a Mexican hat dance over a segment, you can be sure that the associated winding is shorted.

Field coils

After armatures, the next most likely source of trouble is the field coils. Meld hookups vary. The majority are connected in a simple series circuit, although you will encounter starters with *split fields*, each pair feeding off its own insulated brush. Field resistance values are not as a rule supplied in shop manuals, although a persistent mechanic can obtain this and other valuable test data by contacting the starter manufacturer. Be sure to include the starter model and serial number.

In the absence of a resistance test, which would detect intracoil shorts, the only tests possible are to check field continuity (an ordinary ohmmeter will do) and to check for shorts between the windings and frame. Connect a lamp or megger to the fields and touch the other probe to the frame, as illustrated in Fig. 11-17. Individual fields can be isolated by snipping their leads.

The fields are supported by the pole shoes, which in turn, are secured to the frame by screws. The screws more often than not will be found to have rusted to the frame. A bit of persuasion will be needed, in the form of penetrating oil and elbow grease. Support the frame on a bench fixture and, with a heavy hammer, strike the screwdriver exactly as if you were driving a spike. If this does not work, remove the screw with a cape chisel. However, you might (depending on the starter make and your supply of junk parts) have difficulty in matching the screw thread and head fillet.

Coat the screw heads with Loctite before installation and torque securely. Be sure that the pole shoe and coil clears the armature and that the leads are tucked out of the way.

Bearings

The great majority of starters employ sintered bronze bushings. In time these bushings wear and must be replaced to ensure proper teeth mesh at the flywheel and prevent armature drag. In extreme cases the bushings can wear down to their

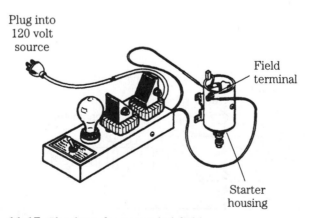

11-17 Checking for grounded fields. Tecumseh Products Co.

bosses so that the shaft rides on the aluminum or steel end plates. The old bushings are pressed out and new ones pressed in. Tools are available to make this job easier, particularly at the blind boss on the commutator end plate.

Without these tools, the bushing can be removed by carefully ridging and collapsing it inward, or by means of hydraulic pressure. Obtain a rod that matches shaft diameter. Pack the bushing with grease and hammer the rod into the boss. Because the grease cannot easily escape between the rod and bushing, it will lift the bushing up and out. Because of the blind boss, new bushings are not reamed.

Final tests

The tests described thus far have been *static* tests. If a starter fails a static test, it will not perform properly, but passing does not guarantee the starter is faultless. The only sure way to test a starter—or, for that matter, any electrical machine—is to measure its performance under known conditions, and against the manufacturer's specifications or a known-good motor.

No-load performance is checked by mounting the starter in a vise and monitoring voltage, current, and rpm. Figure 11-18 shows the layout. The voltage is the reference for the test and is held to 12V. Current drain and rpm are of course variable, depending on the resistance of the windings, their configuration, and whether or not speed-limiting coils are provided. You can expect speeds of 4000–7000 rpm and draws of 60–100A. In this test we are looking for low rpm and excessive current consumption.

The locked-rotor and stall test requires a scale to accept the pinion gear (Fig. 11-19). It must be made quickly, before the insulation melts. Mount the motor securely and lock the pinion. Typically the voltage will drop to half the normal value. Draw might approach or, with the larger starters, exceed 100A.

Solenoids

Almost all diesel starters are engaged by means of a solenoid mounted on the frame. At the same time the solenoid plunger moves the pinion, it closes a pair of contacts to complete the circuit to the starter motor. In other words the component

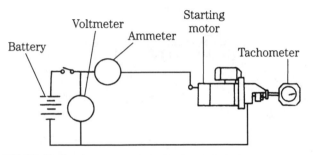

11-18 No-load starter test. Marine Engine Div., Chrysler Corp.

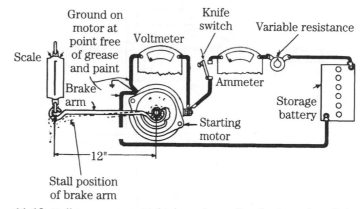

11-19 Stall torque test. Multiply scale reading by lever length in feet to obtain torque output. <small>Onan</small>

consists of a solenoid or *linear motor* and a relay (Fig. 11-20). Some circuits feature a second remotely mounted relay, as shown in Figs. 11-20 and 11-21. The circuit depicted in the earlier illustration was designed for marine use.

The starter switch might be 30 ft or more from the engine. To cut wiring losses a second relay is installed, which energizes the piggyback solenoid. The additional relay (2nd relay in Fig. 11-21) also gives overspeed protection should the switch remain closed after the engine fires. It functions in conjunction with a direct current generator. Voltage on the relay windings is the difference between generator output and battery terminal voltage. When the engine is cranking, generator output is functionally zero. The relay closes and completes the circuit to the solenoid. When the engine comes up to speed, generator output bucks battery voltage and the relay opens, automatically disengaging the starter. As admirable as such a device is, it should not be used continually, because some starter overspeeding will still occur, with detrimental effects to the bearings.

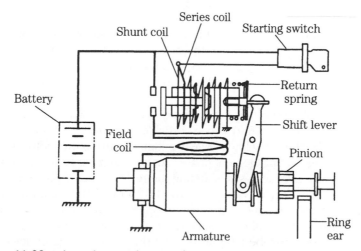

11-20 Solenoid internal wiring diagram.

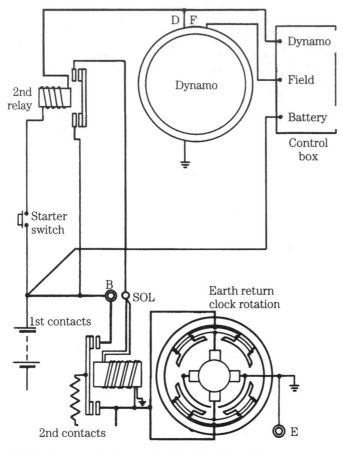

11-21 Overspeed relay and solenoid wiring diagram. GM Bedford Diesel

Should the generator circuit open, the starter will be inoperative because the return path for the overspeed relay is through the generator. In an emergency the engine can be started by bridging the overspeed terminals. Of course, the transmission or other loads must be disengaged and, in a vehicle, the handbrake must be engaged.

In addition, you will notice that the solenoid depicted in Fig. 11-21 has two sets of contacts. One set closes first and allows a trickle of current to flow through the resistor (represented by the wavy line above the contact arm) to the motor. The armature barely turns during the engagement phase. But once engaged a trigger is released and the second set of contacts closes, shunting the resistor and applying full battery current to the motor. This circuit, developed by CAV, represents a real improvement over the brutal spin-and-hit action of solenoid-operated and inertia clutches and should result in longer life for all components, including the flywheel ring gear.

Solenoids and relays can best be tested by bridging the large contacts. If the starter works, you know that the component has failed. Relays are sealed units and are not repairable. But most solenoids are at least amenable to inspection. Repairs to the series or shunt windings (illustrated in Figs. 11-20 and 11-21) are out of the question unless the circuit has opened at the leads. Contacts can be burnished, and some designs have provision for reversing the copper switch element.

Starter drives

Most diesel motors feature positive engagement drives energized by the solenoid. The solenoid is usually mounted piggyback on the frame and the pinion moved by means of a pivoted yoke (Fig. 11-20).

Regardless of mechanical differences between types, all starter drives have these functions:

- The pinion must be moved laterally on the shaft to engage the flywheel.
- The pinion must be allowed to disengage when flywheel rpm exceeds pinion rpm.
- The pinion must be retracted clear of the flywheel when the starter switch is opened.

Figure 11-22 illustrates a typical drive assembly. The pinion moves on a helical thread. Engagement is facilitated by a bevel on the pinion and the ring gear teeth of the flywheel. Extreme wear on either or both profiles will lock the pinion.

The clutch shown employs ramps and rollers. During the motor drive phase the rollers are wedged into the ramps (refer to Fig. 11-23). When the engine catches, the rollers are freed and the clutch overruns. Other clutches employ balls or, in a few cases, ratchets. The spring retracts the drive when the solenoid is deactivated.

Inspect the pinion teeth for excessive wear, and chipping. Some battering is normal and does not affect starter operation. The clutch mechanism should be disassembled (if possible), cleaned, and inspected. Inspect the moving parts for wear or deformation, with particular attention to the ratchet teeth and the ramps. Lubricate with Aero Shell 6B or the equivalent. Sealed drives should be wiped with a solvent-wetted rag. Do not allow solvent to enter the mechanism, because it will dilute the lubricant and cause premature failure. Test the clutch for engagement in one direction of pinion rotation and for disengagement in the other.

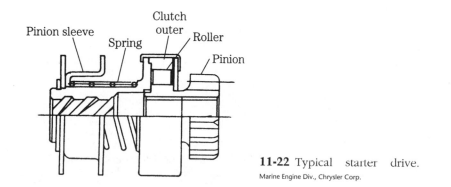

11-22 Typical starter drive.
Marine Engine Div., Chrysler Corp.

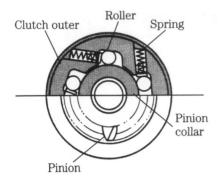

11-23 Overrunning clutch. Marine
Engine Div., Chrysler Corp.

Adjustment of the *pinion throw* is important to ensure complete and full mesh at the flywheel. Throw is measured between the pinion and the stop ring as shown in Fig. 11-24. Adjustment is by adding or subtracting shims at the solenoid housing (Fig. 11-25), moving the solenoid mounting bolts in their elongated slots, or turning the yoke pin eccentric.

Charging systems

The charging system restores the energy depleted from the battery during cranking and provides power to operate lights and other accessories. It consists of two major components: an alternator and a regulator. The circuit can be monitored by an ammeter or a lamp and is usually fused to protect the generator windings.

Alternators generate a high frequency 3-phase AC voltage, which is rectified (i. e., converted to direct current) by internal diodes. Wrong polarity will destroy the diodes and can damage the wiring harness. Observe the polarity when installing a new battery or when using jumper cables. Before connecting a charger to the battery, disconnect the cables. Should the engine be started with the charger in the circuit the regulator might be damaged. Isolate the charging system before any arc welding is done. Do not disconnect the battery or any other wiring while the alternator is turning. And, finally, do not attempt to polarize an alternator. The exercise is fruitless and can destroy diodes.

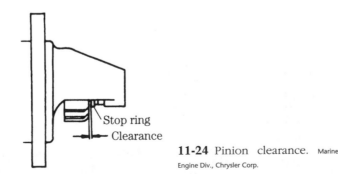

11-24 Pinion clearance. Marine
Engine Div., Chrysler Corp.

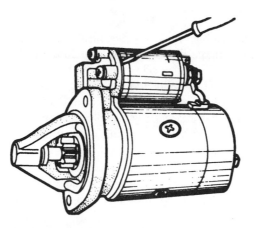

11-25 Shims between solenoid body and starter determine the pinion Clearance.
Marine Engine Div., Chrysler Corp.

Initial tests

The charge light should be on with the switch on and the engine stopped. Failure to light indicates an open connection in the bulb itself or in the associated wiring. Most charging-lamp circuits operate by a relay under the voltage regulator cover. Lucas systems employ a separate relay that responds to heat. The easiest way to check either type is to insert a 0–100A ammeter in series with the charging circuit. If the meter shows current and the relay does not close, one can safely assume that it has failed and should be replaced. The Lucas relay can be tested as shown in Fig. 11-26. You will need a voltage divider and 2.2W lamp. Connect clip A to the 12V terminal. The lamp should come on. Leaving the 12V connection in place, connect clip B to the 6V tap. The bulb should burn for 5 seconds or so and go out. Move B to the 12V post and hold for no more than 10 seconds. Then move it to the 2V (single cell) tap. The bulb should come on within 5 seconds. These units do not have computerlike precision, and some variation can be expected between them. But the test results should roughly correlate with the test procedure. Do not attempt to repair a suspect relay.

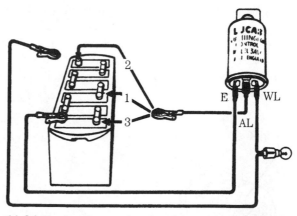

11-26 Testing Lucas charging lamp relay. To distinguish these relays from turn-signal flashers, Lucas has coded them green. GM Bedford Diesel

Test the alternator output against the meter on the engine or by inserting a test meter in series between the B terminal and battery. Voltage is monitored with a meter in parallel with the charging circuit. Discharge the battery by switching on the lights and other accessories. Connect a rheostat or carbon pile across the battery for a controlled discharge. (Without this tool you will be reduced to guessing about alternator condition.) With the load set at zero, start the engine and operate at approximately midthrottle. Apply the load until the alternator produces its full rated output. If necessary open the throttle wider. An output 2–6A below rating often means an open diode. Ten amps or so below rating usually means a shorted diode. The alternator might give further evidence of a diode failure by whining like a wounded banshee.

The voltage should be 18–20V above the nominal battery voltage under normal service conditions. It might be higher by virtue of automatic temperature compensation in cold weather.

Assuming that the output is below specs, the next step is to isolate the alternator from the regulator. Disconnect the field (F or FD) terminal from the regulator and ground it to the block. Load the circuit with a carbon pile to limit the voltage out-put. Run the engine at idle. In this test we have dispensed with the regulator and are protecting the alternator windings with carbon pile. No appreciable output difference between this and the previous tests means that the regulator is doing its job. A large difference would indicate that the regulator was defective.

Late-production alternators often have integrated regulators built into the slip ring end of the unit. Most have a provision for segregating alternator output from the regulator so that "raw" outputs can be measured. The Delcotron features a shorting tab. A screwdriver is inserted into an access hole in the back of the housing (Fig. 11-27); contact between the housing and the tab shorts the fields.

Note: The tab is within 3/4 in. of the casting. Do not insert a screwdriver more than 1 in. into the casting.

Bench testing

Disconnect the battery and remove the alternator from the engine at the pivot and belt-tensioning bracket. Three typical alternators are shown in exploded view in Figs. 11-28 through 11-30.

Remove the drive pulley. A special tool might be needed on some of the automotive derivations. Hold the fan with a screwdriver and turn the fan nut counterclockwise. Tap the sheave and fan off the shaft with a mallet. Remove the throughbolts and separate the end shields.

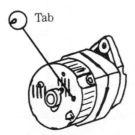

11-27 Location of Delcotron shorting tab.

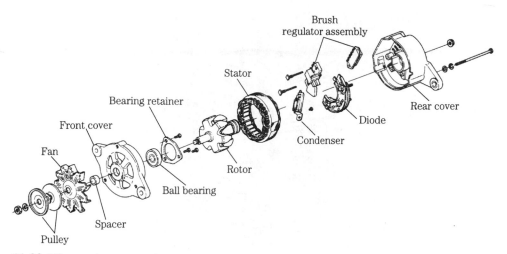

11-28 US generic pattern alternator.

Inspect the brushes for wear. Some manufacturers thoughtfully provide a wear limit line on the brushes (Fig. 11-31). Clean the holders with Freon or some other non-petroleum-based solvent and check the brushes for ease of movement. File lightly if they appear to bind. The slip rings should be miked for wear and eccentricity. Ten to twelve thousandths should be considered the limit (Fig. 11-32). Slip rings are usually, but not always, integral with the rotor. Removable rings are chiseled off and new ones pressed into place. Fixed rings can be restored to concentricity with light machining.

Determine the condition of the rotor insulation with a 120V test lamp (Fig. 11-33). The slip rings and their associated windings should be insulated from the shaft and pole pieces. If you have access to an accurate, low-range ohmmeter, test for continuity between slip rings. The resistance might lead one to suspect a partial open; less could mean an intracoil short.

The stator consists of three distinct and independent windings whose outputs are 120° apart. It is possible for one winding to fail without noticeably affecting the others. Peak alternator output will, of course, be reduced by one third. Disconnect the three leads going to the stator windings. Many European machines have these leads soldered, while American designs generally have terminal lugs. When unsoldering, be extremely careful not to overheat the diodes. Exposure to more than 300°F will upset their crystalline structure. Test each winding for resistance. Connect a low-range ohmmeter between the natural lead and each of three winding leads as shown in Fig. 11-34. Resistance will be quite low—on the order of 5 or 6 Ω—and becomes critical when one group of windings gives a different reading than the others.

Test the stator insulation with a 120V lamp connected as shown in Fig. 11-35. There should be no continuity between the laminations and windings.

The next step is to check the diodes (Fig. 11-36). You might already have had some evidence of diode trouble in the form of alternator whine or blackened varnish on the stator coils. The diodes must be tested with an ohmmeter or a test lamp of the same voltage as generator output.

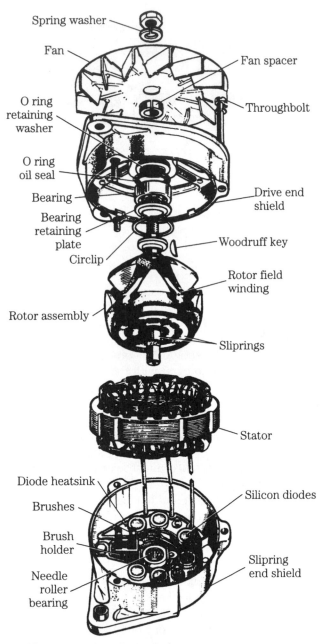

Spring washer

Fan

Fan spacer

O ring
retaining
washer

Throughbolt

O ring
oil seal

Bearing

Drive end
shield

Bearing
retaining
plate

Circlip

Woodruff key

Rotor field
winding

Rotor assembly

Sliprings

Stator

Diode heatsink

Brushes

Silicon diodes

Brush
holder

Needle
roller
bearing

Slipring
end shield

11-29 Lucas 10-AC or 11-AC alternator. GM Bedford Diesel

Test each diode by connecting the test leads and then reversing their polarity. The lamp should light in one polarity and go out in the other. Failure to light at all means an open diode; continuous burning means the diode has shorted. In either case it must be replaced. You might use a low-voltage ohmmeter in lieu of a lamp.

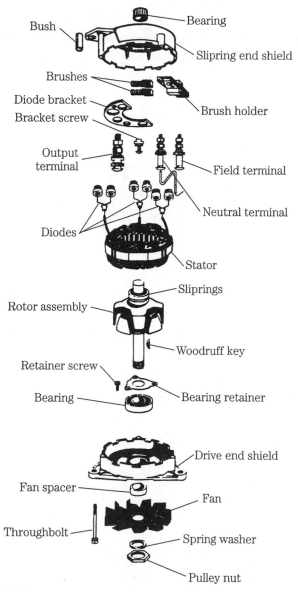

Bush

Bearing

Slipring end shield

Brushes

Diode bracket

Bracket screw

Brush holder

Output terminal

Field terminal

Neutral terminal

Diodes

Stator

Sliprings

Rotor assembly

Woodruff key

Retainer screw

Bearing

Bearing retainer

Drive end shield

Fan spacer

Fan

Throughbolt

Spring washer

Pulley nut

11-30 Prestolite CAB-1235 or CAB-1245 alternator. GM
Bedford Diesel

Expect high (but not infinite) resistance with one connection, and low (but not zero) resistance when the two leads are reversed.

To simply service and limit the need for special tools, some manufacturers package mounting brackets with their diodes. The bracket is a heat sink and must be in intimate contact with the diode case. Other manufacturers take the more traditional approach and supply individual diodes, which must be pressed (not hammered) into

Wear limit line

11-31 Brush showing wear limit line. Chrysler Corp.

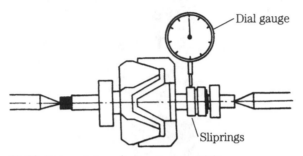

Dial gauge

Sliprings

11-32 Determining slipring concentricity. Marine Engine Div., Chrysler Corp.

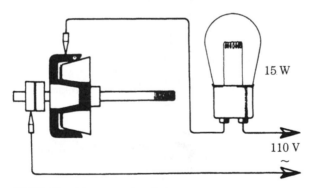

15 W

110 V
~

11-33 Checking rotor insulation. GM Bedford Diesel

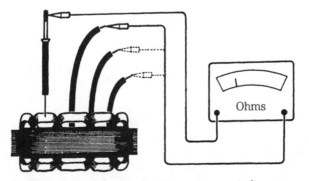

Ohms

11-34 Comparison test between stator windings. GM Bedford Diesel

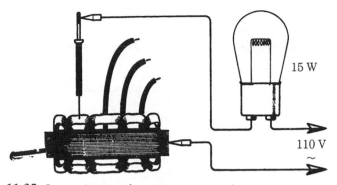

11-35 Comparison test between stator windings. GM Bed Ford Diesel

their sinks. K-D Tools makes a complete line of diode removal and installation aids, including heat sink supports and diode arbors of various diameters. Figure 11-37 shows a typical installation with an unsupported heat sink. Other designs might require support.

Soldering the connections is very critical. Should the internal temperature reach 300°F the diode will be ruined. Use a 150W or smaller iron and place a thermal shunt between the soldered joint and the diode (Fig. 11-38). The shunt might be in the form of a pair of needle-nosed pliers or copper alligator clips. In some instances there might not be room to shunt the heat load between the diode and joint (Fig. 11-38) It is only necessary to twist the leads enough to hold them while the solder is liquid. Work quickly and use only enough solder to flow between the leads. More solder merely increases the thermal load and increases the chances that the diode will be ruined.

Alternator bearings are sealed needle and ball types. They are not to be disturbed unless noisy or rough. Then bearings are pressed off and new ones installed with the numbered end toward the arbor.

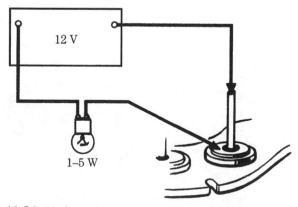

11-36 Diode testing. GM Bedford Diesel

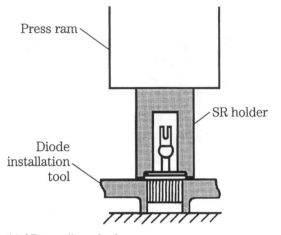

11-37 Installing diode. Marine Engine Div., Chrysler Corp.

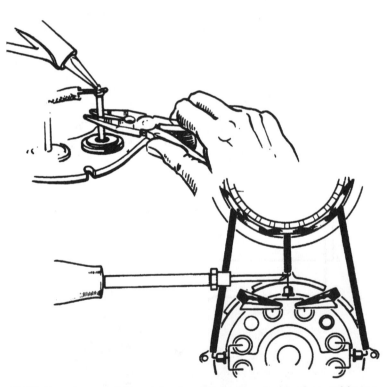

11-38 Two forms of heatsinks to protect the diode when soldering.
GM Bedford Diesel

Voltage regulation

Most alternator-based charging circuits employ voltage regulation (as opposed to voltage and current regulation). The regulator can be external to the alternator or integral with it. External regulators can be mechanical or solid state.

External regulators

Figure 11-39 is a schematic of a typical mechanical voltage regulator. The voltage-sensing winding is in parallel with the output and drives NC (normally closed) contacts. As the generator comes up to speed, voltage increases until the winding develops a strong enough field to open the contacts. Rotor output then passes through dropping resistor RF.

Depending on the make and model, voltage adjustment is accomplished by bending the stationary contact, moving the hinges in elongated mounting holes, or by screw (Fig. 11-40). In theory, the correct point gap should correspond with an output voltage of approximately 15V at 68°F or 28V for 24V systems. In practice, better results are had by measuring alternator output voltage at the battery terminals. Assuming that specifications are available, core and yoke gap adjustments also can be made.

Clean oxidized contacts with a riffle file or a diamond-faced abrasive strip. Do not use sandpaper or emery cloth. Inspect the dropping resistor (often found on the underside of the unit), springs, and contact tips for evidence of overheating. Check the regulator ground connection.

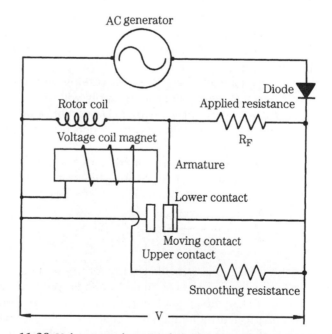

11-39 Voltage regulator in alternator circuit. Marine Engine Div., Chrysler Corp.

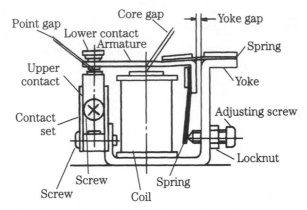

11-40 A typical relay and its adjustment points. Marine
Engine Div., Chrysler Corp.

Before discarding a defective regulator, attempt to discover why it failed. Burnt points or heat-discolored springs mean high resistance in the charging circuit or a bad regulator ground.

Solid-state regulators

Transistorized regulators are capable of exceedingly fine regulation, partially because there are no moving parts. Durability is exceptional. On the other hand any internal malfunction generally means that the unit must be replaced. As a rule, no repairs are possible. Failure can occur because of manufacturing error (this usually shows up in the first few hours of operation and is covered by warranty) high curent draws, and voltage spikes.

The mechanic must be particularly alert when working with transistorized circuits. The cautions that apply to alternator diodes apply with more force to regulators if only because regulators are more expensive. Do not introduce stray voltages, cross connections, reverse battery polarity, or open connections while the engine is running.

The regulator might be integral with the alternator or might be contained in a separate box. In general, no adjustment is possible; however, the Lucas 4TR has a voltage adjustment on its bottom, hidden under a dab of sealant.

Batteries

The battery has three functions: provide energy for the starting motor; stabilize voltages in the charging system; and, for limited periods, provide energy for the accessories in the event of charging-circuit failure. Because it is in a constant state of chemical activity and is affected by temperature changes, aging humidity and current demands, the battery requires more attention than any other component in the electrical system.

Battery ratings

Starting a diesel engine puts a heavy drain on the battery, especially in cold climates. One should purchase the best quality and the largest capacity practical. The physical size of the battery is coded by its *group number*. The group has only an indirect bearing on electrical capacity but does ensure that replacements will fit the original brackets. In some instances a larger capacity battery might require going to another group number. Expect to modify the bracket and possibly to replace one or more cables.

The traditional measure of a battery's ability to do work is its *ampere-hour* (A-hr) *capacity*. The battery is discharged at a constant rate for 20 hr. so that the potential of each cell drops to 1.75V. A battery that will deliver 6A over the 20-hr period is rated at 120 A-hr. (6A × 20hr.). You will find this rating stamped on replacement batteries or in the specifications.

Like all rating systems, the ampere-hour rating is best thought of as a yardstick for comparison between batteries. It has absolute validity only in terms of the orginally test. For example, a 120 A-hr. battery will not deliver 120A for 1 hr., nor will it deliver 1200A for 6 minutes.

Cranking-power tests are more meaningful because they take into account the power loss that lead-acid batteries suffer in cold weather. At room temperature the battery develops its best power; power output falls off dramatically around 0°F. At the same time, the engine becomes progressively more difficult to crank and more reluctant to start. Several cranking-power tests are in use.

Zero cranking power is a hybrid measurement expressed in volts and minutes. The battery is chilled to 0°F; depending on battery size, a 150 or 300 A load is applied. After 5 seconds the voltage is read for the first part of the rating. Discharge continues until the terminal voltage drops to 5V. The time in minutes between full charge and effective exhaustion is the second digit in the rating. The higher these two numbers are for batteries in the same load class, the better.

The *cold cranking performance* rating is determined by lowering the battery temperature to 0°F (or, in some instances, 20°F) and discharging for 30 seconds at such a rate that the voltage drops below 1.2V per cell. This is the most accepted of all cold weather ratings and has become standard in specification sheets.

Battery tests

As the battery discharges, some of the sulfuric acid in the electrolyte decomposes into water. The strength of the electrolyte in the individual cells is a reliable index of the state of charge. There are several ways to determine acidity, but long ago technicians fixed on the measurement of specific gravity as the simplest and most reliable.

The instrument used is called a *hydrometer*. It consists of a rubber bulb, a barrel, and a float with a graduated tang. The graduations are in terms of specific gravity. Water is assigned a specific gravity of 1. Pure sulfuric acid is 1.83 times heavier than water and thus has a specific gravity of 1.83. The height of the float tang above the liquid level is a function of fluid density, or specific gravity. The battery is said to be fully charged when the specific gravity is between 1.250 and 1.280.

An accurate hydrometer test takes some doing. The battery should be tested prior to starting and after the engine has run on its normal cycle. For example, if the engine is shut down overnight, the test should be made in the morning, before the first start. Water should be added several operating days before the test to ensure good mixing. Otherwise the readings can be deceptively low.

Use a hydrometer reserved for battery testing. Specifically, do not use one that has been used as an antifreeze tester. Trace quantities of ethylene glycol will shorten the battery's life.

Place the hydrometer tip above the plates. Contact with them can distort the plates enough to short the cell. Draw in a generous supply of electrolyte and hold the hydrometer vertically. You might have to tap the side of the barrel with your fingernail to jar the float loose. Holding the hydrometer at eye level, take a reading across the fluid level. Do not be misled by the meniscus (concave surface, Fig. 11-41) of the fluid.

American hydrometers are calibrated to be accurate at 80°F. For each 10°F above 80°F, add 4 points (0.004) to the reading; conversely, for each 10°F below the standard, subtract 4 points. The standard temperature for European and Japanese hydrometers is 20°C, or 68°F. For each 10°C increase add 7 points (0.007); subtract a like amount for each 10°C decrease. The more elaborate hydrometers have a built-in thermometer and correction scale.

All cells should read within 50 points (0.050) of each other. Greater variation is a sign of abnormality and might be grounds for discarding the battery. The relationship between specific gravity and state of charge is shown in Fig. 11-42.

The hydrometer test is important, but by no means definitive. The state of charge is only indirectly related to the actual output of the battery. Chemically the battery might have full potential, but unless this potential passes through the straps and terminals, it is of little use.

Perhaps the single most reliable test is to load the battery with a rheostat or carbon pile while monitoring the terminal voltage. The battery should be brought up to full charge before the test. The current draw should be adjusted to equal three

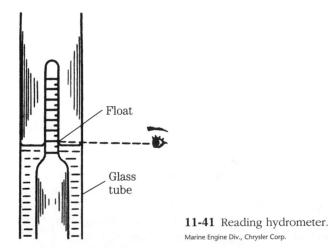

Float

Glass
tube

11-41 Reading hydrometer.
Marine Engine Div., Chrysler Corp.

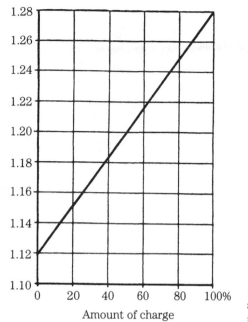

11-42 Relationship between state of charge and specific gravity. <small>Marine Engine Div., Chrysler Corp.</small>

times the ampere-hour rating. Thus, a 120 A-hr. battery would be discharged at a rate of 360A. Continue the test for 15 seconds and observe the terminal voltage. At no time should the voltage drop below 9.5V.

In this test, sometimes called the *battery capacity* test, we used voltage as the telltale. But without a load, terminal voltage is meaningless. The voltage remains almost constant from full charge to exhaustion.

Battery maintenance

The first order of business is to keep the electrolyte level above the plates and well into the reserve space below the filler cap recesses. Use distilled water. Tap water might be harmful, particularly if it has iron in it.

Inspect the case for cracks and acid seepage. Periodically remove the cable clamps and scrape them and the battery terminals. Look closely at the bond between the cables and clamp. The best and most reliable cables have forged clamps, solder-dipped for conductivity. Replace spring clip and other clever designs with standard bolt-up clamps sweated to the cable ends. After scraping and tightening, coat the terminals and clamps with grease to provide some protection from oxidation.

The battery case should be wiped clean with a damp rag. Dirt, spilled battery acid, and water are conductive and promote self-discharge. Accumulated deposits can be cleaned and neutralized with a solution of baking soda, water, and detergent. Do not allow any of the solution to enter the cells, where it would dilute the electrolyte. Rinse with clear water and wipe dry.

Charging

Any type of battery charger can be used—selenium rectifier, tungar rectifier, or, reaching way back, mercury arc rectifier. Current and voltage should be monitored and there should be a provision for control. When charging multiple batteries from a single output, connect the batteries in series as shown in Fig. 11-43.

Batteries give off hydrogen gas, particularly as they approach full charge. When mixed with oxygen, hydrogen is explosive. Observe these safety precautions:

- Remove all filler caps (to prevent pressure rise should the caps be clogged).
- Charge in a well-ventilated place remove from open flames or heat.
- Connect the charger leads before turning the machine on. Switch the machine off before disconnecting the leads.

In no case should the electrolyte temperature be allowed to exceed 115°F. If your charger does not have a thermostatic control, keep track of the temperature with an ordinary thermometer.

Batteries can be charged by any of three methods. *Constant-current* charging is by far the most popular. The charging current is limited to one-tenth of the ampere-hour rating of the battery. Thus a 120 A-hr. battery would be charged at 12 A. Specific gravity and no-load terminal voltage are checked at 30-minute intervals. The battery might be said to be fully charged when both values peak (specific gravity 1.127–1.129, voltage 15–16.2V) and hold constant for three cranking intervals.

A *quick charge*, also known as a booster or hotshot, can bring a battery back to life in a few minutes. The procedure is not recommended in any situation short of an emergency, because the high-power boost will raise the electrolyte temperature and might cause the plates to buckle. Disconnect the battery cables to isolate the generator or alternator if such a charge is given to the battery while it is in place.

A constant-voltage charge can be thought of as a compromise between the hotshot and the leisurely constant-current charge. The idea is to apply a charge by keeping charger voltage 2.2–2.4 V higher than terminal voltage. Initially the rate of charge is quite high; it tapers off as the battery approaches capacity.

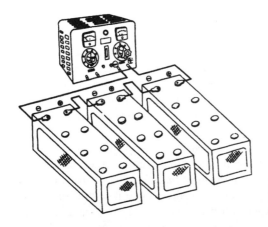

11-43 When charging multiple batteries, connect in series.
Marine Engine Div., Chrysler Corp.

Battery hookups

One of the most frequently undertaken field modifications is the use of additional batteries. The additional capacity makes starts easier in extremely cold weather and adds reliability to the system.

To increase capacity add one or more batteries in parallel—negative post connected to negative post, positive post to positive post. The voltage will not be affected, but the capacity will be the sum of all parallel batteries.

Connecting in series—negative to positive, positive to negative—adds voltage without changing capacity. Two 6V batteries can be connected in series to make a physically large 12V battery.

Cable size is critical because the length and cross-sectional area determine resistance. Engineers at International Harvester have developed the following recommendations, which can be applied to most small, high-speed diesel engines:

- Use cables with integral terminal lugs.
- Use only rosin or other noncorrosive-flux solder.
- Terminal lugs must be stacked squarely on the terminals. Haphazard stacking of lugs should be avoided.
- Where the frame is used as a ground return, it must be measured and this distance added to the cable length to determine the total length of the system. Each point of connection with the frame must be scraped clean and tinned with solder. There should be no point of resistance in the frame such as a riveted joint. Such joints should be bridged with a heavy copper strap.
- Pay particular attention to engine-frame grounds. If the engine is mounted in rubber it is, of course, electrically isolated from the frame.
- Check the resistance of the total circuit by the voltage drop method or by means of Ohm's law ($R = E/I$).
- Use this table as an approximate guide to cable size for standard-duty cranking motors.

System voltage	Maximum resistance	Cable size & length
12V	0.0012Ω	Less than 105 in., No. 0
		105 to 132 in., No. 00
		132 to 162 in., No. 000
		162 to 212 in., No. 0000 in parallel.
24V	0.0020Ω	Less than 188 in., No. 0
		188 to 237 in., No. 00
		237 to 300 in., No. 0000
		300 to 380 in., No. 0000, or two No. 0 in parallel.

- Suggested battery capacity varies with system voltage (the higher the voltage, the less capacity needed for any given application), engine displacement, compression ratio, ambient air temperature, and degree of exposure. Generalizations are difficult to make, but typically a 300 CID engine with a 12V standard-duty starter requires a 700A battery for winter operation. This figure is based on SAE J-5371 specifications and refers to the 30-second output of a chilled battery. At 0°F capacity should be increased to 900A. The International Harvester 414 CID engine requires 1150, or 1400A at 0°F. Battery capacity needs roughly parallel engine displacement figures, with some flattening of the curve for the larger and easier-to-start units. In extremely cold weather—below −10°F—capacity should be increased by 50% or the batteries heated.

- RV and boat owners often install a second battery to support accessory loads. Arranging matters so that there is always power available for starting requires special hardware; otherwise, the state of charge of either battery is the average of the two. Figure 11-44 illustrates this proposition.

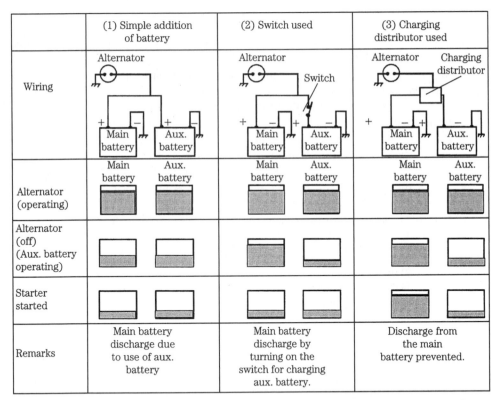

	(1) Simple addition of battery	(2) Switch used	(3) Charging distributor used
Wiring			
Alternator (operating)			
Alternator (off) (Aux. battery operating)			
Starter started			
Remarks	Main battery discharge due to use of aux. battery	Main battery discharge by turning on the switch for charging aux. battery.	Discharge from the main battery prevented.

11-44 Hard wiring batteries in parallel divides the load equally between both. Installing a manual switch in the B+ line confines the load to the auxiliary battery *until the switch is closed for charging*. When this happens, the main battery promptly discharges into the auxiliary. The best solution is to purchase a charging distributor. To avoid problems, furnish the vendor with a complete schematic and alternator characteristics.

12
CHAPTER

Cooling systems

Roughly 30% of the heat energy produced by combustion must be dissipated by the cooling system. In addition, the cooling system can be required to absorb heat from aftercoolers, engine and transmission oil coolers, hydraulic oil coolers, and other sources.

Air cooling

The great advantage of air cooling is its simplicity. There are no radiators, pumps, or hoses to add dead weight and eventually fail. On the other hand, the cooling fins and aluminum castings that promote heat transfer also transmit sound. Air-cooled engines are noisy. Nor does air cooling provide the precise temperature control necessary for good efficiency and low exhaust emissions. The centrifugal fan behaves like a turbocharger, pumping out too much air at high speeds and too little when the engine bogs under load.

Most air-cooled engines are small single- and twin-cylinder units developing less than 40 hp. But the concept is also applied to larger engines. Deutz builds a range of modular air-cooled engines of up to 500 hp, some of which are even used in marine applications. Most of the world's armored vehicles are powered by air-cooled diesels.

Figure 12-1 shows the cooling arrangement for small utility engines. The flywheel-mounted fan generates air flow that is directed over the cylinder by means of shrouding. Few of these engines are equipped with temperature gauges, which is a serious oversight. The operator can measure oil temperature with a thermometer placed in the dipstick boss to a depth of about 5/16 in. below the end of the stick, but well clear of metal surfaces. Maximum permissible oil temperature is a judgment call, with one manufacturer suggesting that it should not exceed 210°F (99°C) above ambient.

The thermostatically controlled shutter used by Onan gives an idea of the normal range of air-outlet temperatures (Fig. 12-2). The thermostat, mounted in the air outlet, begins to open the shutter when temperature reaches 120°F (49°C) and extends fully at 140°F (60°C). A second thermostat shuts off fuel delivery when

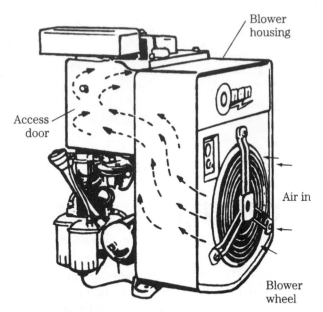

12-1 Air-flow for an air-cooled engine. Courtesy Onan

outlet-air temperature reaches 250°F (121°C). It is fair to say that outlet-air temperature should not exceed 45°F above ambient under load.

Chronic overheating means that the engine is undersized for the application. Other than periodically cleaning the fins and checking the fit of the tin work, there is little a mechanic can do to improve cooling. A very light coat of dull black paint applied over bare metal on exposed surfaces makes a marginal improvement in radiation.

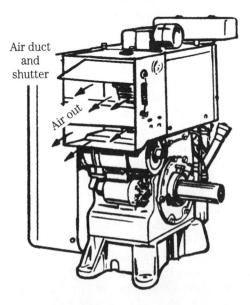

12-2 Thermostatically controlled shutter provides improved temperature control. Onan

Liquid cooling

Early engines were cooled by water in a hopper above the cylinders that was refilled as the water boiled off. Modern practice is to employ closed systems with a radiator or other form of heat exchanger, one or more circulation pumps, and a thermostat. Figure 12-3 illustrates the basic system. When the engine is cold, the thermostat closes to confine most of the coolant within the water jacket. A small

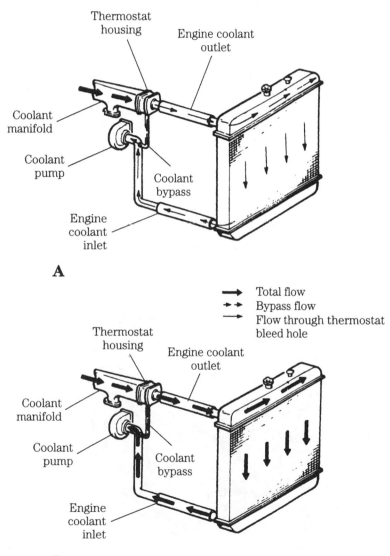

12-3 Coolant flow in system without a provision for air purging. Thermostat closed (A) and open (B). International

fraction of the coolant makes its way to the radiator through an internal bleed port in the thermostat. As shown in the drawings, the bypass often takes the form of a small-diameter hose running from below the thermostat housing to the pump. Some circulation is necessary to prevent local hot-spots in the water jacket.

As water temperature rises—a typical figure is 190°F (87.8°C)—the thermostat opens. Flow then passes from the water jacket through the top hose to the radiator and out of the lower hose to the water pump.

Most cooling systems include one or two additional hoses that function to vent air to the radiator or expansion tank. The system depicted in Fig. 12-4 employs two lines that convey aerated coolant to the radiator header. Air enters from splash

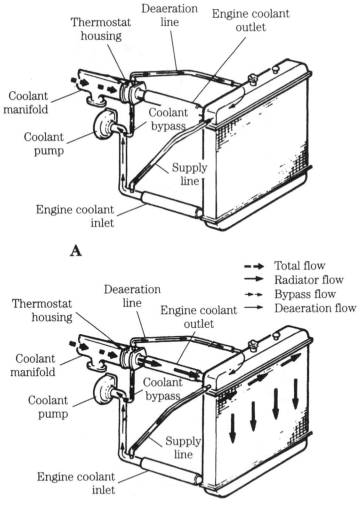

12-4 Coolant flow in a self-purging system. Thermostat closed (A) and open (B). International

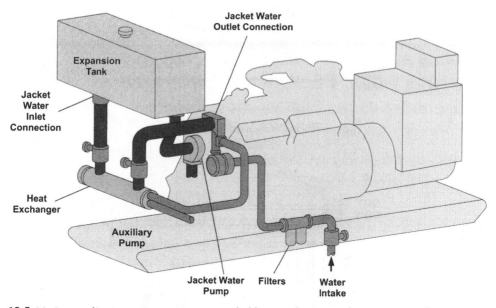

12-5 Marine application using a seawater-cooled heat exchanger and expansion tank. Caterpillar Inc.

entrapment (as, for example, when the radiator is filled from a bucket), past the water-pump seal, and by way of compression leaks across the head gasket. Besides reducing heat transfer—air removes heat 3500 times less efficiently than water—air increases the tendency of the water pump to lose pressure through cavitation.

Modern cooling systems employ an expansion tank, connected to the radiator-cap overflow and, for stationary engines, discharging into the radiator-return line. This expansion, surge, or degassing tank vents entrapped air and exhaust gases, collects overflow, and provides convenient means of replenishing the coolant. If it is to vent gases, the tank must be located above the thermostat housing, which is normally at the highest wetted point on the engine block.

Figures 12-5 and 12-6 illustrate two fresh-water marine systems, one using a sea-water-cooled heat exchanger and the other a keel cooler. Expansion tanks are clearly shown.

As illustrated in Fig. 12-7, standard practice is to distribute coolant to the cylinder liners through an external manifold. Many engines have an oil cooler in series with the main cooling circuit, and multiple liquid-cooled accessories, such as an aftercooler, turbocharger, torque converter, cab heater, and vehicle brakes, plumbed in parallel.

Coolant

Diesel manufacturers recommend a 50-50 mixture of low-silicate ethylene glycol and distilled water. This mixture gives –34°F (–37°C) protection against freezing and raises the boiling point to 226°F (107°C). Pressurization raises the boiling point further.

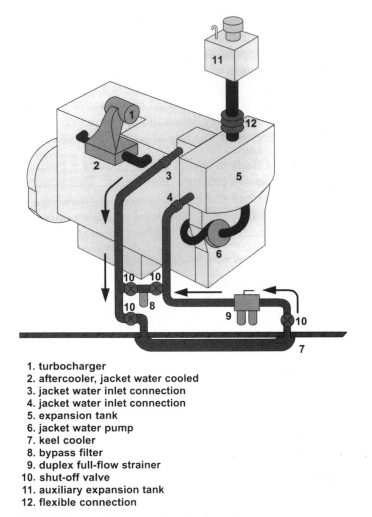

1. turbocharger
2. aftercooler, jacket water cooled
3. jacket water inlet connection
4. jacket water inlet connection
5. expansion tank
6. jacket water pump
7. keel cooler
8. bypass filter
9. duplex full-flow strainer
10. shut-off valve
11. auxiliary expansion tank
12. flexible connection

12-6 Marine application with a keel cooler. Courtesy Caterpillar Inc.

Some diesel manufacturers permit the use of standard automotive antifreezes that meet ASTM D3306 specifications. Others insist upon the more stringent D4985 or D6210 specification. D4985 has an inhibitor life of 3000 operating hours or one year, and D6210 is good for 3000 hours or three years. The additive package in Caterpillar ELC (Extended Life Coolant), which is incompatible with other antifreezes, provides protection for 12,000 hours or six years. Depending upon the type of antifreeze used, long-term protection may require that additive package be periodically replenished. Check with your dealer.

Soluble oils and methyl alcohol have no place in modern engines.

While there is no substitute for distilled or ionized water, there are times when operators are forced to use whatever is available. The water should meet these minimum specifications:

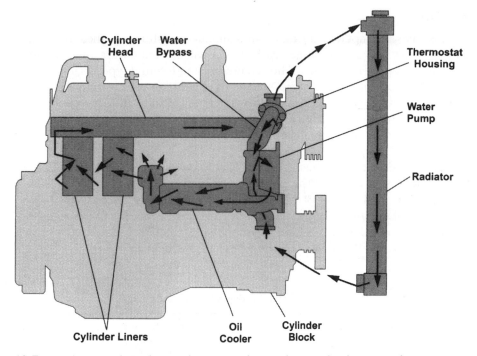

12-7 A radiator, oil cooler, and water jacket make up the basic coolant circuit. Caterpillar Inc.

Chloride (Cl)	40 mg/L (2.4 grains/US gal)
Sulfate (SO$_4$)	100 mg/L (5.9 grains/US gal)
Total hardness	170 mg/L (10.0 grains/US gal)
Total solids	340 mg/L (20.0 grains/US gal)
Acidity	pH 5.5–9.0

The local water utility or agricultural agent can provide data on water quality.

CAUTION: do not overfill radiators. Topping off the radiator or expansion tank when the engine is cold merely wastes coolant, since the surplus goes out the overflow when the engine warms. Continued overfilling dilutes the antifreeze.

A major drawback associated with ethylene glycol is the way it reacts with lube oil. Should a leak develop at the head gasket or oil cooler, shut the engine down immediately and make the necessary repairs. Assuming that the engine will still turn over, you might be able to avert a complete teardown by flushing the lubrication system with Butyl Cellosolve or an equivalent product. But check with the factory first, since the procedure was developed for Detroit two-cycles and could damage more heavily loaded engines.

The procedure consists of replacing the contaminated lube oil with a 50-50 mix of Cellosolve and SAE 10 oil. The engine is run at 1200 no-load rpm for one hour, then for another 15 minutes with pure SAE 10 in the crankcase. If the bearings survive, that is, if oil pressure remains steady and no knocks are heard, the operation was successful and the engine may be returned to service.

Overheating

Most diesel temperature gauges are calibrated to show overheating at around 226°F (108°C), but panel gauges cannot be trusted. Verify with an accurate thermometer, and preferably one of the remote-sensing infrared types that can be used to detect hot spots in the radiator. Other signs of possible overheating include steam from the radiator overflow after shutdown, coolant leaks, and low coolant levels. Low coolant levels can be both a cause of overheating and its result, as coolant boils off and escapes out the vented cap or expansion tank. In any event, refill the radiator and see if the problem recurs.

Severe overheating results in blown head gaskets, warped or cracked cylinder heads, and scored pistons. The pistons for the aft cylinders usually fail first. As a point of interest, the style of combustion chamber influences how pistons seize. Pistons for IDI engines generally bind at the area just below the upper ring land. DI pistons are more likely to seize on their thrust faces.

Table 12-1 lists the most common causes of overheating.

Table 12-1. Overheating problems and causes

Problem	Possible causes
Low coolant level	External leaks at hoses, radiator, and radiator cap
	Internal leaks at head gasket, cylinder head, aftercooler, torque-converter cooler, etc.
Low heat transfer	Clogged radiator
	Scale accumulations in water jacket
	Marine growth on keel cooler or insufficient flow of raw water through heat exchanger
Insufficient coolant flow	Thermostat stuck closed
	Loose water-pump belts
	Water-pump failure
	Clogged radiator core
Insufficient coolant pressure	Failed radiator pressure cap
	Failed coolant-pressure relief valve (when fitted)
Coolant overflow	Combustion gases entering system—loose or cracked cylinder head, leaks at head gasket, precombustion chamber, cylinder liner
	Entrapped air in cooling system
Insufficient air flow through radiator	Low or no fan speed—electric or hydraulic fan drive failure, loose belts, worn pulleys
	Fan installed backwards
	Shutter not opening
	Shrouding not installed properly
High inlet air temperature or restriction	Clogged air cleaner
	Clogged aftercooler
	Turbocharger failure
Exhaust restriction	Turbocharger failure
	Water in muffler or loose muffler baffle
	Clogged particulate trap/catalytic converter (when fitted)

Coolant leaks

Loss of coolant results from intrusion of combustion gases into the water jacket, a failed radiator cap or cap seal, and leaks. Most leaks are obvious, but look for telltale rust stains at gasketed joints with special attention to the block/head interface. Freeze plugs rust from the inside out. Very small leaks can be found by coloring the coolant with dye and pressurizing the system with a hand pump.

Corrosion and scale

A fairly accurate idea of the corrosion present in the water jacket can be had by removing the radiator hose from the thermostat housing. A film of light rust that wipes off with rag may be present on new engines and has no significance. Layered rust is cause for concern.

Most corrosion can be blamed on poor maintenance. Ethylene glycol antifreeze is safe so long as its additive package remains active. When in doubt about the condition of the antifreeze, change it. Antifreeze test kits for nitrite and phosphate inhibitors quickly pay for themselves in fleet service.

The pH level of the coolant should be neutral, that is between 7.0 and 10.5. Below 7.0 the coolant is acidic and dissolves iron; above 10.5 it turns alkaline and attacks aluminum, solder, and other nonferrous metals. The test apparatus is available from chemical supply houses.

Another, and sometimes difficult to correct, source of corrosion is electrolysis. When two dissimilar metals are subject to voltage, the least "noble" of these metals erodes. The coolant acts as the conductor and aluminum becomes the sacrificial metal, eroding about twice as rapidly as cast iron under the same voltage. Check for loose, dirty, or rusted ground connections, missing engine-to-ground straps, and for the presence of sacrificial anodes on marine applications. Using a digital ohmmeter, check the resistance between each grounded electrical component and the battery negative post. Resistance should be less than 0.03Ω. A higher reading means that the component is not properly grounded.

Exhaust gas seeping into the coolant also contributes to corrosion. Make up a hose connection to a sensor boss on the thermostat housing and collect the coolant in a glass container. Run the engine at normal temperature under load. An occasional bubble is acceptable, but a steady stream of bubbles indicates an exhaust leak.

Scale consists of calcium carbonate and sulfate, iron, copper, silica, and trace metals. According to General Motors, 1/16 in. of scale is the thermal equivalent of 4½ in. of cast iron. The primary source of scale is hard water that does not meet the minimum specifications listed earlier.

Cleaning

Reverse flushing removes loose rust and silt from the radiator and block. To flush the radiator, disconnect both hoses and insert a flushing gun, connected to a 100-psi air line, into the lower hose (Fig. 12-8 A). Fill the radiator and inject air in short bursts. Continue until the water flows freely. With the thermostat removed, do the same for the block (Fig. 12-8 B).

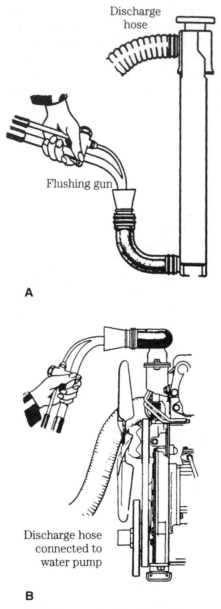

Discharge hose

Flushing gun

A

Discharge hose connected to water pump

B

12-8 Reverse flushing the radiator (A) and the block (B). GM Bedford Diesel

Two-part cleaners, consisting of oxalic acid and a neutralizing compound, can remove light rust and scale deposits without taking the engine out of service. Any of the commercial products work, or you can mix your own acid by adding 2 lb of sodium bisulfate with 10 US gal of water. The neutralizer consists of ½ lb of sodium carbonate crystals per 10 gal of water.

The presence of oil in the coolant can usually be traced to a leaking oil or transmission cooler, although other sources cannot be dismissed. A leak can develop across any interface that separates pressurized lube oil from coolant. Oil leaks result in local overheating that does not register on the temperature gauge. According to Detroit Diesel 1.25% of oil by volume in the coolant increases firedeck temperature by 15%. Once the leak is found and repaired, drain the cooling system and refill with water and two cups of non-foaming dishwasher detergent. Run the engine for 20 minutes or so, adding more detergent as necessary to emulsify the oil. Drain the system, flush with water to remove all traces of detergent, and refill with approved coolant.

Cavitation erosion

Wet cylinder liners flex under combustion pressures and pull away from the surrounding coolant. Air bubbles form in the void, attach themselves to the cylinder liners and implode, leaving tiny pits in their wake. The action is progressive and, at some point, the pitting breaks through to admit coolant into the cylinder (Fig. 12-9).

New liners have protection in the form of an oxide coating. But the oxide eventually pits and one must rely on additives, such as nitrates, to inhibit bubble formation and slow the rate of metal loss. The operator should avoid lugging the engine, which increases the amplitude of vibration, and do what he can to minimize piston slap. The fewer cold starts, the better. But cavitation erosion will be with us as long as diesel engines use wet liners.

Overcooling

Standard practice is to size the cooling system to yield a top radiator-tank temperature of 210°F (99°C) at a maximum ambient temperature of 110°F (43°C). A properly calibrated thermostat and the use of radiator shutters in cold climates should prevent overcooling and the resulting gum and carbon deposits.

12-9 Cavitation erosion results in coolant leaks into the cylinders. Caterpillar Inc.

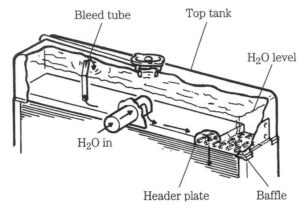

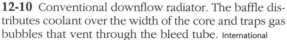

12-10 Conventional downflow radiator. The baffle distributes coolant over the width of the core and traps gas bubbles that vent through the bleed tube. International

Radiators

Heavy-duty radiators have large frontal areas and free-flowing cores with no more than 10 or 11 fins/in. These radiators are square, or nearly square, so that a single fan can pull or push air over most of the core surface. As shown in Fig. 12-10, coolant flows down from the header tank, usually made of brass, across a baffle, and through the vertical copper tubes. In the better examples, the tubes are brazed to the header tank. The Modine Beta Weld process, originally developed for offroad-vehicle radiators, produces a stronger and more reliable joint than soldering.

Radiators for industrial/commercial applications have a built-in safety factor known as the drawdown capability. Standards vary somewhat with the engine maker, but a properly sized system can lose between 10% or 15% of its coolant without overheating. Figure 12-11 charts this relationship.

Passenger-car radiators come out of a different tradition. Low hood lines result in elongated radiators, which often require two fans to cool. Since the early 1990s, aluminum has been specified for the core, and header tanks are often made of plastic. Designers compensate for loss of heat transfer—aluminum conducts heat only about half as well as copper—by narrowing the fin spacing and making the cores thicker. These almost solid radiators require large amounts of fan power to cool.

Automotive cooling systems must be accepted as they are, since the engineering that goes into them and underhood space limitations preclude much by way of modification. But cooling systems for stationary applications can often be improved with a bit of judicious tinkering. For example, most have simple box shrouds that can be replaced by much more efficient venturi shrouds. Other modifications are discussed below under "Fans."

As far as routine maintenance goes, green slime (chromium hydroxide) and sediment can be removed by reverse flushing the core or with oxalic acid. When overheating persists, farm out the radiator to a specialist for chemical cleaning.

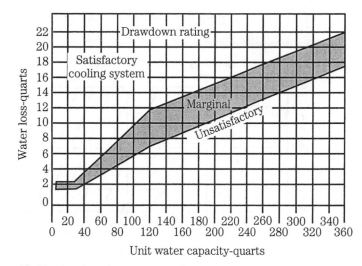

12-11 The drawdown rating refers to the system's ability to function with reduced coolant volume. Acceptable drawdown capacities are shown here for systems with capacities of up to 360 quarts. International

Fins, especially closely spaced fins, tend to clog and should periodically be cleaned with compressed air, high-pressure water, or steam. Some of the worst offenders are off-road vehicles that splatter their radiators with mud and fork-lift trucks that handle cotton or other fibrous products.

Leaks in brass and copper radiators can often be repaired on site. The secret lies in the soldering, or more exactly, in the preparation for soldering. Remove all traces of grease and oil from the damaged area. Then, using fine emery paper, sand down the area to bright metal. Apply generous amount of flux (a mixture of muriatic acid and zinc powder) and solder. Hard, 60/40 solder makes a stronger joint than the softer grades. If the solder bubbles and skates, the surface is still not clean. The solder should sink into the base metal and harden with a mirror-like glaze. Once the repair is made, wipe off all traces of flux.

Pressure caps

Pressure, generated by the radiator cap, raises the boiling point of the coolant (Fig. 12-12) and reduces the tendency of the water pump to cavitate. The cap includes two check valves, one that opens under vacuum (1) and a second, spring-loaded valve (2) that regulates system pressure (Fig. 12-13). Pressures range from 3 or 3 psi to more than 15 psi for engines operating at high altitudes.

A pressure cap that fails to hold pressure can cause boiling and loss of coolant, which ultimately results in overheating. A cap that fails to vent after shutdown, when the volume of coolant shrinks, often collapses the radiator hoses.

And finally, a word of caution. Radiator caps cam open in two stages. The first stage releases pressure and diverts coolant downward, away from the mechanic's

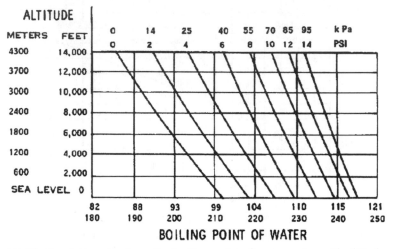

12-12 Graph showing how the boiling point of water varies with altitude and pressure. Caterpillar Inc.

hands. The second stage frees the cap from the radiator neck. All manufacturers warn that the engine should be allowed to cool before removing the cap. Risk increases with temperature and altitude. Releasing pressure on an overheated engine releases a geyser of boiling coolant and superheated steam that, while directed downward, rebounds off adjacent surfaces.

Fans

Fans for stationary applications operate continuously with power transmitted by one or more v-belts from the engine crankshaft. At 35 mph or so, road-going vehicles generate enough ram air velocity to dispense with the fan. Most

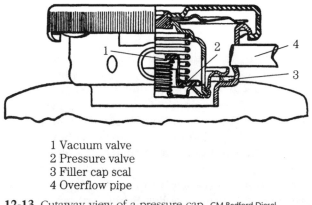

1 Vacuum valve
2 Pressure valve
3 Filler cap scal
4 Overflow pipe

12-13 Cutaway view of a pressure cap. GM Bedford Diesel

Table 12-2. Recommended v-belt tension*

Belt width	Tension	
	New	**Used**
3/8	445 +/− 22N (100 +/− 5 lb)	400 +/− 22N (90 +/− 10 lb)
1/2	534 +/− 22N (120 +/− 5 lb)	400 +/− 44N (90 +/− 10 lb)
5V	534 +/− 22N (120 +/− 5 lb)	400 +/− 44N (90 +/− 10 lb)
11/16	534 +/− 22N (120 +/− 5 lb)	400 +/− 44N (90 +/− 10 lb)
3/4	534 +/− 22N (120 +/− 5 lb)	400 +/− 44N (90 +/− 10 lb)
15/16	534 +/− 22N (120 +/− 5 lb)	400 +/− 44N (90 +/− 10 lb)
6PK	667 +/− 22N (120 +/− 5 lb)	467 +/− 44N (105 +/− 10 lb)
8K	800 +/− 22N (120 +/− 5 lb)	489 +/− 44N (110 +/− 10 lb)

*Courtesy Caterpillar Inc.

passenger cars and light trucks use electric fans that cycle on and off in response to radiator header-tank temperature. Some earlier vehicles employed belt-driven fans with viscous clutches that used silicone as the working fluid. Clutch action was less than positive, allowing the fan to turn at 700—1000 rpm when it should have been disengaged. And when fully engaged, the clutch slipped, reducing fan speed by about 5%.

Belt-driven on/off fans represent more recent thinking. Fans for light trucks employ electromagnetic clutches, similar to those used on air-conditioning compressors. Pneumatic clutches are favored for larger trucks with air brakes. The engine control unit (ECU) controls fan's on-off time in response to coolant temperature and other variables.

The primary maintenance requirement for belt-driven fans is to periodically check belt tension with a Borroughs or equivalent gauge (Table 12-2). You can get a rough idea of tension by applying thumb pressure to the longest belt run between pulleys. A half-inch of "give" is about right, but err on the loose side since belts are less expensive than bearings. Examine belts for oil damage, heat checking, and wear. Severe belt or pulley wear causes the belts to ride on the bottom of the pulley grooves. Paired belts should be replaced as a set: the stretch of the worn belt cannot be adjusted for without over-tensioning the replacement.

Check bearings by removing the belts and turning driven components by hand. "Hard spots" or perceptible side play means that the associated bearing should be replaced. Viscous-clutch units sometimes fail to reach speed as they age. The only way this fault can be detected is to measure airflow velocity with a Caterpillar 8T-2700 or equivalent tool.

There is little that can be done to improve the performance of clutched fans, which have been, or should have been, engineered for the application. But fixed-speed fans can often benefit for creative tinkering. Modifications include:
- Removing restrictions to air flow in front of the radiator and aft of the fan.
- Substituting a venturi-type shroud for the box shroud fitted to many industrial engines. Blade tips should come within 0.5 in. or less of the shroud.

- Increasing the radiator area swept by the fan. This can be done by specifying a larger diameter fan and, on suction fans, by displacing the fan at least on blade width behind the radiator.
- Investing in a high-efficiency fan and adjusting the drive ratio to turn it at the recommended blade-tip velocity.

Thermostats

A thermostat is a heat-sensitive valve that opens to admit coolant to the radiator when the temperature of coolant in the cylinder head reaches a predetermined temperature. Most thermostats open at 190°F (87.8°C).

Figure 12-14 shows a wax-pellet thermostat in the closed and open positions. When closed, a small amount of coolant passes through a bleed port to the radiator, but most returns to the pump through the bypass valve (8 in the drawing). As the coolant warms, the wax pellet (6) liquifies and expands to open the valve (2) to the radiator. Action is progressive. In the partially open position, coolant flow splits between the bypass valve and the radiator. At full open, all flow goes to the radiator. As the engine warms, radiator header-tank and cylinder temperatures should equalize.

Brass frames tend to break, and the metal bellows (rather than the telescoping tubes shown in the drawing) develop cracks. To check temperature response, heat the thermostat in water, supporting it away from the metal sides of the container (Fig. 12-15). The temperature rating, almost always stamped on the frame, refers to the temperature at which the thermostat cracks open. Raise the water temperature to about 25°F (15°C) above the temperature rating, adding antifreeze as necessary to prevent boiling. Ten minutes should be enough time for the unit to open fully.

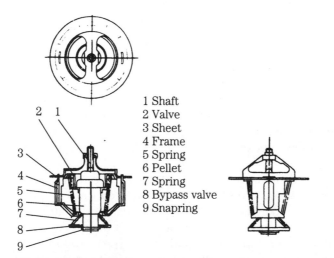

1 Shaft
2 Valve
3 Sheet
4 Frame
5 Spring
6 Pellet
7 Spring
8 Bypass valve
9 Snapring

12-14 Pill-type thermostat. Marine Engine Div. Chrysler Corp.

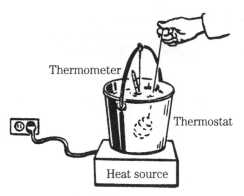

12-15 Testing a thermostat. International

Water pumps

As shown in Fig. 12-16, the water pump consists of a cast-iron or aluminum body, bearing assembly, ceramic face seal, shaft, and pressed-on impeller. Light-duty pumps employ prelubricated bearings; the better types are plumbed into the

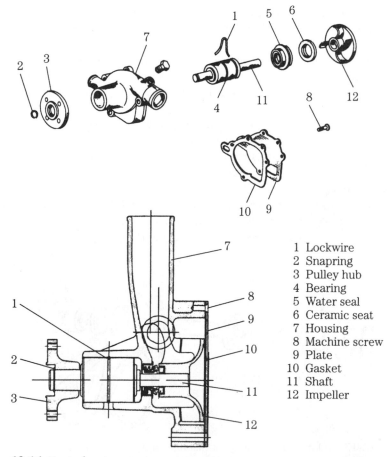

1 Lockwire
2 Snapring
3 Pulley hub
4 Bearing
5 Water seal
6 Ceramic seat
7 Housing
8 Machine screw
9 Plate
10 Gasket
11 Shaft
12 Impeller

12-16 Typical water pump. Marine Engine Div., Chrysler

engine oiling circuit. All have a bleed port outboard of the seal on the underside of the body casting. The port must remain open to provide an escape route for coolant that gets past seal faces and, on pressure-lubricated pumps, to prevent coolant contamination of the lube oil.

Pump failure is signaled by:

- Leaks from the bleed port. Some moisture is normal, but coolant streaks in this area mean seal and possible bearing failure.
- Perceptible side play or "hard spots" as the shaft is turned by hand. In severe cases, bearing wear allows the impeller to make rubbing contact with the pump body.
- Low pressure. Some engines have test ports on the inlet and outlet sides of the pump so that the pressure rise—usually in the order of 12 psi—can be measured. In the absence of these ports, about all one can do is intuit pump performance from the volume of coolant going into the radiator.

Water pumps for nonpassenger car applications have factory support in the form of replacement seals and rebuild kits (Fig. 12-17). Most pumps require the use of an arbor press to remove and install the impeller and bearing pack.

Raw-water pumps assist the engine water pump in marine applications and in applications with remotely mounted radiators. The Jabsco unit in Fig. 12-18 employs prelubricated bearings and a mechanical face seal. The symmetrical impeller enables the pump to be driven in either direction by the expedient of swapping inlet and outlet port connections. Some models include a drain cock; for the unit illustrated, cover screws must be loosened in preparation for freezing temperatures.

To disassemble, remove the cover-retaining screws, end cover (2), and impeller (5). If the pump has much time on it, the impeller will be stuck. Remove it with a suitable puller or by spreading plier jaws under the impeller and tapping the shaft with a brass drift. Remove the cam screw and the cam (6). Replace the wear plate (7) as a precautionary measure.

Support the pump body in a vise and loosen the pinch bolt (24) that secures the body to the bearing housing (15). Remove the slinger (14) and pry off the inner seal from the pump body casting. Press the shaft (25) and bearings out of the casting.

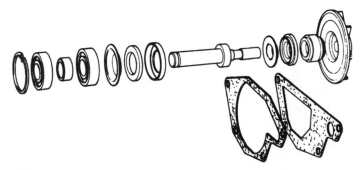

12-17 Perkins rebuild kit. Most manufactures can supply the seal separately.

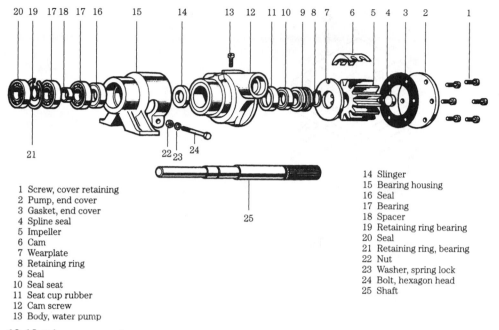

1 Screw, cover retaining
2 Pump, end cover
3 Gasket, end cover
4 Spline seal
5 Impeller
6 Cam
7 Wearplate
8 Retaining ring
9 Seal
10 Seal seat
11 Seat cup rubber
12 Cam screw
13 Body, water pump

14 Slinger
15 Bearing housing
16 Seal
17 Bearing
18 Spacer
19 Retaining ring bearing
20 Seal
21 Retaining ring, bearing
22 Nut
23 Washer, spring lock
24 Bolt, hexagon head
25 Shaft

12-18 Jabsco raw-water pump. GM Bedford Diesel

Working from the impeller side, drive out the inner seal. All that remains is to separate the bearing pack from the shaft by removing the circlip (8).

Hoses

Hose condition is best checked by feel. Replace hoses that feel soft after the engine has shut down and pressure in the system has dissipated. A loose inner liner on the lower radiator hose creates an overheating problem that is nearly impossible to diagnose without removing the hose for inspection. Change all hoses every three years or 4000 hours. Close heater-hose valves during summer months to reduce the potential for leaks.

Most mechanics prefer to use stainless-steel clamps with a worm gear that engages serrations on the ribbon. It is doubtful whether these clamps provide more security than OEM types. Whatever style of clamp is used, avoid over-tightening. The clamp should compress the hose sheathing and not cut into it.

Accessories

As mentioned previously, cooling systems may incorporate turbo aftercoolers, oil coolers, and various other heat exchangers. Design variations make generalizations difficult. Most of these devices, such as the seawater aftercooler shown in Fig. 12-19, can be disassembled for cleaning and, with the proper fixtures, pressurized to detect leaks.

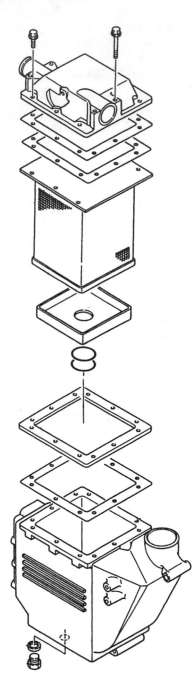

12-19 Yanmar aftercooler uses seawater as the medium of heat transfer.

13
CHAPTER

Greener diesels

This chapter examines ways in which people all over the world are working to make diesel engines more socially useful and less harmful to the environment. The intensity of the effort is reminiscent of the 1960s and '70s when some of the best minds dropped out of college to live on pizza and play with computers. Many of these enthusiasts work within the confines of corporations, firms such as Bosch, Denso, Stanadyne, Magneti Marelli. Others pursue their dreams on their own. Go by a London fish-and-chips joint early in the morning and, chances are, you will find one of these pioneers collecting waste cooking oil to use as diesel fuel.

Brazil

In the summer of 2006, a group of students and technicians arrived at Vila Soledade[1], a remote village built along the banks of the Amazon. Like hundreds of other communities in the region, Vila Soledade is remote from the national grid and must generate its own electrical power. In theory, the 700 or so townspeople could have electricity for five or six hours a day. But diesel fuel is an expensive luxury for farmers living barely above the subsistence level and their antique gen-set frequently broke down. Sometimes they waited weeks for parts. In terms of the government's Human Development Index, the people in Vila Soledade are among the most disadvantaged in the country.

The team brought with them a Brazilian-built MWD TD229EC-6 generator powered by six-cylinder, turbocharged engine, modified to run on unprocessed palm oil. This oil is local product that costs nothing except for the labor involved in its extraction.

The modified engine circulates coolant through a holding tank to preheat the oil to 65°C (Fig. 13-1). A transfer pump then moves the warm oil through a filter to a second tank where it is heated to 85°C for injection. With heat, the palm oil moves from the viscous esterarine phase to the more pumpable oleine phase. Sensors

---

[1]"Energy from vegetable oil in diesel generator—results of a test unit at Amazon region," Dr. Suani Teixeira Coelho, M. Sc. Orlando Christiano de Silva, et al., Brazilian Reference Center on Biomass, COMEC Nov., 2004.

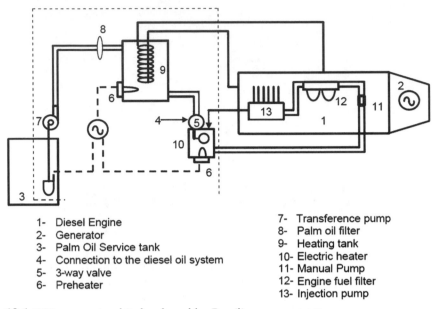

1- Diesel Engine
2- Generator
3- Palm Oil Service tank
4- Connection to the diesel oil system
5- 3-way valve
6- Preheater

7- Transference pump
8- Palm oil filter
9- Heating tank
10- Electric heater
11- Manual Pump
12- Engine fuel filter
13- Injection pump

13-1 SVO conversion kit developed by Brazilian government.

report fuel pressure and ambient, fuel, and exhaust temperatures, which are recorded daily by the operator.

Vila Soledade now has electricity for six hours every day. To prolong filter life the engine is run for a half hour on diesel fuel immediately after starting and just before shut down. The rest of the time it operates on raw palm oil.

Stack emissions on diesel fuel and palm oil were found to be almost identical, except that the vegetable oil produced no sulfur oxide. An analysis of the lube oil allowed change periods to be extended to 200 hours, which was surprising, since straight vegetable oil (SVO) has a reputation for contaminating lube oil. Misfiring developed after 600 hours due to dirty injectors. The injectors were changed and at 800 hours the head was lifted to decarbonize the combustion chambers. The major problem was fuel filters that gummed over and required frequent changes.

Reliable and affordable electricity has transformed the community. With the money saved on fuel, the villagers purchased equipment to process acaí, one of the major crops of the region, and many families have purchased lamps, television sets, refrigerators, and freezers. Most adults now attend night school.

While the Vila Soledade project impacts less than a 1000 people, it is part of a national commitment to wean the nation from fossil fuels. According to Luis Inácio Lula de Silva, the President of Brazil, "In the next 10 to 15 years, Brazil will become the most important country concerning renewable energies. . . . No one will have the ability to compete with us."

Lula's confidence seems well placed. According to the government, 43.8% of Brazil's current energy needs are met by renewable fuels, compared to 13.8% for the rest of the world. Ethanol has cut gasoline consumption almost in half, and a major push is underway to do something similar for diesel fuel.

The state oil company Petrobras squeezes as much as 20% more diesel fuel from conventional petroleum stocks by blending lighter hydrocarbons into the mix during refining. The tradeoff is a small reduction in heat value and a slightly lower flash point. Even so, the country still imports 10% of the diesel fuel it uses, which is a major draw on hard currency.

At the new century dawned, the government acted to reduce imports and, hopefully, to improve conditions in the agricultural sector. In only 12 months it set the stage for a massive influx of biodiesel by organizing production, establishing a regulatory framework, and arranging lines of credit.

In 2003, Petrobras introduced B2, a 2% blend of various vegetable oils and conventional diesel into most areas of the country and, by 2010, B5 will be universal. The program follows traditional practice, in that the raw vegetable oil undergoes transesterfication (reaction with methanol and caustic soda) for conversion to a fatty acid methyl ester, or FAME. This process reduces viscosity to something over 4 mm^2/s, which is in line with the European standard of 3.50–5.00 mm^2/s. B2 has a density of 879 kg/m^3 compared with 860–900 kg/m^3 for European diesel.

B2 will yield an annual saving of 800 million liters of fuel and US $160 million that would be spent on imports. Several government agencies are working with Peugeot, International, and various parts suppliers to make B100 practical for agricultural machinery and portable power plants. Locomotives will, it is predicted, soon operate on B25 and, with some modification, heavy trucks should be able to use B10.

In addition, the Brazilians hope to export FAME to Europe, where legislation mandates 5.75% renewables by 2010. Europe does not have the agricultural base to support such a commitment.

In a parallel program, Petrobras has converted three refineries and is building several more to process vegetable oil by hydrogenation. According to the company, H-bio blends have superior ignition characteristics and have shown themselves to be harmless to fuel systems.

Researchers at the University of Brasilia have explored pyrolysis as a means of converting soybean, palm, and castor oils to diesel-compatible fuel.[2] The technology is not new: pyrolysis was widely used in China during the Second World War to convert ming oil into feed stocks, which were then used to produce gasoline and diesel.

The university researchers used a stainless-steel reactor to distill vegetable oil into four fractions, at 80°C, 80°–140°C, 140°–200°C, and >200°C. These moderate temperatures left a heavy residue accounting for about 2% wt of the oil in the apparatus, which was discarded. As shown in Table 13-1, the distilled fractions have all of the salient characteristics of diesel fuel.

U.S. and Europe

Mixtures of FAME (vegetable oils converted to fatty acid methyl esters) and conventional diesel fuel are a fact of life in Europe and the United States. B2 is widely

---

[2]Daniela G Limá, et al, "Diesel-like fuel obtained by pyrolysis of vegetable oils," Journal of Analytical and Applied Pyrolysis, available on the Internet at www.sciencedirect.com.

Table 13-1. Physical and chemical characteristics of distilled vegetable oil

Properties	Soy	Palm	Castor	Brazilian diesel fuel
Density @ 20° C (kg/m3)	844	818	822	820–880
Cetane	50	53	31	45
Sulfur (% wt.)	0.008	0.010	0.013	0.20
Acid index	116	133	208	NA

available in this country and provides the assurance of adequate lubrication that may be needed with low-sulfur (15 ppm) fuel. Europe is moving to a B5 standard and may adopt a B20 standard in the near future.

Renewable fuels appeal to something deep in human nature and, unlike petroleum fuels, do not contribute to global warming. Rather than collecting in the atmosphere, the CO_2 released by burning vegetable products is recycled back into plant growth.

A joint statement issued by Denso, Bosch, Dephi, Siemens VDO, and Stanadyne approves the use of B5 if, the FAME content conforms to the EURO EN 14214 standard and the base petroleum fuel to EURO EN590. In contrast to ASTM (The American Society for Testing Materials) standards for fuels sold in this country, EURO standards include resistance to oxidation. FAME is biodegradable, which means that it oxidizes rapidly in the presence of heat, water, and various metallic ions. By-products of oxidation include solids and highly corrosive formic and acetic acids.

Other potential problems that fuel-system makers identify with FAME are listed in Table 13-2.

Table 13-2. Possible effects of FAME on fuel systems

FAME characteristics	Failure mode	Effects
Fatty acid methyl esters in general	Swelling, softening, and cracking of elastomers, especially nitrile rubbers	Fuel leaks Filter clogging
Free methanol	Aluminum and zinc corrosion Low flash point	Fuel system corrosion Safety concerns
Free water	Corrosion Reversion of FAME to fatty acids and methanol Bacterial growth	Fuel system corrosion Filter clogging
Free methanol	Corrosion of nonferrous parts	Fuel system corrosion
Free glycerin	Sediments Lacquer buildup	Filter clogging Injector coking
High viscosity at low temperatures	Excessive heat in rotary distributor pumps Higher stresses on pump-drive systems	Premature failure of pumps and related parts Poor nozzle atomization
Solids	Reduced lubrication	Premature fuel system failure

Consequently, the manufacturers say they "accept no legal liability attributable to operating their produces with fuels for which their products were not designed, and no warranties or representations are made as to the possible effects of running their produces with such fuels." In other words, American B2 users are on their own, without warranty protection.

Yet many of the same corporations produce fuel-system components used with Brazilian B2 that soon will be replaced with fuels of higher FAME percentages. According to the government, all fuel-system warranties will be honored. Brazil does not ascribe to the International Monetary Fund/World Bank model of development and, consequently, has a great deal of leverage over multinational corporations doing business in the country.

DIY FAME

Several companies market DIY transesterfication plants to convert vegetable oil into fatty acid methyl ester. There is no guarantee that the resulting FAME conforms to the ASTM D6751 standard for the commercial product, but the critical parameter—vicosity—is reduced and most solids and water are eliminated. By using waste cooking oil as the feedstock, the cost of the fuel comes out to less than $1 a gallon. Restaurants usually give the oil away, since they would otherwise must pay to have it removed. Nor is there any shortage of the product: according to one estimate, three billion gallons of cooking oil are discarded every year in the United States.

Extreme Biodiesel, based in Orange, California, markets the Eliminator, consisting of two large plastic tanks, pumps, filters, and associated plumbing. The apparatus can be set up in a garage and will, its makers say, produce 100 gal of diesel fuel in 12 hours.

The process begins with pumping the oil through a primary stage of filtration to remove large solids. This is followed by titration, a simple chemical test to determine the amount of methanol and caustic required for conversion. The user then mixes the chemicals with the oil, which is heated to speed the process. Several hours later, glycerin, the by-product of transesterfication, drops out.

Impurities that remain are removed by water spray, passed through a water separator and a 15 μm filter. An alternative to the water spray is to treat the fuel with Magnesol, a decontaminate widely used in the food processing industry. At this point, the FAME is ready to be mixed with conventional diesel fuel.

Straight vegetable oil

It is possible to operate diesel engines on straight vegetable oil (SVO) on the model of the Vila Soledade palm-oil project. Rudolf Diesel's first commercial engines, one of which was demonstrated at the 1900 Paris World Fair, ran on straight peanut oil. The inventor, who had no love for oil companies, wanted farmers to be able to produce their own fuel. Even after it became clear that SVO could not compete with cheap petroleum, Diesel struggled for years to adapt his engines to run on coal dust.

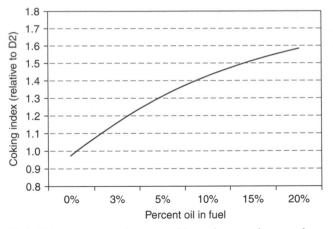

13-2 Coking, that is, how quickly carbon is deposited on internal engine parts, increases with the percentage of SVO added to conventional fuel.

Rapid increases in oil prices during the 1970s sparked a new interest in SVOs in Europe. A leading figure in this movement was Ludvid Elsbett (1913–2003), an engineer of remarkable talent who held more than 400 patents. In 1937 he was hired as a department manager for Junkers Motorenwerke AG, with responsibilities for the gasoline direct injection system used in the Jumo 211 engine that powered the JU-87, JU-88, and other military aircrafts. As a point of interest, the DI system consisted of 1575 individual parts, many of them precision lapped.[3] After the war he was employed by the truck manufacturer MAN and, during the 1970s led an abortive attempt to adapt alternative fuels to diesels in Brazil.

Subsequently, Elsbett and his sons opened a shop in Bavaria that became the first, if not the only, commercial source of SVO automobile engines.

Thousands of enthusiasts, working from information on the Internet, have converted their cars and light trucks to burn SVO. The type of oil used is determined by price and availability. Rape-seed oil is favored in Europe and sunflower-seed oil in this country. Many of these enthusiasts run their cars on waste cooking oil, unprocessed except for filtration.

But SVO is not without problems. One of the most intractable is the way vegetable oil carbons over injectors and combustion chambers. Figure 13-2 graphs the coking effects of sunflower oil against No. 2 diesel. Note that carbon deposits increase almost linearly with the amount of SVO mixed with diesel. This imposes a fairly severe maintenance requirement and, even with frequent de-carbonization, does not obviate the possibility of reduced engine life due to skewed nozzle spray patterns and contaminated lube oil.

---

[3] "A penny for your thoughts, Miss Shilling!," P.M. Green, Aeroplane Monthly, Feb., 1997, p. 39.

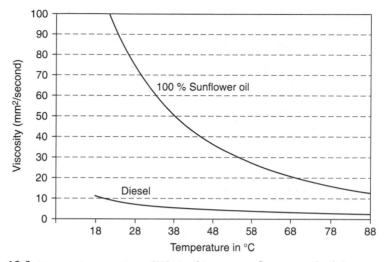

13-3 At room temperature, SVO, in this case sunflower-seed oil, has an order of magnitude greater viscosity than No. 2 diesel.

Straight vegetable oil also presents a viscosity problem at low temperatures. As shown in Fig. 13-3 sunflower oil is an order of magnitude more viscous than No. 2 diesel at 18°C. The high viscosity of SVO is enough to make it an outlaw fuel, unrecognized by diesel manufacturers and government agencies. People who use the fuel mitigate the viscosity with heat and by mixing the SVO with diesel, alcohol, and even gasoline. SVO also attacks seals, hoses, and other elastomer parts.

The best candidates for conversion are older, IDI engines with Bosch PE-type inline pumps and pintle injectors. The pumps are nearly indestructible and the injectors can be easily disassembled for cleaning. Distributor-type injector pumps should be avoided, since lubrication is entirely fuel dependent and electronic versions of these pumps cut off fuel delivery during coastdown. There have been reports that the alcohol present in FAME de-laminates the internal timing sensor (the sensor that tracks pump-cam position relative to the engine camshaft) on Bosch PV44 radial-piston pumps.

Most SVO conversions use conventional diesel fuel for starting and have a second tank, heated by engine coolant, for vegetable oil. A thermostatically controlled resistance heater at the suction side of injector pump takes some of the strain off the pump. Manually controlled glow plugs can remain energized until engine temperatures stabilize.

Dual-fuel engines

With some tradeoff in performance, diesel engines can be made to run on a mixture of diesel fuel and compressed natural gas. Benefits are reduced NOx and PM emissions, lower fuel costs, and for the five Caterpillar 3170B trucks operated by

Pima Gro Systems in California, full warranty coverage. The gas, injected into the manifold at 125 psi, is the primary fuel, ignited by small amounts of No. 2 diesel. A dual-fuel ECU adjusts pulse width for both sets of injectors with respect to manifold pressure, charge-air and gas temperatures, and fuel mapping. At low rpm the engines "skip fire," with three cylinders receiving diesel fuel and no fuel to the others. As loads and speeds increase, progressively greater amounts of natural gas are injected and the remaining three cylinders come on line. Skip firing reduces emissions and provides the surplus air to ease the transition from one fuel to the other. These trucks run about 350 miles a day without major problems.

Conventional fuel

Serious efforts are underway to ameliorate some of the worst effects associated with conventional fuels. In accordance with the Kyoto Protocols, the EU has agreed to reduce emissions of CO_2, which are believed to be the primary cause of global warming. Because diesels produce less carbon dioxide than SI engines, European governments promote their use with tax breaks and diesel-friendly emissions regulations (Table 13-3). Many experts believe that, by 2008, half of the new cars registered in Western Europe will be diesel powered.

Some translation is needed:

- PM (particulate matter) refers to the solids in diesel exhaust that are sometimes visible as soot.
- NOx (nitrogen oxides) is a blanket term for various oxides of nitrogen, such as NO and NO_2.
- HC (hydrocarbons) come about as the result of incomplete combustion of fuel and lube oil.
- CO (carbon monoxide) is an odorless gas, lethal at high concentrations.

Table 13-4 lists the U.S. Environmental Protection Agency (EPA) 50-state standards for light vehicles that were phased in during 2004 with full compliance scheduled for 2010. These standards apply both to diesel- and gasoline-fueled vehicles. Heavy pickup trucks, such as the Silverado, and large SUVs are partially grandfathered until 2008, when they will have to meet the same standards as other vehicles.

What the EPA calls NMHC (nonmethane hydrocarbons) is known as HC. HCHO, better known as formaldehyde, is a powerful carcinogen.

Table 13-3. **Current and pending EURO exhaust emissions regulations**

	Euro 4 Jan, 2005	Euro 5 Jan, 2010*	Euro 6 > 2013
PM	0.025 (g/km)	0.005 (g/km)	0.0025 (g/km)
NOx + HC	0.30 (g/km)	0.250 (g/km)	N/A
NOx	0.25 (g/km)	0.200 (g/km)	0.08 (g/km)
HC	NA	0.10 (g/km)	0.05 (g/km)
CO	0.50 (g/km)	1.00 (g/km)	0.50 (g/km)

* Recommendation of the European Commission.

Table 13-4. EPA Tier 2 light-duty vehicle emissions standards

	Mileage	Bin 8	Bin 7	Bin 6	Bin 5	Bin 4	Bin 3	Bin 2
NMHC	50k	0.100	0.075	0.075	0.075	0.070	0.055	0.010
	120k	0.125	0.090	0.090	0.090			
CO	50k	3.4	3.4	3.4	3.4	2.1	2.1	2.1
	120k	4.2	4.2	4.2	4.2			
NOx	50k	0.14	0.11	0.08	0.05	0.04	0.03	0.02
	120k	0.20	0.15	0.10	0.07			
PM	50k	0.02	0.02	0.02	0.01	0.01	0.01	0.01
	120k							
HCHO	50k	0.015	0.015	0.015	0.015	0.011	0.011	0.004
	120k	0.018	0.018	0.018	0.018			

In the EPA scheme of things, manufacturers are free to classify their engines into any of the seven bins shown in the table. But the fleet average, that is, emissions from all engines the manufacturer sells, must conform to Bin 5 standards after 2010. Rather than juggle the mix by counter-balancing relatively dirty Bin 7 or 8 engines with super-clean Bin 2 or 3 models, most manufacturers aim at producing Bin 5 engines across the board.

NOx and PM

Bin 5 limits NOx to 0.07 g/mile and PM to 0.01 g/mile, which are, respectively, one-sixth and one-half of EURO 4 limits. The PM standard approaches that of rubber dust from vehicle tires. Some explanation is required to understand why the EPA has taken these draconian measures, which affect diesel engines far more than their SI counterparts.

Both NOx and PM pose serious health risks, especially for urban populations. NOx reacts with sunlight to produce smog and is the pollutant most responsible for low-level ozone. The health effects of long-term exposure to ozone include chronic respiratory problems, reduced potential for exercise, and more frequent emergency-room visits. Ironically some of the most severe ozone exposure occurs in suburbs, downwind of city traffic.

PM, sometimes visible as the sooty component of diesel exhaust and felt as a stinging sensation on the face, presents an array of health problems, one of which is lung cancer. According to the American Lung Association, 50,000 Americans die each year because of exposure to PM, taken for purposes of the study as particles of less than 10 μm in diameter. No one has identified the biological mechanism involved, nor does the association try to distinguish between diesel PM, dust, and tobacco smoke. But there is a strong statistical correlation between inhaling PM and dying early. Perhaps a more telling statistic is that between 1980 and 2000, asthma increased 160% in American children under the age of four. African-American children, who live in disproportionate numbers in urban areas with high levels of diesel PM, die from asthma at the rate 11.5 per million, or more than four times the rate for white children.

Diesel engines are relatively clean in terms of HC and CO, but produce large amounts of NOx and PM. By insisting that all engines, regardless of fuel used, meet the same emissions standards, the EPA has moved to close what it identifies as the "health gap" between CI and SI.

Unfortunately, the two most toxic pollutants in diesel exhaust are the most difficult to control. The same high combustion-chamber temperatures that burn off HC and convert CO to CO_2 generate NOx and PM.

Designers were able to meet EPA Tier 1 and EURO 4 NOx and PM standards with in-cylinder controls, such as the use of exhaust gas recirculation and retarded injection to reduce flame temperatures. However, Tier 2 and pending European standards cannot be met by engine modifications alone. For that, one needs exhaust aftertreatment.

NOx aftertreatment

The high level of air in diesel exhaust makes the three-way catalytic converters used on SI automobiles impractical for CI engines. But it is possible to engineer around the problem, although at the price of some complexity.

Figure 13-4 illustrates the system developed by Houston Industrial Silencing for standby generators and other stationary applications. It covers all bases with a highly efficient muffler, two catalytic converters and a PM trap. The microprocessor-controlled injection system delivers the correct amount of ammonia (NH_3), to the first-stage converter. Ammonia reacts with NOx to produce free nitrogen and water.

$$NH_3 + NOx = N_2 + H_2O$$

Stated more precisely to account for the NO and NO_2 components of NOx, the reaction goes.

$$4\,NO + 4\,NH_3 + O_2 = 4\,N_2 + 6\,H_2O \quad 2\,NO_2 + 4\,NH_3 + O_2 = 3\,N_3 + 6\,H_2O$$

A second converter, downstream of the first, causes HC and CO to react with surplus O_2 in the exhaust to yield carbon dioxide and water:

$$HC + CO + O_2 = CO_2 + H_2O$$

A solution of 75% water and 25% ammonia is normally used as the oxidant. The dry (anhydrous) form of ammonia can also be injected directly into the exhaust or mixed with steam prior to injection. In any event, the rate of injection must be precisely calibrated for each installation by sampling NOx concentrations at the exhaust outlet.

When using platinum as the catalyst, the system is said to reduce 90% or more of the NOx present and at least 95% of the CO. HC conversion rates are comparable to those for CO. Platinum operates in a temperature band of 460°–540°F, attainable by a well-engineered installation operating under constant load.

NOx-reduction systems for motor vehicles pose more formidable problems (Fig. 13–5). These systems must be compact, capable of operating under variable loads and, when used on heavy trucks, must function 290,000 miles without maintenance. For automobiles, the zero-maintenance requirement is 100,000 miles.

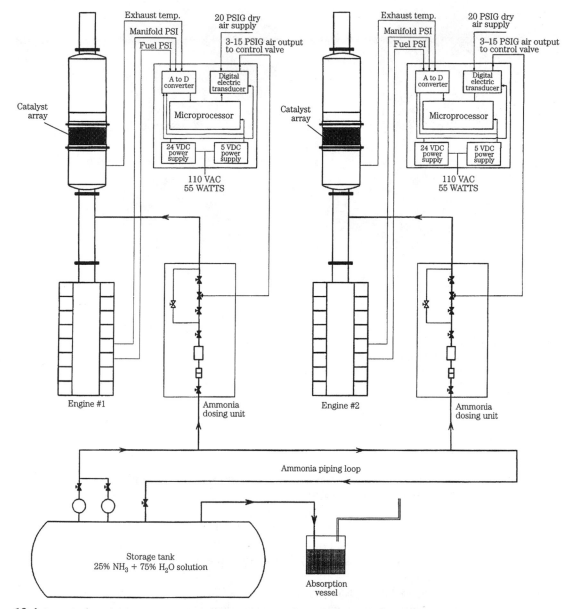

13-4 A typical stationary engine installation that combines ammonia dosing with a PM trap and efficient sound deadening.

Ford and Bosch employ a solution of urea $(NH_2)_2CO$ and water as the ammonia carrier. Urea breaks down into ammonia (NH_3) in the catalytic converter, which reacts with NOx as described above to produce water and nitrogen gas. Under ideal temperature conditions, these systems are claimed to oxidize more than 90% of available NOx. The urea component amounts to about 5% of diesel fuel consumption.

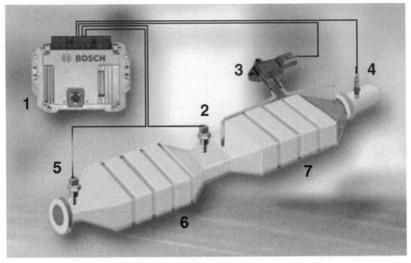

13-5 Bosch aftertreatment system for motor vehicles. The particulate trap (7) converts NOx to NO_2, which then oxidizes the carboniferous components of soot into CO_2. Those components that escape oxidation collect in the filter element. A differential pressure sensor (3) monitors the pressure drop across the trap. When backpressure becomes excessive, the ECU initiates regeneration by briefly supplying more fuel than the engine can use. The surplus fuel ignites in the particulate trap to burn off the soot and return backpressure to normal levels. Sensors (2 and 5) monitor exhaust temperatures going into and leaving the catalytic converter (6).

A delivery system filters the urea, atomizes it with compressed air, and adjusts delivery to engine load, speed and exhaust temperature. One or more oxygen sensors continuously sample the exhaust to provide feedback.

AdBlue, the Bosch marketing name for urea (actually a common fertilizer), is available at service stations in Germany. Ford and Mercedes-Benz have received EPA approval to introduce these converters in the U.S. although the agency would prefer a true zero-maintenance system.

Honda's 2.2 L i-CTDi, scheduled for introduction before 2010 when Tier 2, Bin 5 comes into effect, does away with urea injection. The catalytic converter is coated with two absorbent layers. The outer layer traps NOx during lean-burn operation. At intervals, the engine goes rich. Hydrogen, obtained from the unburnt fuel in the exhaust, reacts with the captive NOx to produce ammonia (NH_3). The second layer of absorbent material stores NH_3 until the engine returns to its normal, lean-burn operation. At this point, the ammonia reacts with NOx to yield water and nitrogen gas.

The system should have very little effect upon fuel consumption: according to a Honda spokesman, when running at 60 mph the engine operates lean for three minutes, then rich for five seconds. Amazing.

PM trap

The trap, also called a diesel particulate filter (DPF), mounts upstream of the catalytic converter. Because of the high temperatures involved, filter elements are made of porous ceramic or sintered metal. When the filter clogs, the computer senses the restriction as excessive backpressure, and orders injection late during the expansion stroke. Raw fuel floods the trap and spontaneously ignites, burning off carbon and soot accumulations. Periodically, the trap must be disassembled for removal of ash and other hard deposits.

Peugeot-Citroen employs a honeycomb silicon carbide PM trap that is cleaned with a combination of fuel and a proprietary additive, known as Eolys. Since Eolys is not the sort of stuff you buy at the corner gas station, the EPA will probably not approve it.

Ultra-low sulfur diesel

Sulfur clogs PM traps and poisons catalytic converters, and is in itself a source of PM. Consequently, the EPA ruled that the sulfur content of Nos.1 and 2 diesel be reduced from 500 ppm to 15 ppm. ULSD (ultra-low sulfur diesel) may be used in any vehicle engine, and is mandatory for 2007 and later models.

The changeover, which begun in mid-2006, put at least one small refinery out of business, but has generally progressed smoothly. One concern was that hydrogenation, the process refiners use to remove sulfur, also lowers the lubricity of ULSD. Additives bring the fuel up to 500 ppm lubricity levels. Aside from scattered reports, difficult to document, there seems to have been no serious problems in this regard.

Because diesel engines live so long, it will be decades before the full effect of Tier 2 standards and ultra-low sulfur fuel make themselves felt. If EPA projections are correct, our grandchildren should see a 2.6 million ton reduction in annual emissions of NOx and a 100,000 ton reduction in PM.

Meanwhile, the regulatory climate in the U.S. is changing. The reality of global warming has now been accepted by most opinion leaders, including many of the same conservatives who led the fight against ratification of the Kyoto Accords. Effects of other emissions are, for the most part, invisible and detected only by statistics on hospital admissions and longevity. Signs of catastrophic climate change are more difficult to ignore.

Environmentalists hailed the recent Supreme Court decision that empowers the EPA to treat carbon dioxide emissions under its general mandate to protect the environment. Other pollutants can be converted chemically into less noxious compounds. But no conversion technology, no magic bullet, exists for CO_2. The only way to reduce carbon dioxide emissions is to burn less fuel. As the most efficient form of internal combustion, the diesel engine has a bright future.

Index